The Workbench Book

The Workbench Book

Scott Landis

The Taunton Press

All photos by the author, except where noted below the individual
photos or on p. 244.

First printing: October 1987
Second printing: September 1988
Third printing: April 1989
Fourth printing: March 1991
International Standard Book Number: 0-918804-76-0
Library of Congress Catalog Card Number: 86-51321
Printed in Italy

A FINE WOODWORKING Book

FINE WOODWORKING® is a trademark of The Taunton Press, Inc.,
registered in the U.S. Patent and Trademark Office.

The Taunton Press, Inc.
63 South Main Street
Box 5506
Newtown, Connecticut 06470-5506

To Gerald Bannatyne, who uses no workbench at all. With wisdom and grace, he taught me more than he'll ever know.

I owe thanks to a great many people and institutions for their unselfish contributions. My first and greatest debt is to the craftsmen themselves, many of whom appear in these pages. Your insights were always welcome, and greatly enriched my own appreciation of the subject.

I particularly want to thank Jay Gaynor, Richard Starr, Tage Frid and Ian Kirby for allowing me to include their written material in sidebars. Thanks also to: Jerry Grant, Rob Tarule and G.L. Gilmore for granting me access to their unpublished manuscripts; Sam Manning for the use of his drawings; Greg McAvoy for demonstrating his Emmert vise in the photos on p. 139; Paul Kebabian for allowing me to photograph parts of his tool collection; and Charles Hummel and The Winterthur Museum for access to the Dominy and Greber collections. Photos have come from a number of sources and these are credited where they appear throughout the book and on p. 244.

I greatly appreciate the efforts of the following individuals, who read and commented on portions of the text: Michael Fortune, Ron Hickman, Frank Klausz, Nina Maurer, Tom Nelson, Toshio Odate, Richard Schneider, Peter Shapiro and Simon Watts. Special thanks to John Alexander, who not only read several chapters, but vigorously challenged my assumptions and avidly shared my discoveries.

The staffs of several museums, schools and companies went out of their way to facilitate my research. I particularly thank the people at: The Winterthur Museum; Mercer Museum; Hancock Shaker Village; Fruitlands Museums; Mt. Lebanon Shaker Village; Shaker Village at Chatham, New York; Colonial Williamsburg; The J. Paul Getty Museum; Mystic Seaport Museum; Hordamuseet of Fana, Norway; Skoklosters slott at Baalsta, Sweden; The Rockport Apprenticeshop; Robert Larson Co.; C.F. Martin Co.; Record Marples Ltd.; The Paramo Tools Group Ltd.; and The Black & Decker Mfg. Co. Thanks also to Valerie Oakley at the Southbury Public Library in Southbury, Connecticut, for her research assistance.

Finally, I would like to recognize the vital role played by three other friends and colleagues: Mark Kara, who rendered all the color and thumbnail drawings; Heather Brine Lambert, who built many of the benches on paper and whose good eye is responsible for gracefully tucking an unruly tangle of material between covers; and especially Roger Holmes, who helped me find order in the chaos. His good ear and fine humor always encouraged me and helped keep my feet on the trail.

If ever a book was written with the help of many, this is it. Hundreds of people, in several different countries, welcomed me into their workshops, their homes and their lives. They freely shared their experience and my excitement—indeed, they fueled my momentum with their own enthusiasm. I hope I have done them justice. Regardless of what we all may have learned in the process about the workbench and our craft, these many friends remain my greatest reward. You have made the journey well worth taking.

Contents

Introduction

Why a book about workbenches?

This book grew, in part, out of my personal attachment to tools. I'm one of those fanatics who can spend more time building a canoe than paddling it. Regarding most subjects, I relish the journey at least as much as the arrival. The workbench is no exception. It is the foundation tool of the woodworking trade, upon which all handwork is performed and without which we would have difficulty completing a single project. Although the bench takes different forms, it is perhaps the only tool that is common to every branch of the craft—from country chairmaking to urban cabinetmaking. That it is ubiquitous is partly the cause of its neglect. Library shelves are crammed with books about furniture, and to a lesser extent about the tools that are used to make it. But none attach more than a passing interest to the workbench.

Still, the bench had been a source of inspiration to my own work, and I was convinced that other craftsmen felt the same way. The workbench is to the dedicated woodworker what an instrument is to the virtuoso musician. In the hands of a master, the bench can be made to produce works of brilliance; it can be 'played' with an almost audible clarity. Like a musical instrument, however, the bench is no better than the person using it. At the same time, even the most skilled craftsman with the finest tools will be limited by a poorly made or ill-conceived workbench.

This intimate relationship—between the maker, his tools and his bench—suffuses our working life. Experienced craftsmen can navigate around their benches (and the inner recesses of their shops) blindfolded as easily as some of us can find our way around the inside of a dark refrigerator.

Such rapport does not come overnight, and is not easily severed. I've heard several stories that illustrate the bond that can develop between craftsman and bench. According to one story, an aircraft pattern shop preserves its benches even though they have been made obsolete by modern machinery. The shop retains one workbench per man, paying homage to the hand-tool tradition, although the workers now only sit upon their benches to eat their lunch. Another tale about a deceased Lithuanian cabinetmaker in New York City tells how his cremated ashes are stored in a box on the shelf beneath his bench. And every Christmas his former co-workers fondly toast his memory by dumping a shot of vodka into the box.

It's no coincidence that antique benches are finding their way into homes around the country, as sideboards, kitchen counters or simply for display. In a turbulent modern world, the workbench is a reminder (real or imagined) of a time when men were measured by, in the words of the merchant John Wanamaker, "...the plumb of honor, the level of truth and the square of integrity, education, courtesy and mutuality."

"The whole earth is full of monuments to nameless inventors."
—Otis T. Mason, *The Origins of Invention*

On another level, this book was inspired by a failed business. Several years ago I was a partner in a bench-building venture in Toronto, Ontario. We set forth (our naiveté matching our enthusiasm) to design and manufacture a workbench that would take into account all that a workbench should be. (The making of this bench is described in Chapter 5.) We built into our benches the quality and detail you would expect to find in a bench you built for yourself.

I trucked a demonstration model to trade shows around the Northeast to gain exposure. It drew many compliments but few sales. In one instance, I eavesdropped on a revealing conversation that took place between a woodworker and his wife at what they thought was a respectable distance away from the booth: "Nice bench," said he. "You can make that," said she. "Pretty pricey," he admitted. "It's only wood," she added. He returned later for a brochure and to check a few dimensions, even crawling underneath to mentally record the vise mechanics. (It would have been unseemly to pull out a tape measure.) I wasn't surprised to see the same fellow cross the aisle and plunk down a thousand dollars on a gleaming tablesaw.

I gained two lessons from that scenario. The first is that the bench is taken for granted. There's a common misconception that it is somehow less of a tool than the tablesaw or jointer. To many woodworkers, metal is an alien material that calls for a specialist; wood is our birthright. We hobble along, using a pair of rickety sawhorses or a glue-spattered plywood table, while we demand accuracy of our tablesaw within 0.001 in.

The corollary is that, because the bench is made of wood and built like a piece of furniture, the woodworker knows he can build it himself. What's more, many of us feel an obligation to do so. This notion is rooted in a centuries-old European tradition that has served the trade well. Bench-building has long been a testing ground for the old-world novice. In modern workshops around the country, however, I found that woodworkers have been tripping over stacks of bench lumber and rusting hardware for years, waiting to find the time and the inspiration to get started.

I began researching this book by looking at benches— the old and new, the big and small, the elaborate and simple. I was on the trail of the custom-built (more than the commercial) bench, and I was looking for quality—either of design or of construction. The benches I found were often unique. Over the space of a year, I traveled the country to interview bench-makers and bench users, writing about and photographing the results. I spent days touring New England, plumbing the innards of Shaker benches, hunting for signatures, dates and salient features. In Northern California, I spent a comfortable night on top of a sturdy plywood and angle-iron bench built into a 1958 Chevy step van that had been converted to a rolling workshop. My travels took me back to Canada and once, briefly, to Great Britain. My guides and hosts were stockbrokers, farmers, retired motorcycle mechanics, antique dealers—most of them woodworkers, both amateurs and professionals.

With regard to their workbenches, the craftsmen I met were of two distinct schools. For some, the bench was a lifeline to a cherished tradition. They copied its design from one they had learned on, or failing that, from an adopted traditional form. Members of the second group usually prefaced our first meeting with a disclaimer: "The bench works for me. . . ." The implication being that it might not work for anyone else.

In either case, I found that craftsmen often were surprised by my interest. After all, they protested, the bench was built to do a job. Comparing their own bench to some imagined paragon of 'benchness,' they assumed theirs was too crude or too ordinary to be of interest. They were usually wrong, but sometimes it happened that the workbench I went to see was not the one that intrigued me. I might stumble upon the most interesting detail elsewhere in the same shop or perhaps across town in the mind of another craftsman. Word of mouth led me from bench to bench and shop to shop, from one end of the country to the other.

What began as a search for fine benches soon became much more. Craftsmen are used to talking about their work, their design aesthetic, even their planes and chisels, but not about their bench. Once encouraged, however, they embraced the subject passionately, as though discussing their own children. They were glad to share their knowledge, but they were just as eager to find out what others had to offer. Like the traveling 'drummer' of another era, I was pumped for information about the treasures I'd seen and the people I'd met along the way. At these moments I felt most richly rewarded. Beyond anything this book might contribute to the design, construction and lore of the workbench, I was delighted to find that the process also could bring together the craftsmen who build and use them. The story that unfolds in these pages is as much about the people as about the benches themselves.

When any extended project nears completion, I suppose it's natural to regret the things left unsaid. Faced with the inevitable limitations of time and space, I have had to omit some fine examples of creativity, and skip lightly over others. Even as friends marvel at the range of benches I've been able to muster, I receive another call or letter to remind me that there is so much left to discover. There are old benches that raise (and perhaps answer) important questions about the history of tools, habits of work and patterns of trade and cultural migration. There is much to be gleaned from a broader study of tool and bench use around the world. And I've surely missed many new benches that have been built to reflect the specific needs of their makers or the continuing evolution of the woodworking trades.

While not being the last word on the subject, and certainly not the first, this book will, I hope, inspire a renewed interest in and a broader appreciation of the workbench. If you decide to build your own—and I hope you will—I encourage you to look forward and back in the process and to enjoy it. Keep in mind that a workbench should be custom-fitted to its user. Accordingly, the drawings that appear throughout the text are intended to show how things go together—not to be reproduced verbatim. If you already have a bench, you will discover more accessories and ways to improve it than you ever dreamed of. And if you have no interest in building a bench at all, but you enjoy the makers' stories as much as I did, you might want to skim the technical details and get on with the fun.

Early in my journey someone suggested that the workbench is not a project for beginners. It's far too complicated, he argued, and how is the beginner to know what he needs until he's got enough experience to understand what a bench can and should do for him?

Fair enough, I thought, but how can you get that experience without a workbench? I also recalled that building my own first bench about 15 years ago provided me with a short but intense course in modern woodworking. In the space of a few weeks making the bench in a friend's shop, I gained an understanding of (and an immense respect for) the jointer, thickness planer, tablesaw and drill press—all tools I had previously kept at arm's length. On later benches, I was introduced to the hollow-chisel and horizontal mortiser and the router. I also learned that truing a maple benchtop is one of the best hand-planing exercises around. At least as important as the tool handling skills was the foundation I laid for my understanding of clamping and gluing, dovetail and mortise-and-tenon joinery, and basic wood technology. These lessons could be applied to building houses as well as to building cabinets. I was truly being educated by my bench—what other project could offer as much?

If a novice were to ask me now, "Should I build the workbench of my dreams?", my response would be one of qualified encouragement—qualified in only two respects. It will take much longer than you expect, and by the time it is built, you may have different dreams. A workbench need not be forever. It can be modified, sold or given to a friend. As your needs and skills change, so may your workbench—I know one fellow who has survived five generations of benches.

Finally, remember that a good workbench won't make you a better woodworker, but it sure helps.

Workbenches for all seasons. Top right: Michael Fortune's modern bench. Above: Turn-of-the-century Swedish benches; No. 2 in a series of 11 educational posters, distributed by Gleerup's University Bookshop, Lund, Sweden.

This 1816 painting by G. Forster depicts the interior of an English woodworking shop. The vises and benches resemble those described by Peter Nicholson in his Mechanical Exercises (see p. 12), published four years earlier.

The Evolution of the Workbench

Chapter 1

I began writing this book in celebration of the modern workbench. At the time, it seemed there were sufficient examples of weird and wonderful contemporary devices to keep me occupied and to delight the most jaded woodworker. More than a year after I hit the trail, the yards of file folders and notebooks crammed with photos and drawings confirmed the accuracy of my initial perception. If anything, I underestimated the range and sheer volume of what was out there.

What I hadn't fully appreciated was the role played by tradition in the design of the modern workbench. Sure, I knew that dovetailed tail vises and mortise-and-tenoned legs and stretchers had been around for a while. I also knew a few people who owned and worked on 19th-century or early 20th-century benches. But I was surprised to discover the wealth of effective workholding devices that were invented long before the contemporary woodworking revival. The origins of the tail vise can be traced nearly 500 years. Some of the most unusual leg vises were built in the 18th and early 19th centuries. And the panoply of iron vises that flourished around the turn of this century puts to shame any recent tool catalog.

I have spent a lot of time on my back this past year, brushing away cobwebs and accumulated debris to peer with the aid of a flashlight at some of the workbench handiwork of our forebears. Staring up at the underside of some particularly well-built tail vise, or at some rough and ready shaving horse, I frequently found myself reflecting on the evolutionary nature of the workbench.

It wasn't until I'd seen a lot of benches that I began to appreciate the extent of cross-pollination that occurs between a workbench, the tools and technology of its period and the purpose for which it is built. I found this relationship reflected in virtually every bench I saw—from the traditional European and North American cabinetmaker's and joiner's benches and their contemporary derivatives, to Japanese beams and trestles, country woodworker's brakes and horses, and benches used by boatbuilders, luthiers and carvers.

Although the lineage of the workbench reaches back to prehistory, the most familiar woodworking bench—that used by the traditional cabinetmaker—is a relatively recent development. It acquired its distinctive characteristics only in the last three or four centuries. As an indispensable jig for producing flat, square stock with a hand plane, the modern bench, like all the other, less common benches to be found in this book, grew out of the most basic, portable workholding system: the human hand was the original vise, the ground or the body our first worksurface.

"The one grand stage where [the carpenter] enacted all his various parts so manifold, was his vice-bench; a long rude ponderous table furnished with several vices, of different sizes, and both of iron and of wood."
—Herman Melville, *Moby Dick*

Gerald Bannatyne, an Ojibway Indian from northern Ontario, uses a traditional crooked knife to make snowshoes. Bannatyne's lap is his workbench, his left hand a vise.

Japanese foot vise

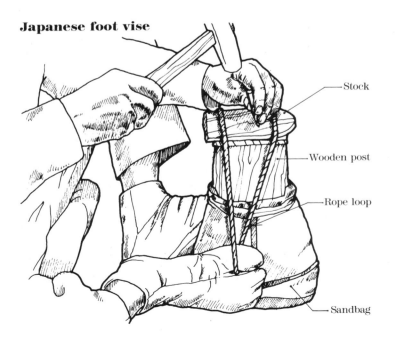

Stock

Wooden post

Rope loop

Sandbag

This basic method survives among a few craftsmen even today. A dwindling handful of native North Americans rely upon a crooked knife, or canoe knife, to make everything from canoes and snowshoes to flat boards. Wielded with one hand, the work held fast in the other, the crooked knife is at once the nomad's drawknife and scrub plane. It is enabled by its curved blade to sculpt three-dimensional surfaces or remove wood in the middle of a wide board. From a totally different culture, the ancient Japanese foot vise shown in the drawing below allows the craftsman to hold small stock securely using his toes and a loop of rope. Such a device represents an important development in the progress of technology, since it frees both hands for precise work. The creative application of the human body as clamp, vise and worksurface is typical and abundantly practical in nomadic societies, and wherever simple tools are required to perform a wide variety of functions.

Early woodworkers (and rural craftsmen through the ages) were generalists—gatherers and hunters who needed weapons and transportation, or later, farmers who needed shelter and implements. They were craftsmen by necessity. Accordingly, their workholding devices and cutting tools had a variety of applications. An ax, for example, can be used to fell a tree, remove its limbs, split it in half, and then hew it flat or shape it round. A shaving horse and drawknife can be used to rive flat stock like shingles and clapboards, or to shape curved stock like bucket staves and chair legs. The basic hand plane, by comparison, is much more limited. It is used to flatten boards; the standard cabinetmaker's workbench is designed primarily to hold them.

Crooked knives, drawknives and spokeshaves all involve a pulling motion. As the Japanese well know, pulling affords more control than pushing because the tool is drawn closer to the craftsman's center of gravity as the stroke progresses. This is critical when the tool requires extensive manipulation and when the craftsman himself is the workholding device. The push of the craftsman's weight directly on the material or through a lever, as on a shaving horse, is a natural counterpoint to the pull on the tools. Together they make a highly efficient, interdependent system of tool use and workholding.

The nature of the material being worked is also important. Much early woodworking was done (and much country woodworking continues to be done) with green, or wet, wood rather than dry wood. Green wood is easier to work than dry, and thus requires less formidable equipment to hold it. It is split directly from the log, resulting in oddly shaped baulks of timber that are not conveniently chucked in a vise on a flat bench. Its saturated fibers clog such highly developed instruments as a hand plane, so it is more commonly worked with open-bladed tools like the hatchet, adze and drawknife. These conditions make it logical to work green wood from a sitting position, holding the work in hand, by sitting on it or by pushing on some form of lever mechanism. When wood is dry and rectilinear, sedentary work at a shaving horse, or holding the work against your lap or chest, is much more difficult.

In one very simplified view, the history of the workbench is the gradual development of an aid or replacement for the body as holding device. As tools and tasks became more refined and holdfasts, clamps and vises were adopted, work could proceed more quickly without it having to be physically restrained by the craftsman. But specialization has its cost. It enhances certain functions at the sacrifice of others. When a piece of wood is clamped to the flat top of a cabinetmaker's bench, its whole length can be conveniently flattened and squared; fixed in a face vise, its ends can be worked into a complex joint like a dovetail. A whole range of commonplace work, however, becomes more difficult when done at a conventional bench. If you've ever tried to shape a canoe paddle or an ax handle at one, you'll know what I mean. I've built snowshoes at a workbench and had to devise all sorts of contraptions to hold the curved, riven members.

This tradeoff is what David Pye so aptly describes as the distinction between the 'workmanship of risk' and the 'workmanship of certainty' in his book *The Nature and Art of Workmanship*. Simply put, when you work against your thigh or chest with a crooked knife—or even with a drawknife at a shaving horse—there are many more opportunities to go wrong than when you use a plane at the bench. The edge of the knife, like that of the ax and adze, is exposed and unrestrained. Nothing but your own skill controls the depth and direction of each cut. Your body plays an integral role in the process by supporting the work, keeping it from moving, and in manipulating the tool.

As Pye says, "All workmen using the workmanship of risk are constantly devising ways to limit the risk by using such things as jigs and templates." The spokeshave is a perfect example of a 'risky' tool (the drawknife) made more manageable with the addition of a simple guide. What you can accomplish with the modified tool is at once more predictable and more restricted. The cabinetmaker's workbench is one of the most basic flattening jigs at his disposal. It is, however, no more appropriate for many country woodworkers, boatbuilders, or Japanese carpenters than is a drawknife for planing veneer.

The logical extension of the limitation of risk was the invention of machinery—highly refined jigs, really—for cutting and shaping wood. Where a lot of stock had to be removed, it was less efficient to lift the plane to the plank than to carry the wood to the machine. Labor that once might have been performed by the muscle power of an apprentice could be accomplished with the aid of sweatless power. Obviously, this development required ambulatory craftsmen. With their role now redefined, many modern craftsmen only rarely manipulate tool on wood. Instead, they are coordinators, moving materials between an assortment of dedicated machines, sometimes to be finished or detailed at a workbench.

The drive to limit risk parallels that of woodworking specialization. Among ancient Egyptian artisans there were carpenters, wagon builders, sculptors and cabinetmakers. As Josef Greber notes in his book *The History of the Plane*, "The furniture of the common people [in Egypt] differed less in form and construction from the fine pieces of the nobles, merchants and priests than in the quality of the materials and the lack of decoration." In this respect not much has changed since the days of the Pharoahs. Using the ax and adze as primary shaping tools, and a wide variety of solid woods and veneers, the Egyptian craftsman built all sorts of articles—beds, boxes, chairs, folding stools, etc.—fully as wide a range as has the modern designer/craftsman.

Greber goes on to say that "every major change in furniture had been connected with a change in the development of the plane." He credits the Greeks with the invention of the plane, but the Romans with having developed and popularized it. As both a piece of planed furniture and a functional companion to the plane, the workbench underwent a dramatic metamorphosis.

Roman engineering and architectural skills and the vast expanse of the Empire created a social, political and technological climate that allowed woodworking, like other crafts, to flourish for 1,000 years. Greber suggests that it was an anonymous Roman carpenter who "made the first workbenches for holding the stuff efficiently. . . ." No more would a craftsman be constrained to work on the ground, gripping a chunk of wood in one hand while planing or chopping it with the other. By placing the work upon a bench and fixing it with wedges or against a stop, he could hold it securely while freeing both hands to attend to the plane.

Egyptian carpenters use adzes, saws and an early form of 'vise' and 'workbench' to build a shrine. Stucco-work detail from the tomb of Nebanon and Ipuki, ca. 1475 B.C. From a print by Davier.

The drawing of the Roman workbench below was based on a stout oak plank found in 1934 in Saalburg, Germany. The plank (dated ca. 250 B.C.) contains four angled mortises for the splayed legs and an extra mortise in the center of the board near one end, probably for a bench stop. On one edge of the board are two more rectangular mortises, perhaps intended to house pegs that could support long work next to the benchtop, or as a part of some kind of brake or wedging device. Relief carvings show Roman craftsmen either seated at low benches of this type, or standing next to taller ones in order to plane or saw in the same manner as a 20th-century woodworker. Occasionally, Roman benches are depicted with stretchers connecting the splayed legs on each end of the bench, an arrangement that provides a much sturdier foundation for heavy work.

It is difficult to fathom the implications of so apparently simple an innovation. Pliny the Elder (23-79 A.D.) describes extravagant Roman display tables made of carved ivory and elaborate inlay that were valued at more than a million sesterces, or the price of a large estate, "if anyone would pay so much for an estate." It would be an overstatement to claim that such opulent furniture was the result of the workbench, but the existence of a reliable workholding apparatus certainly helped make it possible.

Just as a glacier deposits a moraine, the Romans left their skills as artisans and builders wherever the Empire extended. Aided by the sizable stands of virgin timber north of the Rhine, woodworking all over Northern Europe—principally in what is now Germany and Scandinavia—blossomed. Magnificent wooden palaces were built, replete with wooden furniture and paneling. By the end of the Roman era, the Norse began to assert their own hegemony, largely on the strength of their warriors and the finely crafted wooden ships that bore them. The elegant artifacts found in Norse ship burials at Oseberg and Gokstad represent some of the finest surviving examples of early European woodwork.

As the Roman Empire receded from Europe, many of the cultural and technological advances it engendered began to erode. Greber reports that in the tenth century, Theophilus, a Benedictine monk in Cologne, described the dressing and fitting of altar tables and doors "only with the tools which coopers use." This suggests a remarkable regression in the skills of European craftsmen. It is not so surprising, in this context, to discover that the European joiner's benches employed as recently as the 15th century are virtually indistinguishable from that early Roman form—built more than 1500 years before. (In his book *Woodworking in Estonia*, A. Viires describes benches used in rural Estonia of the 1940s and 1950s that are virtually identical to the Roman example.)

Roman workbench

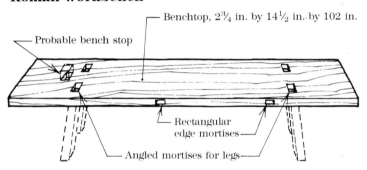

— Benchtop, 2¾ in. by 14½ in. by 102 in.

— Probable bench stop

— Rectangular edge mortises

— Angled mortises for legs

The workbench in this medieval illustration employs a simple system of bench stops to resist the planing action. Its design is no more sophisticated than that of the Roman bench at top, built over 1400 years earlier. From Das Hausbuch der Mendelschen, *Nuremberg, ca. 1400.*

The major political and cultural renewal that began in the 14th century heralded the revitalization of the crafts in general. During the early Renaissance, trade increased between the decentralized states of Europe and the secular guilds were consolidated. Fine furniture was no longer the exclusive domain of the Church. In cities all over the continent, joiners exercised their new status by bolting from the carpenters' guilds to form their own trade organizations. In many places they were even successful in making it illegal for a carpenter or a turner to plane wood or to cut a mortise-and-tenon joint. Offenders continued to be prosecuted under these statutes in some small European towns as recently as the 18th century.

Such jealous control of the craft furthered specialization. Jacques-André Roubo, in his treatise *L'Art du Menuisier* (see Chapter 2), notes, for example, that the leg vise was much preferred by cabinetmakers for its superior workholding ability, but he is perplexed and chagrined by its lack of appeal to joiners. This may be accounted for, at least in part, by the traditional conservatism of the trades. But it also may reflect the simple fact that joiners of the period worked with much larger pieces of solid wood, for which the simple bench stop, bench hook and holdfast were more efficient.

As tools and technology and social patterns continued to develop and diverge, the workbench had to adapt. Workshop scenarios depicted in Renaissance illustrations provide some of the best records of its continuing evolution. These often emphasize its prominence, as is the case with the 16th-century intarsia guild chest shown at the top of the facing page. The bench is the focal point of the joiner's tools. Protruding pegs and wedges on the benchtop, and the cutaway portion on its front edge reflect further tentative advancements in the use of leverage and wedges in workholding technology. (The stepped object in the foreground is an early progenitor of the modern board jack, or bench slave.)

As furnituremakers and carpenters tackled more ambitious projects, their tools were refined to meet the challenge. High-style craftsmen of the Renaissance (and later periods) began to use more and more exotic veneers and ornate moldings. Their work involved sophisticated joinery and unprecedented attention to quality of finish. It was during this period that the 'modern' workbench was developed. As Greber notes, "The Golden Age of furniture made tremendous demands on the skills of the joiner. It soon became apparent that the old tools would not do the job and many improvements were made. The workbench received a screw vice; veneer and fret saws appeared...while the number of plane types especially increased."

The earliest record I have found of a tail vise and face vise is on the unique German workbench shown at center right. This drawing (also from Greber's collection) is dated 1505, predating by more than a century the next recorded appearance of a tail vise on a workbench. But it is also noteworthy for several unusual characteristics of its construction. The enclosed tail vise, in which the dog moves rather than the vise block itself, is very complex, even when compared with much later, more common L-shaped designs. In fact, I have seen only two such vises in use on modern benches (one of these is shown on pp. 130-131). The benchdogs are similar to those documented by Roubo almost 300 years later in Paris. But the flexibility of being able to reposition them anywhere along the benchtop was a feature apparently unknown even to Roubo. The twin-screw face vise is also the forerunner of face and leg vises commonly employed for the next few hundred years. Near the right front corner of the bench is a notch that appears to be a vestige of the wedging or levering device shown on the more primitive intarsia guild bench, above.

By the late 17th or early 18th century, the traditional cabinetmaker's bench—replete with tail vise and front vise (also called a chop)—was firmly established in Northern Europe. The Swiss bench below, dated ca. 1700, is an early example of a fully equipped workbench. It is remarkable, not only for its ornament, but also because (apart from the carved cabinets and turned legs) it is identical to modern, commercially made benches.

The workbench is the center of attention in this 16th-century intarsia guild chest. The notch in the front edge of the bench may have been used as a brake or wedging device. Note the bench slave in the foreground and the hand ax and chopping block; the latter were probably used to prepare stock directly from the log.

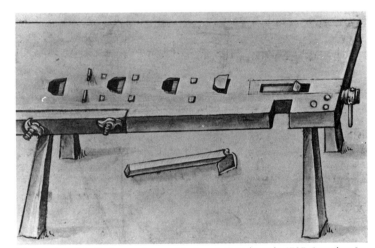

The enclosed tail vise and movable benchdogs in this 1505 drawing by Nuremberg engineer Löffelholz are highly developed for the period.

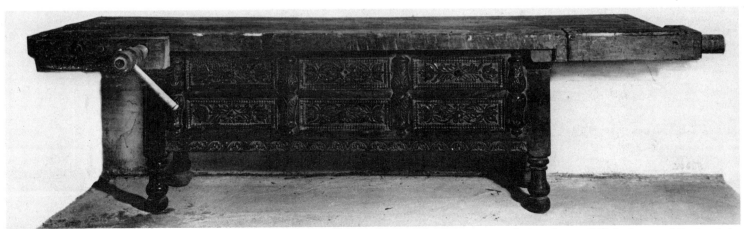

This carved beech joiner's bench was built ca. 1700 in Ilanz, Grisons, Switzerland. The top measures $27\frac{1}{2}$ in. wide by $106\frac{1}{4}$ in. long and the bench is $29\frac{1}{2}$ in. high. In their construction, the tail vise and the shoulder vise are identical to vises used on modern 20th-century benches.

Skokloster bench

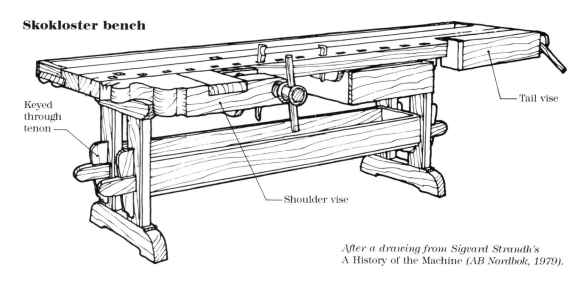

Keyed through tenon

Shoulder vise

Tail vise

After a drawing from Sigvard Strandh's
A History of the Machine *(AB Nordbok, 1979).*

Félibien's bench has a single planing stop (B) and an iron holdfast (C), with alternate holdfast holes in the top and both front legs. From Des Principes de L'Architecture *(Paris, 1676).*

A similar bench of about the same period is in the castle museum at Skokloster in the Swedish province of Uppland. This workbench, shown above, has the same tail vise and shoulder vise, but has a tool tray in the top. It also uses a more common keyed through-tenon construction for the base. The most interesting feature of the Skokloster bench is the shoulder vise on the left front corner, which is so elegantly shaped. (There is also a clamping board in front of the screw, to distribute pressure over a wider area.) It is possible that, like the Swiss bench on p. 9, the design of the shoulder reflects the creative expression of the craftsman who built it. But it seems more likely that, in casting about for a hefty chunk of wood for the vise, the maker seized upon a discarded corbel, the remnant of some architectural project, and decided to incorporate it in his bench, thus serving both a practical and a decorative function.

The tendency to mix and match parts salvaged from other projects is not uncommon in old benches. As a piece of utilitarian shop furniture, the workbench was often built out of available materials. When subsequent owners set about repairing or modifying an old bench, they probably felt even less constrained to maintain whatever integrity may have existed in the original. Old benchtops are frequently mounted on newer bases. Vises are added or removed. Open frame bases may be enclosed at a later date with clapboards taken off the side of a house. It's not unusual to find hardware and materials from several centuries mixed in one workbench. This curious amalgamation of disparate elements can present a minefield of obstacles for anyone attempting to assign provenance to an old workbench. But it also provides tantalizing clues about the various owners and lifetimes that a bench has survived, as noted by Jay Gaynor in "Dating workbenches" on the facing page.

During the 18th and 19th centuries, highly developed benches seemed to be unique to Swiss, German and Scandinavian craftsmen. The benches depicted in French and English encyclopedias of the same period are rather pedestrian by comparison. In one of the earliest such examples, André Félibien's *Des Principes de L'architecture* (Paris, 1676), the workbench could hardly be more prosaic. Félibien shows only a single stop and a holdfast used to secure work to the bench—the same devices described by Roubo nearly a century later. (The 'Stent panel,' pictured in W.L. Goodman's *The History of Woodworking Tools,* is another fine example in low-relief woodcarving of this type of simple, 17th-century joiner's bench.)

Dating workbenches

Determining the age of an old bench can be tricky. Here, Jay Gaynor outlines a sensible process. Former partner in The Jamestown Tool Company of Jamestown, North Carolina, Gaynor is currently Curator of Mechanical Arts at Colonial Williamsburg.

Dating an old workbench begins with asking its owner what he knows of its history. After all the stories are told, however, you'll probably find yourself still wondering how old the bench really is. Answering that question can be a relatively simple research task, or it can be a long, involved mystery.

Start by trying to determine whether your bench was commercially made or built by a woodworker for his own use. If it has a company name stamped on it or cast into a piece of metal hardware, the answer is usually easy. Several firms sold workbenches during the 19th and early 20th centuries. By comparing your bench with catalog illustrations (a number of old catalogs have been reprinted), you can usually come up with a good estimate of its date. If you can't locate a catalog, you might at least determine when the company was in business by looking it up in commercial directories of the city in which it was located.

If the bench is unmarked, looking at the old catalogs is still a good place to start. They will show you what commercially made benches of the period looked like. If your bench resembles closely those in the catalogs, chances are good that it was factory-made, no earlier than the mid-19th century.

What if everything points to the bench being made by a woodworker for his own use? The top is made of a solid slab or two. The joinery looks like one-off handwork. Here's where the fun begins.

Dating benches is similar to dating old furniture except that, while furniture styles often evolved quickly, bench design was extremely conservative. It is quite possible that a craftsman-made bench dating from the late 19th century might vary little in basic form from one

made a century or more before. Therefore we have to turn to details of construction.

The best place to begin is with that hobgoblin of small minds—consistency. Old benches usually have had hard lives, and over the years everything from tops to legs to vises may have been replaced. Stand back and look at the assembled bench. Does everything seem to fit together well, or do some things appear to be missing or modified? Look for empty holes or mortises. Keeping in mind that benches often were made out of materials on hand, is the general style and quality of workmanship consistent? Note what you find and then disassemble the bench.

Now concentrate on the details. Do the same tool marks show up all over the bench, or were some pieces sawn by a vertical saw while others were cut with a circular saw? Were all the holes bored with the same kind of bits? Are some of the primary worksurfaces rough while others are carefully smoothed? Are some pieces chamfered, but not others? Do components fit together well? Is the hardware all of the same type? Is the wear pattern predictable, or are some parts that would have been worn in use amazingly crisp while others show the ravages of time and dropped tools? Is the color of the wood consistent, or do some parts look less oxidized than others? Are paint splatters and stains consistent?

At this point, you can start drawing conclusions. An important one to begin with is whether or not all bench parts started out life together. Expect some inconsistencies. A part here or there may have been salvaged from another project. Big holes, like those for a holdfast, may have been bored with an auger of different design than the bits used to bore smaller holes. The lumber for the legs may have been bought from a local mill, while the top was hand-hewn from the log. But, if everything appears consistent or the inconsistencies can be accounted for reasonably, you probably have a piece that 'retains its integrity.'

The following details will help you date a bench. If the bench seems unaltered, they can allow you to assign it an approximate date. If you are analyzing components of a bench that has been altered, they can be a guide to determining how and when the work was done.

• **Nails:** Forged nails, identified by their hammered, irregular heads and shanks that taper on all four sides, were common through the first quarter of the 19th century. After about 1800, machine-cut nails were available. These can be identified by shanks that taper on only two sides. Early cut nails have hand-forged heads, but since the 1820s cut nails have been made with machine-formed heads that are usually flat and rectangular. Modern wire nails were introduced in quantity about 1890.

• **Wood screws:** Most wood screws made prior to about 1850 have blunt ends. The earliest were hand-filed, while later ones had machine-cut threads. Both types usually have coarsely finished shanks, but hand-filed threads tend to be deeper. If screws that appear to be original look like modern, gimlet-pointed ones, your bench probably dates after the mid-19th century.

• **Saw marks:** Hand-saw marks do not tell you much. Neither do the vertical marks left by 'up-and-down' sawmills, which were used in North America as early as the 17th century, and, in some parts of the country, as late as the 20th century. But if you see circular saw marks, the component on which they appear, although it might date as early as 1820, is more likely to have been cut after 1850. A combination of up-and-down and circular saw marks is cause to suspect modification.

• **Planer marks:** Machine planer marks, identified by a closely spaced pattern of smooth, parallel marks across the grain, indicate that the piece was made after about 1830. The marks left by some

early planers are gently curved rather than straight.

• **Drill bits:** Spiral augers were introduced late in the 18th century, and a bench made using them is likely to date after 1800. The older bit types—shell, spoon, nose and center—continued to be used throughout the 19th century.

• **Hardware:** Most early benches had a minimum of hardware. Hand-forged fittings might indicate an early date, while cast-iron ones usually appear on benches built in the late 19th century or later.

• **Vise screws:** The common use of iron vise screws dates to the same time as cast-iron fittings. During the late 19th century, hardware firms sold both wooden and iron screws for those who wanted to build their own vises. Wooden screws are often difficult to date by their threads. But if the head and handle look to be one-off turned examples, they may support other, early dating details.

These dating techniques may permit you to determine only the earliest date your bench could have been made. Innovations came at different times to different parts of the country. A bench made in the backwoods in 1920, for example, may have the same features as an urban one made 100 years earlier. Ultimately, the precise dating of any bench is possible only if it is marked with the actual date or other datable circumstances of its manufacture, or if it has a specific, reliable history.

After you have determined the probable history of your bench, you've got another important task. Figure out how to own it responsibly. Benches dating before 1840 or 1850 are very rare. If you own one, you own an important document that, by characteristics as subtle as stains or wear patterns, might have much to tell us about early benches and early woodworking methods. Think twice before you refinish it, replace any broken parts, or plane the top. Maybe it should be left to tell its story.

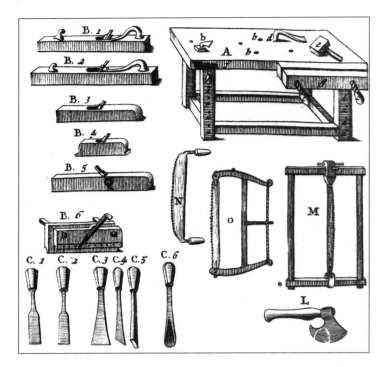

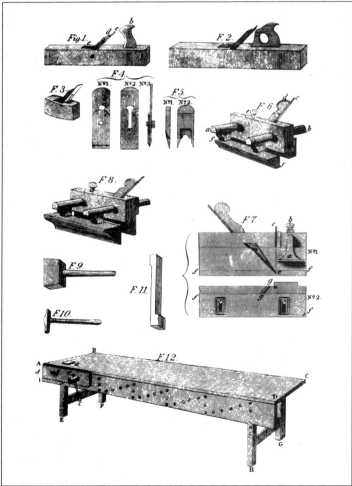

The detail from Joseph Moxon's Mechanick Exercises *(London, 1703), at top, shows a bench with a twin-screw face vise and an adjustable peg that screws into holes tapped in the left front leg. The face vise in the detail from Peter Nicholson's* Mechanical Exercises *(London, 1812), at bottom, is mounted at the left corner of the bench. A wooden guide bar (d) keeps the jaw level with the benchtop and the front apron provides a variety of positions for supporting one end of a board.*

Several years later, Joseph Moxon's *Mechanick Exercises* (London, 1703) portrays an almost identical workbench, with the addition of a twin-screw vise and a wooden hook mounted on the front of the bench. Moxon's work was the first technical introduction to the crafts published in English.

The wooden hook (so called by Roubo), which Moxon has drawn on the bench's left front edge, is used to jam one end of a board while its edge is planed. Below the hook (and barely visible in the engraving) is a single wooden screw, which according to Moxon was used in conjunction with support pins in the other leg to hold boards for edge-planing. The position of the large twin-screw vise, which Moxon says is "sometimes fixed to the side of the Bench," may reflect the artist's imagination. Such a vise mounted on the right corner would make it almost impossible to use either the single stop or the hook.

By 1812, the English workbench had undergone some modification, as shown by the engraving in Peter Nicholson's *Mechanical Exercises.* Nicholson begins his preface by paying homage to Moxon, but in the next sentence asserts that "the progress of science, and the changes in matters of art have rendered the work obsolete and useless." On Nicholson's bench, a single-screw vise is affixed to the left corner of the top, directly in front of the stop (which Nicholson calls a bench hook). A plank may be clamped in the vise and supported along its length by pegs inserted in a series of holes drilled along the bench's wide front apron and the far leg.

Perhaps the most interesting feature of Nicholson's bench is the horizontal piece, which is mortised into the vise jaw and supports its left end. About this guide, Nicholson says only that its purpose is to keep the vise jaw from turning around. This guide is more commonly associated with the vertical leg vise; when drilled with a series of holes into which a pin is inserted, it serves to keep the vise jaws parallel. On Nicholson's face vise, the guide also may have been used to support work clamped horizontally in the vise, or perhaps to hold a small block the same thickness as the work (thereby keeping the jaws parallel).

Of all the characteristics of these historic benches, there is one in particular that bears further comment. To a bench, they are all much lower than the average modern cabinetmaker's workbench. Early Roman benches hover around knee height. Nicholson's bench is 32 in. high. Roubo's is about 31¾ in. Other 18th- and 19th-century American benches typically vary in height between about 28 in. and 33 in.

As Charles Hummel suggests in *With Hammer in Hand,* such low bench heights are probably related to the intensive use of the hand plane by early craftsmen. These benches were, after all, planing benches. The lower they were, the more pressure could be applied to the plane. Many modern woodworkers, by contrast, are as likely to use small, hand-held power tools as they are a plane, and these require control and accuracy more than pressure.

But before jumping to conclusions about workbench height, I should mention a comment made to me by Ralph Salaman, English author and tool collector, regarding the small village workshops he visited in Great Britain in the 1940s. Salaman also observed that "the benches seemed so low down for comfortable working." He soon discovered, however, "that this was because the floors had not been cleaned for many years and there was 9 in. to 12 in. of compacted waste underfoot." So it would seem, at least from Salaman's experience, that all old benches may not have been as low as we think.

Having of necessity been confined to books for information on early European workbenches, I looked forward to tackling North American workbench history. Only 300 years old, this history still lives in museums and in a surprising number of private shops and collections—I was looking forward to seeing the old benches and contemporary reproductions in person. One of my first stops was the Dominy workshops, reconstructed at the Henry Francis du Pont Winterthur Museum, in Delaware, home of several of the oldest documented workbenches in North America.

Between about 1760 and 1840, at least three generations of Dominys worked wood in the Long Island town of East Hampton. Nathaniel Dominy IV (1737-1812), a descendant of English immigrants who arrived in East Hampton in 1648, established the workshop and was succeeded there by his son, Nathaniel V, and grandson Felix. Nathaniel IV probably built the earliest of the three surviving Dominy benches sometime after 1760. The benches were in continual use until about 1840, when the last sale was posted in the workshop account book. The 18th-century demeanor of the shop remained essentially unaltered through the 1940s, when it was reconstructed at Winterthur.

Although there are some slight differences among them, the three Dominy workbenches are all strongly derivative of the English tradition exemplified by Moxon and Nicholson. The Dominys were American born, but like many of their colonial contemporaries, the tools and techniques they used were drawn from the dominant culture in which they practiced their craft. Indeed, until well into the 19th century, most of the tools purchased in North America came from England.

All three benches have a twin-screw face vise mounted on the left corner of the bench and a mortise for a single wooden bench stop behind it, as shown below. They reflect none of the advancements of the Northern European benches described earlier—there are no shoulder vises, tail vises, benchdogs or cabinets. The Dominys were country craftsmen, and like their rural counterparts of every period, they were woodworking generalists. At various times, the family worked as wheelwrights, millwrights, coopers, boatbuilders, clockmakers and furnituremakers. Their simple vises, stops and holdfasts would have proved more adaptable to such diverse activities than highly developed tail vises and shoulder vises. As Charles Hummel, whose book *With Hammer in Hand* documents the lives and work of the Dominys, told me, they probably were aware of the latest technology of the day, but were "conservative by choice."

Despite their simplicity, the Dominy benches exhibit several interesting features and a couple of curious anomalies. The three benches line opposite walls of the narrow, rectangular workshop. The largest one is 28¼ in. wide by 12 ft. 4½ in. long and is centered against the east wall beneath a large window. Two shorter benches are positioned end-to-end along the opposite wall. The workmanship on all three is rather crude, particularly in comparison with Dominy furniture, which demonstrates their capacity for refined work. Many of the twisted wooden base members were obviously riven, and are joined with undersize mortises and single-shoulder tenons, roughly fitted to one another and drawn together with large wooden pins.

The Dominy benches, built in the late 18th century in East Hampton, Long Island, reflect the design of English benches from the same period. Their main features are the twin-screw face vise, the sliding board jack and the single planing stop behind the vise.

The rear legs on two of the benches are splayed slightly, providing a somewhat more stable footprint than parallel legs. I was surprised to find that these benches have no rear stretcher connecting the rear legs, which would have made them much more liable to rock during heavy planing. (The third bench, which is slightly more than half the length of the longest bench, has a rear stretcher that may not be original.) The benches easily could have been braced to the wall, but I was not able to find evidence of this.

Each of the two longer benches has a sliding board jack, which supports the free end of a plank while the other end is clamped in the vise. Variations of this sliding support can be found on many 19th-century benches and are much more flexible than the stationary peg holes of Moxon's and Nicholson's benches. The Dominy benches also have three additional, rectangular board supports that appear to have been designed to pull out from mortises in the front edge of the top. All but one are jammed in their mortises and cannot be withdrawn.

The two-piece tops consist of a slab of red oak in front for a primary worksurface, and a much thinner board behind. The main portion of the longest top was sliced from a massive bole of a tree, about $5\frac{1}{2}$ in. thick by $17\frac{1}{2}$ in. wide. A small grease cup swivels out from the underside of all three benchtops near

the vise to provide ready lubrication of the vise screws, plane bottoms and sawblades.

The twin-screw vises are similar to the one pictured by Moxon, but have handles to make them easier to turn. (Two of the benches have a threaded hole in the left leg below the horizontal vise jaw, and the longer bench's left front leg is mortised near the bottom for a horizontal beam, evidence of earlier, vertical leg vises.) The two screws are adjusted independently to allow the clamping of tapered as well as square objects. Likewise, a board may be trapped vertically between the screws so that its end can be dovetailed. None of the large screws is fitted with a garter, the narrow key that allows the jaw of the vise to retract as the screw is unwound. Consequently, the vise jaws must be pulled out by hand to enlarge the opening. While detailed photos of the early Swiss bench shown on p. 9) reveal a garter in the tail vise, this feature does not appear to have been used regularly on benches in England or North America until sometime in the 19th century.

The vise screws pass through the floating jaw and are tapped into the benchtop itself. The underside of the top has been relieved, as shown below, so that only the first few inches need be threaded. Two of the benches also have narrow slots angled across the relieved channel, indicating that the female threads

The Dominy screws have no garter securing them to the jaw. While this requires that the jaw be retracted by hand when the screw is unwound, it enables the jaw to clamp tapered stock securely.

This bottom view of the vise screw shows the angled slot that was used to draw the tap through the benchtop in cutting the female threads.

were cut using a simple screw tap, like the one shown at right. In principle, this tool is similar to the early threading tap, designed in the second century B.C. by a Greek engineer named Hero (see p. 123). A starter hole is drilled through the bench and a thin metal guide plate is inserted in the slot. The guide registers in the narrow, spiral groove that wraps around the shaft of the tap, drawing it through the hole and engaging the V-shaped cutter in the bench.

There is much to learn from looking at old benches, and even more to gain from working on them, as Rob Tarule demonstrates in the next chapter. Benches like the Dominys' are a rich source of information about woodworking practices during the early colonial settlement of North America. While the Dominys may have lingered somewhat behind the latest innovations in the European capitals, it would be a mistake to consider their benches primitive, or somehow deficient. Like all traditional tools, they were designed to serve the particular needs of the craftsmen who built them.

Antique screw tap

Cutter

Tap

Sheet-metal plate housed in angled slot

Area being threaded

Cut spiral sawkerf in tap shaft at desired pitch.

Trough

Benchtop underside

Note: *Sheet-metal plate engages kerf, drawing cutter through wood.*

A Federal workbench

Michael Dunbar is a Windsor chairmaker and cabinetmaker who works primarily in the Federal style (late 18th and early 19th centuries). In 1976, when he built the bench at right, Dunbar was employed as a chairmaker at Strawbery Banke, a restored historical village in Portsmouth, New Hampshire. "I didn't have a lot to draw on in the way of models," Dunbar recalls. The model he chose—and copied closely—was a friend's bench of early 19th-century vintage.

Two years before they began joining boards, Dunbar and two friends took down a huge beech. They had it milled and then stored the planks to dry. They turned a bunch of green hickory billets, and borrowed a tap and die to cut about a dozen 1½-in.-dia. screws. The hickory screw handles were also turned green, with a knob at one end turned only slightly larger than the diameter of the hole drilled across the head of the screw. While both screw and handle were still sopping wet, the handle was pounded through the head. (This technique works even better if the handle is dry and the head is wet—only the head will shrink.)

Dunbar's bench has a tail vise on the right corner, but his 39-in.-long, twin-screw, cherry face vise gets most of the action. The screws are set 23 in. apart, which provides plenty of room to clamp a Windsor chair seat while its edge is being shaped.

Dunbar's face vise differs from the Dominy vises in two important respects. The screws are keyed to the jaw with a garter so that the jaw retracts when they are unwound. And instead of tapping the benchtop itself, Dunbar installed two dovetail-shaped nuts in the edge. This would have been convenient for a Federal-period craftsman—as indeed it is for a modern one—for two reasons. Long before industrialization and mass production, it was not uncommon for cabinetmakers to purchase wooden screws and matching nuts from fellow woodturners. What's more, being the most valuable parts of an old workbench, these wooden screws and nuts could be easily removed from one bench for use in another—a savings of time and money.

Like the Dominy benches, Michael Dunbar's reproduction Federal-period bench was built for work, not for show. Dunbar uses the twin-screw face vise to clamp Windsor chair seats. The nuts for the screws are dovetailed into the edge of the top, suggesting that the original screws and nuts may have been purchased separately, or that they may have been intended to be replaceable.

HAMMACHER SCHLEMMER & CO.
NEW YORK — SINCE 1848.

CABINET AND PIANO MAKERS' BENCHES

No. A

This has long been the standard Cabinet Makers' Bench. It is constructed throughout of the best selected maple, and is made in the most careful manner. Strong and rigid, it is, without question, the best Cabinet Makers' Bench made.

No. A. Cabinet Makers', 6 ft. 8 in. long; 2 ft. 1 in. wide; 2 ft. 9 in. high; 7¼ in. recess.................Each, $22.50

Hammacher, Schlemmer & Company was a major supplier of workbenches in the second half of the 19th and early 20th centuries. This catalog, ca. 1920, offers everything from professional cabinetmaker's benches (above) to the combination bench-and-tool cabinet (right), which was aimed at the gentleman woodworker.

Until the late 19th century, workbenches were, perforce, as idiosyncratic as the craftsmen who made them. By the turn of the century, however, American woodworkers were able to purchase wooden or metal benchscrews, or even complete benches, from several mail-order companies. Two of the oldest and best known were the Ohio Tool Company, of Columbus, Ohio, and Hammacher, Schlemmer & Company of New York. European firms, such as Wm. Marples & Sons of Sheffield, England, and Joh. Weiss & Sohn of Vienna, Austria, also sold workbenches during the 19th and 20th centuries. European benchmakers continued to sell in North America long after their U.S. competitors changed over to heavy machinery and food processors. Two of the most popular modern, European-built workbenches are described in Chapter 8.

Most 19th- and early 20th-century catalogs offer a wide variety of workbenches. Many of these are designed for school industrial-arts programs and amateur woodworkers, as well as for manufacturers and professionals of all different stripes. Among several of the antique-tool dealers I spoke with, commercially made 19th- and early 20th-century benches are far more common today than the one-off, craftsman-made variety. As a rule, they are well made, if lightly built, and have the conventional tail-vise/wooden face-vise configuration that has helped to define the current concept of what a workbench is. Antique commercial benches usually have laminated tops and are often built out of standard, dimensional lumber. The better bases are assembled with either wedged tenons or bolts, while some of the lighter models rely on glued dowel joints.

No. 100
COMBINATION BENCH AND TOOL CABINET
Polished Oak Cabinet, brass trimmings and containing simply the best tools made.
Complete, $85.00
(For details see pages 468 to 471)

Most tools are made for us by other tradesmen. Our saws, chisels, and planes are made by specialists. Although the workbench can be purchased, too, part of its appeal is that it can be built by any woodworker familiar with the joinery and techniques used in furniture. A traditional cabinetmaker's bench, for example, involves not only dovetails and mortise-and-tenons, but every manner of planing, rabbeting, drawboring, laminating, wedging and pinning. Most of the tools in the workshop can be used to make it. The fact that one workbench is almost a necessity for making another underscores its importance.

The handmade workbench is much more than a highly functional tool. It is also a piece of furniture, and, as such, it provides an opportunity for the craftsman to strut his stuff— it's the woodworker's answer to the needleworker's sampler. European-trained cabinetmakers like Frank Klausz (Chapter 4) told me that the making of a new workbench was the ideal project to assign to a new employee in the traditional workshop. It served as a test of what a young woodworker could do. If he botched his bench, the master would be loath to set him loose on a piece of furniture.

When it comes to workbenches, it is impossible to generalize about either the 'old-world' precision craftsmanship or the slipshod pragmatism of their makers. Some benches were obviously thrown together hastily from the roughest stock and salvaged lumber, while others evidence carefully selected and joined timbers and painstaking laminations. New machinery and man-made laminates have emerged, which have greatly affected both the construction and the use of many of the benches being built today. Many modern cabinetmaking shops no longer have a cabinetmaker's bench, but rely instead on utility tables and machinery. But where the pre-industrial hand-tool and solid-wood tradition survives, the benches haven't changed very much in the last 200 years.

I've been asked on several occasions whether there is an identifiable 'American' workbench—a thorny question to answer with any certainty. After all, the design of early workbenches on this continent, like that of tools and furniture, was lifted directly from Europe. English immigrants tended to adhere to English traditions, French to French traditions, Germans to German traditions, and so on. And because the workbench is so closely related to tools and furniture, it was effectively girded against radical innovation. But—outside the confines of a few very cloistered communities—these traditions evolved, comingling more liberally on this continent than in the Old World.

For the most part, when people refer to workbench 'design' they are talking about the vise, the most distinguishing feature of any workbench. Perhaps the only real American contribution to vise design is the angled leg vise to be found on many old benches from the mid-Atlantic states (see p. 125). And although I've never heard of any such creature in Europe, I wouldn't be surprised to discover that it too had been imported.

I could fill many chapters with descriptions of exotic, craftsman-built benches. I have included a few of the most striking examples in the following chapters in the hope that they will suggest (and inspire) the creative tradition that has always been a part of workbench design. For, after looking at hundreds of benches, two things have become abundantly clear to me. First, 'tradition' is the operative word: whenever I think I've found something unique in the world of benches, it can almost always be traced to some venerable ancestor. And, second, the evolution of the workbench, like that of living things, is never complete.

This 19th-century tail vise (left) features a double dovetail, also found on benches sold by the Vienna company, Jos. Weiss & Sohn. The heads of both tail- and leg-vise screws (above) are stamped with the maker's name, E.W. Carpenter, of Lancaster, Pennsylvania. Photos by Ronald W. Pearson.

Rob Tarule's reproduction of the Roubo bench provides an active testing ground for woodworking methods of the 18th century.

18th Century: The Roubo Bench

Chapter 2

A s Jacques-André Roubo wrote in his classic four-volume treatise on woodworking, *L'art du Menuisier* (Paris, 1769-1775), "The bench is the first and most necessary of the woodworker's tools." That said, Roubo goes on to explain in crisp technical prose and vivid engravings (shown on p. 21) precisely how to build a typical joiner's bench of the 18th century.

Accustomed as we are to today's benches with their complex vises and involved construction, the Roubo bench is so simple that it's tempting to dwell on what it lacks. There's no tail vise, no regimental line of benchdogs marching across the top, no quick-action front vise, no sled-foot trestle base. Roubo's bench has a massive single-plank top and four stout legs; a single, large bench stop (*boëte*) or hook (*crochet*); a wooden hook (*crochet de bois*) screwed to the left front edge of the bench; and an optional leg vise (*presse*). All workholding is accomplished either by pushing the work against the stop or into the maw of the wooden hook, or by securing it to the top or front of the bench with an iron holdfast (*valet*).

How, you may wonder, could such a primitive contraption serve for work on the refined furniture of the period? How would delicate moldings be held or drawers dovetailed? The answer lies in the division of the woodworking trades at the time, as depicted in Roubo's engraving. The workers shown are architectural joiners, not furnituremakers or cabinet-

makers. Leaning against the wall of the shop are the fruits of their labor: no chairs or tables, no casework, only architectural woodwork—windows, paneling, stair stringers. The distinction is important. It means that these men would have spent all their time performing a few operations with great efficiency. For example, the ability to trap short or irregular bits of wood in a vise, while critical for a cabinetmaker, would have been superfluous for the joiner. In looking at this bench, as well as at the other benches in this book, it is clear that the type of work being done is a major determinant in the design of the bench.

Roubo's description of the bench (see p. 21 for excerpts) must have been the most comprehensive to date. That's no surprise, considering that his was the era of the Enlightenment and its great encyclopedias. As the introduction to Roubo's third volume states, "The intention of the Academy [*Académie royale des sciences,* founded in Paris in 1666, which published the work] is to erect a monument to human industry, so that nothing in the arts is neglected from our epoch to illuminate posterity."

What *is* surprising is that Roubo's description of the workbench continued to stand as the last word on the subject for more than 200 years—and it is still exemplary. What seems to have made Roubo unique is that he approached the subject not as a generalist like Denis Diderot or Joseph Moxon, two of the era's other great encyclopedists of craft and industry, but with the experienced eye of a *compagnon menuisier* (journeyman joiner). Described in a prefatory note to the first volume as *un homme de métier,* or craftsman, Roubo, by the publication of the second volume in the following year, had acquired the distinction of *maitre menuisier,* or master joiner, presumably as a direct result of his work on Vol. I. Thus his writing has an accuracy, detail and professional insight often lacking in encyclopedias painted with a broader brush. The *Biographie Universelle Ancien et Moderne* acknowledges this attribute in a comment that has a familiar ring to modern woodworkers: "Roubo offers the phenomenon, rarer in France than in the rest of Europe, of a workman who was distinguished in his field and never thought of giving it up for a higher profession."

Discovering Roubo's workbench was, for me, a rare treat—akin to unearthing in some long lost voyageur's journal a description of my favorite canoe route. Finding a reproduction of the bench was even better. Its owner and builder, timber framer and erstwhile medieval scholar, Rob Tarule, enjoys the bench at least as much for what it tells him about the 18th century as for what it enables him to do.

In the mid 1970s, while seeking a way to combine his interests in woodworking and history, Tarule was advised by a savvy friend to begin by making a workbench. For inspiration his friend produced a copy of Roubo. Tarule translated the workbench text and liked the scale and utility of the bench, as well as its appeal to his sense of appropriate technology. "I'm a lot more interested in things that make things [than in furniture]," he says. So, he decided to reproduce the bench. "In some way," Tarule adds, "I also made it as an antidote to Tage Frid's

bench [p. 95], which was so bloody complicated with its hundreds of moving pieces." He figured there must have been an easier way. I've measured his Roubo reproduction and found that it yields only a handful of vital statistics. "There aren't very many things to measure," Tarule observes. "That's one of the things I like about it. Four legs, four rails, twelve joints. It couldn't be any simpler."

The Roubo bench is so simple, though, that I couldn't help wondering if Tarule was making a virtue of necessity. After all, the workholding system is the guts of the workbench. Wouldn't the woodworker of the period have been the first to adopt more secure methods of holding the work if they had been available? In answer, Tarule refers to Roubo's plates—for the type of work shown, vises may not have been as efficient as stops and holdfasts. Roubo's holdfasts are one-piece iron bars, hand-forged in the shape of an *L.* The long leg, or shank, of the *L* is inserted in a hole drilled through the benchtop. The bent corner, or head, is struck with a mallet, securing the work beneath the pad at the end of the neck. In the process, the shank is wedged firmly across the hole in the bench, as shown in Roubo's cross-sectional engraving. The time spent engaging and disengaging the screw of a vise would have slowed down a joiner and might have cost him his job. One or two blows on the head of a good holdfast will hold a board securely in almost any position; one quick shot on the side of the shank frees it. And the long, straight stock of the architectural trade need only be pushed against a single stop to resist a planing stroke.

Making the Roubo bench is an exercise in small-scale timber-frame construction.

Making a bench in the 18th century

Jacques-André Roubo's L'art du Menuisier (Paris, 1769-1775) provides one of the most thorough descriptions of a workbench ever written. It belies the notion that early workbench (or furniture) construction was haphazard, and between the lines of Roubo's explicit instructions and engravings we may discover much about the way wood was worked in the 18th century. In preparation for building his reproduction of the bench, Rob Tarule translated Roubo's text (Vol. I, pp. 54-67), an excerpt of which appears below. Tarule notes that the pre-metric French inch (pouce) was 27.07mm and the foot, 12.44 in. Thus the modern equivalents of Roubo's measurements would be slightly larger than those described in the translation.

The Appropriate tools for the working of wood

The bench is the first and most necessary of the woodworker's tools; it consists of a top, four legs, four rails and a bottom. The top is made from a strong plank or slab 5" to 6" thick by 20" to 24" wide. The length varies from 6' to 12', but is most commonly 9'. The plank is elm or beech, but usually the latter, which is very firm and has a tighter grain. It should be drilled with a number of holes, into which are inserted the holdfasts. These holes should be...drilled quite perpendicularly. Their number is not strictly defined, but in general one should avoid more than are necessary....

The front legs are each drilled with three holes, into which are inserted the leg holdfasts. Flush with the edge of the bench and 4" or 5" from the bottom of the legs are joined four rails, 4" or more in depth and 2" in thickness. The bottom of the bench is filled by planks carried on ledgers (*f*, Fig. 4) attached to the rails. The length of the planks must be set with respect to the width of the top so as to provide more strength, as one can see in Fig. 1....

The height of the bench is generally 2½' but since all the workers are not of the same height, it suffices to say that the bench should not be higher than the thigh of the worker, because if it is too high, it deprives him of strength, and he runs the risk of becoming bent in a short time. One should take care also that the heart of the wood in the benchtop is up, for it is harder than the other [sapwood], and because if it becomes distorted, it will not move on that side, whereas it cups on the other.

The holdfasts are the iron tools employed to hold the work on the bench in a firm and stable manner. They are ordinarily 18" to 20" long or even 2' in overall length. Their thickness should be $1\frac{1}{16}$" to $1\frac{5}{16}$", and the curve of their hook 9" to 10" long and in the neighborhood of 6" in height. They must be of very strong iron forged from a single piece so that they are less likely to snap. All their strength needs to be at the front. It should also be observed that from the head *g* up to the pad *k* it becomes imperceptibly thinner in such fashion that their extremity is not much more than $\frac{3}{16}$" thick, for this renders it more elastic and facilitates compression. There should also be care taken shaping the bend so that when they are under tension, they do not pinch at the back of the pad, for if they should bend in the middle, they would pinch more, and would mar the work (Fig. 4)....

One compresses and holds the holdfast in the bench by striking on the head *g* with the mallet, and one releases by striking on the head in the opposite direction, that is to say on the side *i*, when removing, or better on the shank *l*. The holdfasts should not be polished, because they will not hold in the bench, which is remedied with a file. Only the pad should be smoothed so that it does not mar the work.

The holdfasts for the legs do not differ from the others, except that they are short. Their function is to hold the wood edgewise along the length of the bench, where it is held in a stable fashion with the help of the wooden hook *m* (Fig. 1). The hook is attached with screws or spikes in the edge of the benchtop, and is sometimes fitted with iron points; but since this mars the work, it is better to omit them, and to taper it [the hook] as pictured (Fig. 1).

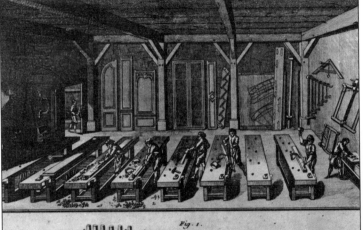

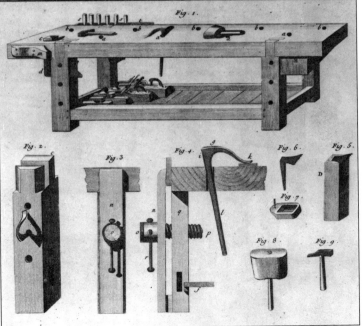

Interior view of an 18th-century joiner's workshop, from Jacques-André Roubo's L'art du Menuisier, Vol. I, plate 11. The hook, stop and holdfasts comprise an efficient workholding system.

Roubo, however, seems at odds with this explanation, at least in his discussion of the leg vise illustrated with the joiner's bench (Figs. 3 and 4 in the plate on p. 21). Introducing the vise as a cabinetmaker's accessory, which holds small work securely and without marring, Roubo is perplexed by the reluctance of his contemporaries to accept the leg vise. "I do not understand," he confesses, "why house carpenters do not adopt this method, which is not only very convenient, but at the same time causes no inconvenience, since one can remove the vise from the bench when one needs to."

One reason could have been the legendary conservatism of the building trades. It's arguable that, in the spirit of the Luddites, tradesmen who feared for their jobs might have resisted a more efficient system. (Sawmills in Great Britain were habitually destroyed throughout the 17th century by pit sawyers who felt threatened by the mills' encroachment, with the result that these establishments became entrenched on the Continent and in North America long before England.) Resistance may have had less inflammatory origins, too. By describing the leg vise as a removable option, Roubo suggests that workers were not yet accustomed to it, which indicates to me that the vise was a relative newcomer to Roubo's Paris. The leg vise shown in Vol. I lacks several improvements shown on leg vises in Vol. III, published five years later, indicating the speed at which vises were developing at the time (see sidebar below) and also, perhaps, the more widespread adoption of the vises.

While my observations are derived from the Roubo text, for Rob Tarule, the bench itself serves as a vehicle for what he calls "reverse detective work." By following Roubo's directions for making the bench and then figuring out how to use it efficiently, he has tried to reconstruct the working method of the 18th century. In the process, Tarule occasionally unearths an otherwise obscure nugget of information, which leads him to make his own educated guesses about the period.

The German cabinetmaker's bench

This so-called German bench, shown in Roubo's Vol. III, introduces several sophisticated variants, apparently unknown to Roubo when he wrote Vol. I. Most conspicuous is the addition of a tail vise to the right corner of the bench and the row of dogholes in the top. Detailed cross-sectional engravings of the tail vise also show a garter, the wooden key that is inserted in a groove turned in the head of the benchscrew to enable the vise to open as the screw is retracted. While a few holdfast holes remain in the top, the large square stop and iron hook are gone.

In place of the optional leg vise and the wooden hook that appeared on Roubo's earlier bench, there are two leg vises—one fixed to the left leg and the other movable along the length of the bench. Roubo explains that this arrangement offers a variety of positions for secure workholding, as well as for gluing. A horizontal beam has been added to the bottoms of both leg vises for accurate and convenient parallel adjustment of the jaws.

In his text, Roubo credits this bench as being of German origin "whether it was invented in Germany," he speculates, "or, as seems more likely, by the German cabinetmakers of whom there are a great number in Paris." In fact, there are examples of Swiss and Swedish benches, at least as sophisticated, that were built one or two centuries before Roubo's publications (see Chapter 1).

The fact that the details of this bench are not disclosed in Roubo's earlier volume, or in any other French or English encyclopedia of the period, indicates that the late 18th century may have been a critical period of transition for the workbench in France and England. Unfortunately, while

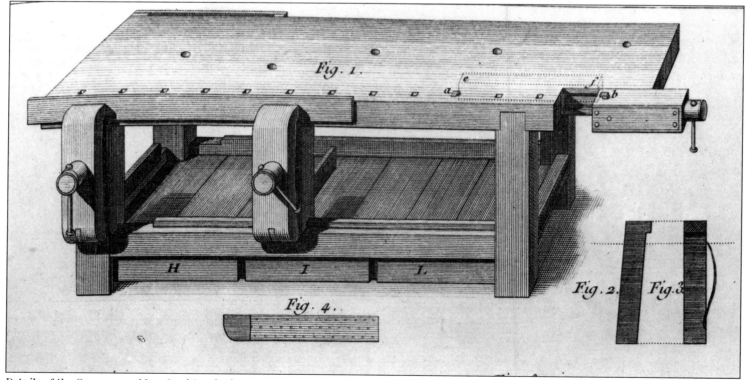

Details of the German workbench with tail-vise developments, from Roubo's Vol. III, plate 279.

Having decided to build a reproduction of the Roubo bench, Tarule first began searching for a top, described by Roubo as a single plank of hardwood, 5 in. to 6 in. thick, 20 in. to 24 in. wide and between 6 ft. and 12 ft. long. Not an easy task in 20th-century New England. "When I copy something," Tarule says, "I try to copy it as accurately as possible." But reality has a way of intervening and, by the time he was through, Tarule had departed from the original Roubo bench several times in building his interpretation. In the first place, nobody stocks dry wood that large, so Tarule knew he'd have to compromise on some of the basic dimensions. (Roubo mentions that the benchtop tends to cup until it is seasoned, which suggests the use of at least partially green wood.) After almost a year of picking over the lumber piles at local Vermont sawmills, Tarule found a mammoth maple plank, which he was able to dress down to a slab almost 5 in. thick, 18 in. wide and 98 in. long.

Roubo's illustrations and description of the tail-vise construction are typically thorough, he doesn't tell us much about its use, except to say that it is particularly handy for holding small work. As much may be gleaned from the root of the French word for cabinetmaker, *ébéniste*, for whom Roubo says this bench is intended. The word *ébène*, or ebony, suggests not only the preciousness of the material commonly employed, but also that it was available in relatively small sizes and applied in thin sheets of veneer. Such material would indeed have required specialized holding and gluing devices, not available on Roubo's original workbench. By contrast, the joiner's workbench would have been more than adequate for the long pieces of straight-grained, solid-wood stock, assembled with mechanical joinery. For these purposes, the simpler bench would have been faster and easier to use than any of its more complicated successors.

Roubo specifies either elm or beech for the top, but Tarule selected hard maple as a reasonable (and available) North American equivalent. Tarule explains that during the 18th century in Europe, elm and beech both grew to very large dimensions, as did oak. While oak might have provided a superior bench wood, he speculates that it would have been in greater demand for the barrel trade. Roubo specifies oak only for the legs and the stop—all require smaller chunks of timber, which may have been offcuts from other projects or less desirable parts of the tree.

Locating a board that size is hard enough, but Roubo goes on to specify that the heartwood should face up. (This made Tarule's task more difficult because the heart of a large maple is often of poor quality—black, punky and knotty—and usually has to be boxed out of the sawlog.) It is harder than the sapwood, Roubo explains, and if the plank warps, it will cup away from the heart. However, since the surface will be distorted on both sides—crowned if the heart is up and bowed if the heart is down—it would have to be flattened in either case. Tarule hypothesized that if the heart is placed up, when the top cups, it will bear more tightly on the shoulders of the leg joints. If the heart were placed down, it could rise off the shoulders. He also noted that it would have been traditional in Roubo's shop to clean and polish the benchtops periodically (and presumably resurface them, if necessary). Aside from enforcing neatness, this practice would have retarded the exchange of moisture between the top and the air, reducing movement. To slow the seasoning process of such a thick balk of wood and to minimize checking, you could wax, paint or otherwise coat the ends of the top until it reaches equilibrium with the moisture in your workshop.

After hand-planing the top flat, Tarule set it aside and turned his attention to other projects. Meanwhile, he says, "the bench was always over my shoulder waiting to be done." It re-emerged about two years later when Tarule became Curator of Mechanick Arts at Plimoth Plantation in Plymouth, Massachusetts. At Plimoth, a restored Pilgrim village of 1627, Tarule spent a lot of time in the woodworking shop "snuffling around in the past," figuring out how tools were used and how things went together. He needed a bench and decided to resurrect the Roubo.

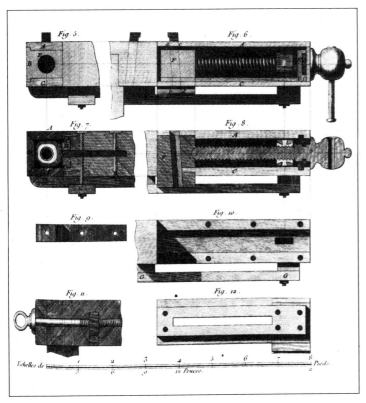

The 4½-in.-thick maple top lends a massive stability to the bench. Roubo specified that the heartwood face up for a tougher work surface.

Tarule/Roubo bench

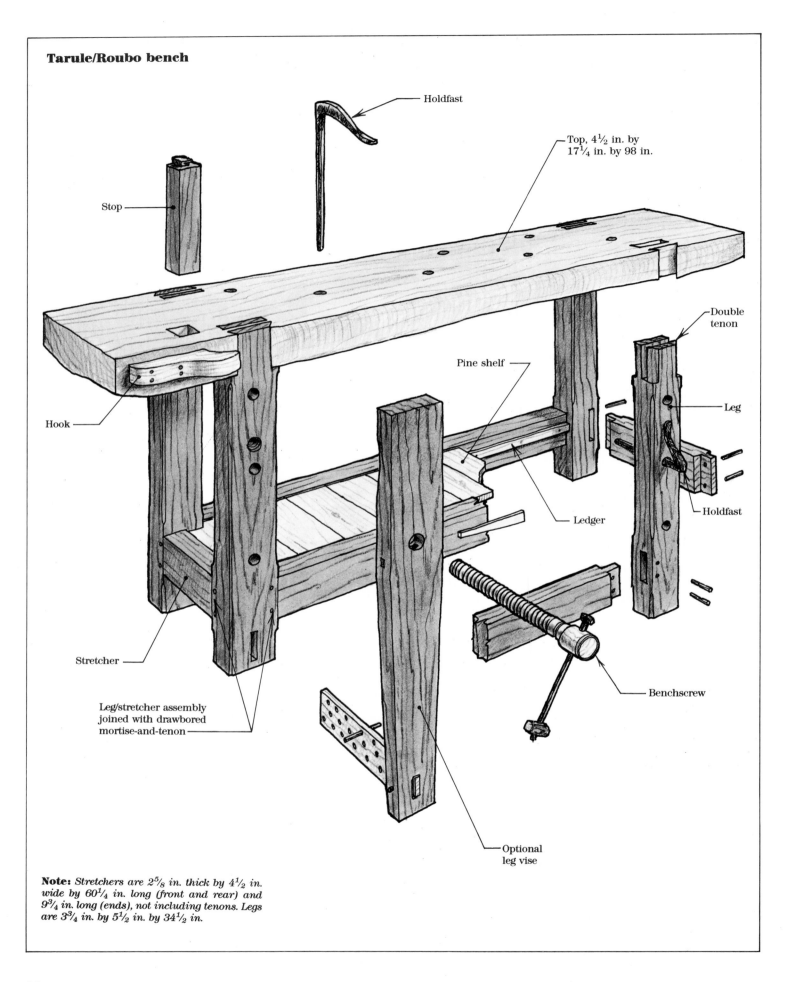

Holdfast

Top, 4½ in. by 17¼ in. by 98 in.

Stop

Double tenon

Pine shelf

Leg

Hook

Holdfast

Ledger

Stretcher

Benchscrew

Leg/stretcher assembly joined with drawbored mortise-and-tenon

Optional leg vise

Note: *Stretchers are 2⅝ in. thick by 4½ in. wide by 60¼ in. long (front and rear) and 9¾ in. long (ends), not including tenons. Legs are 3¾ in. by 5½ in. by 34½ in.*

"I treated the whole thing** as if it were a timber-framing problem," Tarule says, and, with its large members and mortise-and-tenon joinery, it is essentially that. With a few minor exceptions, all surface preparation and joinery was done by hand—he even sawed the ends of the top to length with a two-man crosscut to leave the appropriate tool marks. He left the waney edge and most of the original saw marks on the underside of the top to make the most of the large plank, flattening the underside only in the area of the joints. He cut the legs from scraps of 4x6 red-oak floor-joist material. Although the top was still relatively green (assuming one year of seasoning per inch of thickness), Tarule reasoned that this would allow the top to shrink and seat itself more tightly around the double tenon at the top of each leg. Roubo doesn't mention glue, so Tarule assumed that, as in timber-frame joinery, none was used. Unglued joints can be disassembled, a blessing for Tarule, who has had to move his bench several times.

Roubo's treatment of the double tenon is interesting in what it reveals about his role in recording the workbench material. Rather than drawing the tenon as he sees it, as a historian might, Roubo draws it as he thinks it ought to be. He faithfully reproduces the double tenon, with its front portion strengthened by a dovetail, but chooses to alter the common practice of making the back of the tenon flush with the back of the leg. Instead, he adds a shoulder there, reasoning that doing so enhances the joint's strength and distributes the weight of the top equally to the back and front of the leg.

In the same spirit, Tarule did not reinforce the double tenons with wedges as Roubo recommends. He planned to add them later if the legs loosened up, but wanted to be able to remove the legs to transport the bench. He fitted the joints very carefully and, in apparent vindication of his original theory, as the top dried (and warped about $\frac{1}{4}$ in. over its width), it seated itself more firmly on the legs. So firmly, in fact, that it took the better part of half an hour to drive the top off the four leg joints when Tarule recently moved the bench from Plimoth Plantation to its new home at Acushnet Sawmill near New Bedford, Massachusetts, where he currently works designing timber frames.

Besides being handsome and strong, the exposed end grain of the double tenons also gives Tarule a convenient hard surface that he can use as a small anvil to flatten hardware, or even—perish the thought—as a chopping block. "People did a lot more chopping then than they do now," Tarule says, and he is convinced that the ax was the unsung tool of the early carpenter. To demonstrate, Tarule shaves a peg with a hand ax, using the top of a leg to support it. It's obvious he does this either very carefully or very infrequently (I suspect the latter), because the benchtop shows none of the telltale signs of wayward ax cuts.

Tarule made his stretchers of maple and cut a full-width tenon on each end. The tenon layout was not specified by Roubo, presumably because such construction details were standard and would have been fully understood by craftsmen of the period. The leg-to-stretcher mortise-and-tenons are pinned with two, dry white-oak pegs driven into the marginally wetter red-oak legs—the legs will shrink tight around the pins as the drier pins absorb moisture from the legs. Tarule splits and whittles the pins to a roughly octagonal shape out of green stock to ensure straight grain. They are then set aside to dry, which happens very quickly because of their small size.

The double-tenon joint offers excellent support for the top. Tarule assembled it dry, without glue or wedges, so that the top can be pounded off the legs when the bench is transported.

The whole bench is gathered together like a timber frame with drawbored and pinned mortise-and-tenons.

For strength, it's critical that the tenon shoulders fit tight to the mortised leg. Shrinkage across the width of the leg will open a gap at the shoulder, so less wood between pin and shoulder means less potential shrinkage and a tighter joint. But if the pin is placed too close to the shoulder, you run the risk of weakening the mortise. Tarule placed the pins about ⅝ in. from the edge of the leg (you can safely make these as close as ½ in. from the edge). He drawbored the joint, an old technique whereby the corresponding holes in the tenons are off-set by about 3/32 in. toward the shoulder, as shown in the drawing below. Driving a slightly tapered pin through the holes in the assembled joint pulls the shoulders tight to the leg.

To complete the frame, Tarule chamfered the corners of the legs. ("Roubo didn't say anything about chamfers," Tarule says, "I just put them on.") The base is filled with short lengths of 1-in.-thick pine boards, which rest on ledgers nailed to the insides of the long stretchers. While Roubo's drawing shows this shelf near the middle of the stretchers, Tarule set his only about ⅛ in. below their tops. This enables him to rest one end of his planes on the stretcher, keeping their blades just off the shelf, and the shallow shelf is easier to keep free of chips and shavings than a deeper, four-sided trough would be.

Having scrupulously built the bench only with hand tools, Tarule admitted (somewhat sheepishly) to two lapses into more modern technology. Recently, he used a chain mortiser to excavate a 2¾-in.-square hole through the top for the bench stop, and while the bench was apart he resurfaced the top by running it through Acushnet's thickness planer. "I decided that after six years the wood was going to be hard," he explains. He figures that the top, which lost about ½ in. in width over that time, has finally reached equilibrium and should require only minimal resurfacing in the future. Because of the top's shrinkage and the stable construction of the base frame, we guessed that there would be a gap at the bottoms of the short-stretcher tenons where they enter the legs. Sure enough, we found these joints had all opened slightly, creating an A-frame-like structure. Tarule speculates that this angularity may also contribute to the overall rigidity of the bench. Like the individual joints in a typical timber-frame building, which are each only approximately tight, these work together to contribute to the structural integrity of the whole.

After several resurfacings, the top measures 4½ in. thick, 17¼ in. wide and 98 in. long. The bench is 34½ in. high—

several inches taller than Roubo's specified 31¾ in., or thigh-height work surface (31¾ in. is the modern equivalent of Roubo's pre-metric 2½-ft. measurement). To account for this discrepancy, Tarule points to the considerable variation in recommended bench heights among historians and practitioners. He agrees that a low bench allows for greater pressure when hand-planing, but he prefers a relatively high bench (Tarule is 5 ft. 8 in. tall and the benchtop falls a bit below his elbow), which allows him to plane without having to crouch. The power comes more from his legs than his upper body.

The stop and holdfasts are what make this heavy wooden table into a workbench. The 12-in.-long stop is made from a single chunk of white oak, dried in Tarule's microwave oven. It fits snugly in a square hole in the top and is adjusted, as explained by Roubo, by tapping on its top or bottom with a mallet. In the top of the block, Tarule installed a serrated iron hook (found in a flea market) similar to the one drawn by Roubo. Although Roubo is specific about the shape of the hook and its position in the block, Tarule has been trying different placements of the hook. If the teeth of the hook are allowed to protrude beyond the front edge of the block, as described by Roubo, the block cannot be hammered below the benchtop for an unobstructed work surface. Also, because of the size of the hook's square tang, it is easy to split out the front of the block if the hook is installed too close to that edge. The head of Tarule's hook is thicker than the one illustrated by Roubo, so he has found it convenient to install it in the middle of the block, allowing the head to protrude about ⅜ in. above the block. In this position, the iron hook cannot be engaged if the top of the stop is extended above the bench. This hasn't presented a problem for Tarule so far, however, because most work requires only a slight grip of the teeth at the bottom. Where it is necessary to extend the stop to hold large work, the wooden side of the stop itself is often perfectly adequate. As an alternative, Tarule plans to remount the hook near the front of the stop, as Roubo suggests, and excavate a matching pocket in the benchtop to allow him to tap the hook below the surface of the bench.

Although Roubo describes the placement of holdfasts in his original text, there is considerable variation in the hole patterns of the eight benches illustrated. Tarule bored a minimum of holes in the top and the front legs, enough to accommodate holdfasts in several convenient positions. It's easy to add more

Drawbored mortise-and-tenon

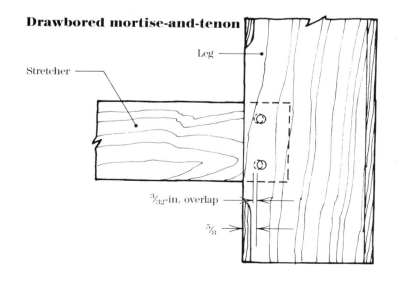

Stretcher

Leg

3/32-in. overlap

⅝

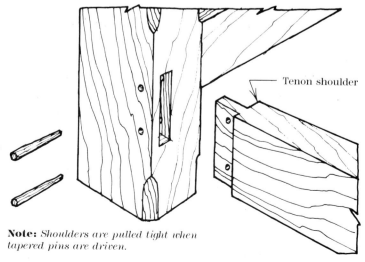

Tenon shoulder

Note: *Shoulders are pulled tight when tapered pins are driven.*

as they are required. After struggling unsuccessfully to get small, commercially made holdfasts ($\frac{5}{8}$-in.-dia. by 8-in.-long shank, 4-in.-long head) to grip, Tarule recently had a pair of hefty iron holdfasts custom-made according to Roubo's description. These are 20 in. and 15 in. long with a $1\frac{1}{16}$-in.-dia. octagonal shank, which holds securely in a $1\frac{1}{4}$-in.-dia. hole bored through the top or front leg of the bench. (The longer holdfast is used in the top, and the shorter one is used in the legs—at 15 in. long it is short enough not to interfere with the rear leg of the bench when it is set.) No hobbyist tools, these hand-forged holdfasts should outlast both Tarule and the bench (the pair was made by a local blacksmith and cost Tarule $130). They are so heavy that they almost set themselves, and they're not inclined to work their way loose with the vibration set up when chopping mortises, or planing.

Roubo's holdfasts are not only large, but they are forged out of a single piece of iron, as opposed to those used on the Dominy benches in Chapter 1. These have a right-angle bend between the shank and neck and are welded from three separate pieces. The elegant curve of the Roubo holdfast tapers from the neck to the pad, adding spring to its grip. While smaller, thinner holdfasts flex both in the shank and the neck, the heavier ones appear to spring only in the neck as the stout shaft is wedged diagonally in the hole. This is shown quite clearly in the drawing at right below, and illustrates how holdfast holes in an old benchtop can become as worn as sandblasted driftwood with repeated use. The long length of shank on Roubo's holdfast makes it possible to fasten thick stock to the bench, or to place a waste block below the work to bring it well above the bench surface. This positions the work at a convenient height for crosscutting while it keeps the sawblade clear of the benchtop, as demonstrated by the joiner in the center of Roubo's engraving.

Tarule discovered that his original commercial holdfasts had to be fitted to holes drilled to within $\frac{1}{16}$ in. of the diameter of their round shank in order to work. The heavier hand-forged holdfasts are much more accommodating of different-size holes. To be sure, however, Tarule drilled several test holes in a piece of scrap the same thickness as the benchtop before riddling the benchtop. (The so-called 'improved bench holdfasts,' which are tightened by a screw and require a steel collar in the benchtop, strike me as a diluted crossbreed of a holdfast and a clamp. They are much less flexible and not nearly as rapid in use as the traditional forged iron holdfast.)

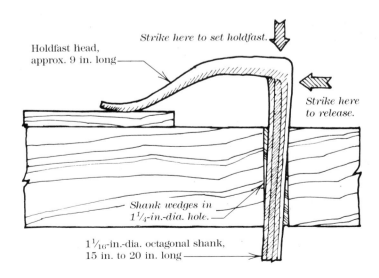

At top, the oak bench stop fits tightly in a matching mortise cut in the benchtop. The teeth of the iron hook protrude just enough to catch the end of a board. Above, Tarule's hand-forged holdfasts greatly enhance the versatility of the bench. The stout shank wedges in one of the holes drilled in the bench and the spring in the gooseneck holds work firmly.

Stops and holdfasts

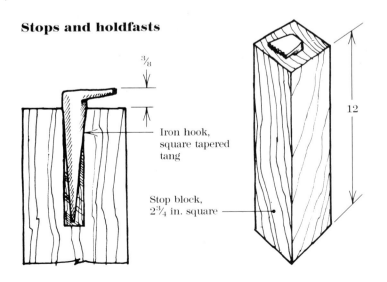

$\frac{3}{8}$

Iron hook, square tapered tang

Stop block, $2\frac{3}{4}$ in. square

12

Strike here to set holdfast.

Holdfast head, approx. 9 in. long

Strike here to release.

Shank wedges in $1\frac{1}{4}$-in.-dia. hole.

$1\frac{1}{16}$-in.-dia. octagonal shank, 15 in. to 20 in. long

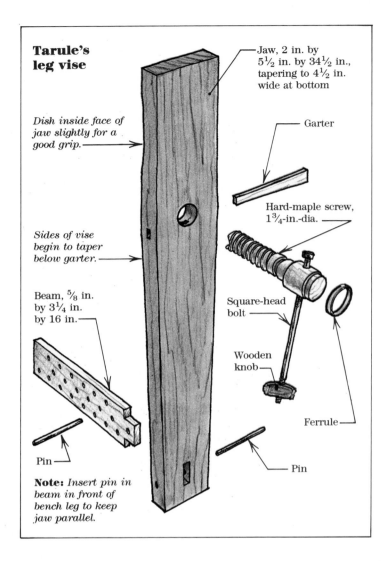

Tarule's leg vise

Jaw, 2 in. by 5½ in. by 34½ in., tapering to 4½ in. wide at bottom

Dish inside face of jaw slightly for a good grip.

Garter

Sides of vise begin to taper below garter.

Hard-maple screw, 1¾-in.-dia.

Beam, ⅝ in. by 3¼ in. by 16 in.

Square-head bolt

Wooden knob

Ferrule

Pin

Pin

Note: Insert pin in beam in front of bench leg to keep jaw parallel.

Tarule modified Roubo's leg vise when he built this one. The garter, which attaches the screw to the jaw, and the horizontal adjustment beam near the bottom make the vise much more effective.

A wooden hook screwed to the left front edge of the bench is used to trap one end of a board for edge-planing (see photo, p. 18). Tarule made his out of a piece of white oak and shaped it to accept stock up to 2 in. thick.

Tarule also built a modified version of Roubo's optional leg vise, shown in Figs. 3 and 4 on p. 21, adding some of the features commonly found on later, 19th-century leg vises. Where Roubo's original vise has no garter, Tarule added one. A slim, tapered oak wedge, the garter fits in a mortise in the vise's side and engages a groove turned below the head of the screw, allowing the jaw to retract as the screw is withdrawn. (In hindsight, Tarule considers his garter design to be "the long way around the barn." A simpler solution would be to bore a round hole in the side of the vise and tap a dowel into the groove.)

To keep the vise jaw parallel to the leg of the bench, Roubo recommends inserting a small block of wood the same size as the work at the bottom between the vise leg and bench leg. Tarule used blocks for years and found them adequate, but just the same, he eventually decided to install the horizontal beam shown in the drawing at left. To determine the hole placement, Tarule laid out a grid of three parallel lines along the length of the beam and then drilled ½-in.-dia. holes at 1½-in. intervals along each line. The holes are staggered by ½ in. on each line to provide maximum flexibility of adjustment. A pin is inserted through the appropriate hole in front of the leg to keep the bottom of the vise from moving toward the leg as the screw is cinched down on the work.

For the screw, Tarule turned a hard-maple cylinder on the lathe and borrowed a friend's screw box to cut the 1¾-in.-dia. threads. It was the largest woodthreading tap and die he could find, and Tarule figures it's plenty big enough. The 2¾-in.-long head could be shortened by an inch without losing anything, however, and you'd be less likely to bump into it. Tarule cut a piece of brass plumbing pipe for the ferrule, but a forge-welded, wrought-iron band would be more appropriate to the period. For the handle, or 'tommy bar,' Tarule used a square-head iron bolt, with a chunk of hardwood roughly twisted onto the threaded end.

As of this writing, the only accessories Tarule hasn't completed are the small drawer, the tool rack and the grease cup. Of the three, Tarule will probably make the grease cup next. He's curious "to see what I'd use it for." (Roubo says that grease is rubbed on the tools "to keep them smooth," and some modern woodworkers, such as Frank Klausz in Chapter 4, use it to lubricate the bottoms of their planes.)

I was surprised that Roubo makes no mention of a finish or, for that matter, of routine maintenance. It may be that, like the joinery details, such information was considered obvious at the time. Tarule prefers an all-purpose 'miracle' finish used at Plimoth Plantation—a mixture of beeswax, turpentine and boiled linseed oil. This homespun recipe calls for about 2 oz. of melted beeswax (roughly an egg-size chunk) cut with a pint of turpentine. Tarule notes that a little wax goes a long way, and too much will just sit on the surface. The linseed oil is added in equal measure to the combined beeswax and turpentine. Tarule applied a coat of this mixture every day for about a week, until the benchtop would absorb no more. You could also apply straight linseed oil first for better penetration, and then top it off with the full blend. The finish doesn't water-stain, it's easily buffed, and the nicest thing according to Tarule is that it's totally repairable—just brush on another coat.

If I've learned anything in the research of this book it is that, on its own, the workbench is only a table—"a top, four legs, four rails and a bottom," as Roubo's understated introduction explains. (In later chapters, I'll describe some workbenches that don't have any legs at all, and a few that don't really have a top.) Like any other creature, when it's removed from its normal context, whether to the dining room or the museum gallery, it loses something. See it in its natural habitat, however, in a working shop, with appropriate tools in the hands of a cunning craftsman, and you can begin to understand it.

On a hot Saturday afternoon in June, I asked Rob Tarule to show me how the bench was used. At one end of the cavernous 50-ft. by 150-ft. timber-frame layout shop at Acushnet Sawmill, Tarule has staked out his territory. He is bordered on three sides by an old lathe, a drafting table and his Roubo workbench. All around are dismembered timber frames, splayed about in varying states of construction.

"Part of the secret of this bench," Tarule began, "is just the mass of the top. The thing is so bloody heavy that it doesn't wobble...you can really pound on it." Although Tarule's bench can tip if he's rough-planing feverishly across the top, it's rock-steady when he works longitudinally, as most of the men in Roubo's engraving do. Tarule points out that the bench Roubo describes was 3 in. to 7 in. wider, which would have stabilized it considerably. The massive four-post leg construction concentrates the weight on four relatively small points. These are much less likely to skitter across the floor than would a sled-foot trestle base.

"When will they invent the vise?" Tarule mutters facetiously while planing against the stop. He continues in a more serious vein: "One of the things that strikes me is the absolute mini-mal amount of holding required." So much can be accomplished quickly and efficiently, he's discovered, using only gravity and friction to hold the work. As I watch Tarule work, the utility of the bench's simplicity is obvious. Wide, long planks can be face-planed securely along the length of the bench held only by the single stop. The weight and surface area of the board creates enough friction for Tarule to take a heavy planing cut near either edge, out of line with the stop, without having the board skid across the top.

Short boards are a different story. Plane strokes can occur only directly against the stop, not off to one side, unless the work is otherwise dogged or held fast. Tarule takes a few off-center swipes with a plane to demonstrate the point, and the stock spins around and off the bench (just missing me).

"Battens are a real big part of the holding system," Tarule says. To gain flexibility, or to plane diagonally across a board, Tarule fastens a batten to the bench with a holdfast. One quick shot with a hammer sets the holdfast. The end of the batten rides against one edge of the board, which is effectively wedged between it and the stop, as shown below. This holds the work securely and enables Tarule to work the entire board from one position. For quicker setups, the holdfast can also ride loosely in a hole, just touching one edge of the stock for a slight lateral support. When one side is done, Tarule flips the board over in an instant with one hand and resumes work on the other side. No need to put down the plane or crank out a tail-vise screw. Tarule's holdfasts are plenty strong enough to fasten work anywhere on the benchtop for chiseling or sawing. A batten can also be used with a holdfast on either end to hold a wide board flat on the bench, as for chopping dovetails. "I don't know what to tell people [about using holdfasts]," Tarule says. "It's limited only by cleverness."

When planing short boards against the stop, Tarule fixes a batten at the opposite corner of the board to hold it in place.

Most conventional modern workbenches have two main operational areas—the tail vise and the front vise. On Tarule's bench, the focal point is the left end of the bench, where the stop and hook and holdfasts convene. "This is where everything happens," he says. To plane the edge of a board, he simply jams one end in the wooden hook on the front edge of the bench. A short board is clamped to the left leg with a single holdfast, or supported by another board below it. A board that spans both front legs can be held in place with a holdfast on each end. While Tarule is busy adjusting boards and holdfasts, I notice an additional feature of the hand-forged, octagonal holdfasts—they sit upright in a waiting position without rolling over in the holes bored in the legs. A board can be positioned easily without his having to hold the holdfast at the same time.

The wooden hook is the one part of the bench Roubo doesn't describe in much detail or picture in action. Tarule wondered what sort of system would have been used to support a board that is too short to span both legs, but too long to get adequate support from just one holdfast on the left leg. He reckoned that a long batten could be attached with holdfasts to both legs, as shown in the photos on the facing page. This could then serve as a convenient platform for a board of almost any length. This arrangement has a lot of advantages. It provides even support over its entire length. Stock can be jammed in the hook for planing, flipped over or end-for-end quickly as it is worked and replaced with another piece of stock of any length—all without any adjustment of the support platform. Unless the stock is unusually warped, it will stand readily on the batten, with only one end wedged in the hook.

As Roubo predicted, Tarule discovered that the leg vise is a convenient accessory to the bench. While the jointing of long boards can be easily accomplished using the hook and holdfasts, short or irregular pieces of wood can be securely gripped in the jaw of the vise. The leg vise can also be used to clamp a board vertically when dovetails or tenons are being cut on one end. This cannot be done using the hook and holdfast combination because the work itself would cover the holdfast holes in the front leg. In lieu of the leg vise, this operation was probably done by clamping the work flat on the bench and working off one edge of the top, as shown on the third bench from the right in Roubo's engraving on p. 21. But it's much easier to cut a neat joint if you can see your layout marks. Tarule also finds the leg vise handy for clamping a plane upside down to shave 2-ft.-long strips of riven wood into thin, round stock, which he cuts short for pinning mortise-and-tenon joints. Short pins are almost impossible to hold, and long, thin strips bend too much to be shaved with a drawknife. By pushing the strips over the bottom of the upside-down plane, and cutting the tapered pegs off the ends, he can get the job done quickly. While he admits that there's no evidence that this method was used by joiners in Roubo's France, Tarule says, "I wouldn't be surprised but what the old-timers did it."

In addition to the stop, hook, holdfasts and vise, Tarule uses a bench hook, both for crosscutting small stock and as a shooting board. The bench hook, which I've found in use by woodworkers all over the country, seems a logical extension of the Roubo bench. A lot of work can be done using only body weight and simple mechanical advantage. (Another early style of bench hook, shown on p. 107, employs two narrow hooks to support both ends of long stock.)

The ease and speed of Tarule's benchwork is a convincing argument on behalf of the simple bench. But is this bench really appropriate for the modern woodworker, who works with a variety of materials and power tools as well as hand tools? I know of only a few woodworkers, such as Ian Kirby (Chapter 6), who reject the tail vise and prefer to work on a bench as simple as Roubo's. "I've done a lot of work on the bench," Tarule says. "For my purposes, it needs to be adaptable to a variety of methods." Tarule grew up using a turn-of-the-century Sears, Roebuck workbench, with a front vise and an end vise. Before he added the workholding devices to the Roubo bench, he kept a Record vise mounted on one end to handle all the miscellaneous small holding tasks of a modern workshop. Tarule unbolted it in honor of my research and discovered that he missed it on several occasions. But now that the stop and the wooden hook are in place and the leg vise is made, he is looking forward to getting to know the bench strictly on its own terms.

These days, inasmuch as Tarule is consumed by timber-frame projects all week, his bench investigation has become a passionate pastime. Ever since his days at Plimoth Plantation, Tarule has been fascinated by the early uses of green wood in woodworking. He's spent time experimenting with riven wood at the bench, and found that with simple tools and even simpler holding devices straight-grained, wet wood works much more easily than dry. How much benchwork (aside from the building of the bench itself) actually utilized green wood is a matter of conjecture. But it seems clear that whatever insights Tarule has already gained into the woodworking world of the 18th century, they won't be his last. "I see myself quite seriously tinkering on this kind of stuff—spending the next ten years figuring out how Roubo used the bench...." Tapping the benchtop, he adds, "If I didn't have to make money, I'd do it all the time."

Bench hook

Cleat

Benchtop

Note: *A bench hook is one of the quickest, most effective ways to secure a board for crosscutting. When you push the work against the top cleat with one hand, the bottom cleat is pushed firmly against the edge of the bench.*

Tarule discovered that a 2x4 batten clamped to the front legs with hold-fasts greatly improves the flexibility of the wooden hook. A board of any length can be rapidly chucked in the hook for edge-jointing while it is fully supported on the batten.

This 12-ft.-long, 19th-century workbench is located in the Tan House at Hancock Shaker Village, Pittsfield, Massachusetts.

19th Century: The Shaker Bench

Chapter 3

**I**n several centuries of new-world furnituremaking, perhaps no other style of furniture has become more identifiably 'American' than that of the Shakers. Their simple ladderback chairs, case pieces and nesting oval boxes are as distinctive and familiar as far more elaborate colonial furniture. The Shakers somehow stand apart from the mainstream of American woodworking while, simultaneously, they set a standard for its design and construction. Even the common dovetail joint and the turned wooden drawer-pull were employed with such skill and frequency by Shaker woodworkers that they have very nearly been usurped as Shaker trademarks. Indeed, as a measure of the success of these craftsmen, the very word Shaker has become synonymous with quality and simplicity, two of the founding tenets of the faith.

Not surprisingly, the Shaker workbench also stands apart from other benches of the period. This became apparent to me through some correspondence I had with a cabinetmaker in Stockholm, Sweden. After describing the European workbenches of his own experience, he added that the most beautiful workbench he had ever seen was at a Shaker museum in New England. It was similar, he said, to an old workbench he had known during his apprenticeship. Although his English was halting and his description vague, I felt certain that the bench he referred to was one of two I had seen at Hancock,

Massachusetts, and, a few miles away over the Taconic hills, at Mount Lebanon, New York. Both Hancock and Mount Lebanon (known as New Lebanon prior to 1861) are former Shaker communities, now museums.

By virtue of their size alone—the Hancock bench is almost 12 ft. long, the Mount Lebanon bench just over 15 ft.—they are both extraordinary. That much is obvious to someone like me who had previously considered a 7-ft. benchtop long. But there is something more special than that. Over 100 years old and in disrepair (the Lebanon bench is incomplete and had been carelessly used and partially cannibalized), these benches possess a certain presence. Every important detail seems intentional, not the haphazard whim of a rough-hewn rural craftsman. The builders of these benches cared. While I am not a Shaker scholar, it seems to me that these benches are distinctly Shaker, at once related to the makers of those delicate chairs and to the monumental built-in cupboards that line the cavernous Shaker dwellings.

Before examining the Shaker workbench, it is important to understand that the Shaker workshop—and everything within it—was a logical extension of the moral and spiritual order that was the touchstone of the faith. The Shakers, officially known as The United Society of Believers in Christ's Second Appearing, grew out of the turmoil and urban poverty of the

"That which has in itself the highest use possesses the greatest beauty."
—Unnamed Shaker

Industrial Revolution. It is an austere, fundamentalist brand of Christianity, with roots in 17th- and 18th-century French and English enthusiastic sects. (The name Shaker is a conjunction of 'Shaking Quaker,' denoting the frenetic 'dancing' worship for which the sect was well known.) The Society's founder and spiritual leader, Ann Lee, a millworker in Manchester, England, in the 18th century, was believed to have been the prophesied reincarnation of the Christ spirit come back to earth to lead her followers into the Millennium.

Acting upon divine revelation, Mother Ann led a loyal band of eight Believers to America in 1774. Preaching a radical gospel of celibacy, equality of the sexes, separation from the world and spiritual purity, Ann Lee launched an evangelizing tour of the Northeast. She faced great hardship in that first decade in America and died in 1784, but not before her efforts had begun to bear fruit. Riding a wave of religious revivalism at the end of the Revolution, working-class people joined the Shakers in droves. Within ten years after Ann Lee's death, ten communities had been established from Maine to New York.

At the height of its popularity, around 1840, the Society of Believers comprised an estimated four to six thousand members in eighteen principal communities and eight states, spreading as far west as Kentucky. The communities were organized into 'Orders,' or 'Families,' each supervised by two Elders and two Eldresses, appointed by other members of the ministry. The role of these four leaders combined parental and ministerial functions, while Deacons and Deaconesses in each Order were responsible for the conduct of the temporal concerns. The income provided by the community's activities—seed and herb sales, broom making, farming, chairmaking, etc.—was more than enough to support the Shakers' largely self-sufficient, Spartan lifestyle, with funds left over to acquire additional property. (The average land holdings of each Shaker community were estimated at about 2700 acres in 1875.)

Consistent with the religion's working-class origins, Shakers considered work a form of worship, their skills a divine gift. Mother Ann said, "Put your hands to work, and your hearts to God…" and the *Millennial Laws*, revised and circulated to the 18 communities in 1845, eventually prescribed in rigorous detail how that was to be accomplished. Every activity of Shaker life was delineated, from the burying of the dead to the folding of ones hands and the segregated climbing of stairs.

Father Joseph Meacham (the Shaker Elder who, along with Mother Lucy Wright, was responsible for molding the first Shaker institutions) helped establish early guidelines for manufacturing: "All work…ought to be faithfully and well done, but plain and without superfluity." Meacham meant *all* work. The Shakers did not apply different standards to furnishings made for workshops and those made for dwellings. It was thus established—almost in scripture—that the Shaker workbench would be as well made as a cupboard or chair.

Equally scrupulous in their relations with the world, brothers and sisters were told by the Millennial Laws not to "manufacture for sale, any article…which would have a tendency to feed the pride and vanity of man, or such as would not be admissible to use among themselves." (Chairs were the only furniture made for sale outside the communities.) Even items deemed acceptable to purchase from the world had to meet certain standards. The early Shakers expected their communities to survive and flourish through the Millennium, so they were not about to be seduced by discount merchandise or slipshod construction. Planned obsolescence would have been blasphemy.

While this attitude was abundantly expressed in the work of later Shaker craftsmen, it is very difficult to distinguish the earliest Shaker woodwork from that of their contemporaries. Many Shaker craftsmen came to the Society after having been apprenticed or established in cabinet shops in 'the world.' This was especially true at the close of the 18th century and in the

The Church Family buildings at Hancock Shaker Village were occupied from 1790 until 1960. In the foreground is the Sisters' Shop. Immediately behind it are the Round Stone Barn and the Tan House, which currently houses the workbench and cabinet shop described in this chapter. Photo by Linda Butler.

first quarter of the 19th, when the young communities were gathering converts. In later years, 'second-generation' woodworkers trained in the Shaker faith and the craft would have been better equipped to conform to the design standards of the community. In addition, when new converts arrived, they brought with them the furnishings of their former existence; what furniture they did require would, out of necessity, have been built quickly and simply.

The Shakers may not have originated the notion that 'form follows function,' but in their craft they unwittingly developed it to a fine art. Stylistically, mature Shaker furniture might be described as stripped-down country furniture of the Federal period. While their worldly counterparts experimented with reeding, beading and rope turning, Shaker cabinetmakers preferred clean, simple lines and delicate, attenuated proportion. They used solid native woods—mainly cherry, maple, birch, beech, pine, butternut and walnut—and studiously avoided imported mahoganies and exotic veneers and inlays. Again, the Millennial Laws were specific: "Beadings, mouldings and cornices, which are merely for fancy may not be made by Believers...." Veneering was considered an adulteration.

It is a common misconception that the Shakers did not paint or finish their work. (Perhaps our perceptions have been distorted by too many years of black-and-white publications.) One look at their multicolored oval boxes or carriers, and even their most utilitarian casework (workbenches included) demonstrates that the Shakers were capable of manipulating color. But even more than applied color, the Shakers were sensitive to the infinite variation in the color and pattern of natural wood. Their fondness for bird's-eye and tiger maple and the more subtle combinations of different woods belies their monochromatic reputation. (The cabinet beneath one workbench at the Shaker Museum in Old Chatham, New York, has walnut frames surrounding tiger-maple panels.)

The reverence that today attends the craftsmanship of the Shakers is in part justified by the consistency with which both quality and simplicity were applied to Shaker work. But only in part. It is also fostered by the mystery and irony that have enshrouded this most successful American Utopian adventure almost since its inception.

In a pragmatic sense, The United Society of Believers formed a bridge between the rural, agrarian society of the 18th century and the industrialized America that followed. It erected a protective envelope in which to launch experiments in mass production and to seek a new balance between hand and machine—all within the context of a medieval system of apprenticeship and the seclusion of walled villages. Any failures or problems the group encountered were diminished by their communal society, their sheer strength in numbers and their agricultural base, which would still put food on the table.

The Civil War and the urban industrialization that followed changed the face of America and perhaps terminally depleted the Shakers' reservoir of converts. The network of Shaker villages has all but disappeared, leaving only a handful of 20th-century Shakers in two communities (Canterbury, New Hampshire, and Sabbathday Lake, Maine), but the Society is admirably survived by the legacy of its hands. Unlike many worldly craftsmen and designers, the Shakers valued their work not for its own sake, but as a reflection of their religious commitment. It is no small irony that this spiritual, non-materialist society is best remembered for its products rather than its beliefs.

The workbench

As in a lot of other Shaker furniture, the distinctive features of a Shaker workbench are not always immediately obvious. As a utilitarian piece of equipment, the Shaker bench has to meet many of the same requirements as a worldly workbench. There is only so much room for variation and development before such a basic tool becomes over-specialized. Though the Shakers, like their contemporaries, distinguished between joiners or carpenters, who made architectural elements, and cabinetmakers, who made furniture and small goods, the workbenches of these craftsmen were probably quite similar. Chairmaking and boxmaking were separate industries with different workholding requirements. Shaker chairs were a production item, mainly comprised of interchangeable turned parts. Thus the lathe was the primary tool and workholding device. Chairs were clamped in a vise like the one shown below while their seats were woven. Shaker boxes were also mass-produced, and they were assembled on benches that were much smaller and less refined than the workbenches used for furnituremaking or joinery.

Sister Sarah Collins weaves a chair bottom at Mount Lebanon, New York, ca. 1930. Photo by William F. Winter, Jr.

The Shaker workbench, like others in the world, has many standard components: a tail vise and dogholes, a front vise, and room for tool storage beneath the top. Likewise, most of the same materials, hand tools and machinery available to the Shakers for workbench making were the same as those used by their worldly counterparts. As a result, similar woods may be found in both Shaker and non-Shaker benches, joined with the same mortise-and-tenon or dovetail joints.

It is unclear exactly when the Shakers began building workbenches. Perhaps a few were brought along when woodworkers joined the fold. (Gideon Turner, an early convert, became a member of New Lebanon in 1788 with "1 Set Carpenters tools & 1 Set Joiners Tools" valued at eight pounds.) Or, more likely, makeshift arrangements may have been employed until permanent workshops could be built and proper benches installed. In any case, journal entries and a couple of dated benches indicate that Shakers were building benches by the first or second quarter of the 19th century. This coincides with the period during which most Shaker furniture was built and the stylis-

tic features that distinguish it today were firmly entrenched. Although Shaker life and work became increasingly codified at the same time, no precise description of the 'proper' workbench or its appropriate usage has yet been discovered. (The idea that such a description might exist is not as farfetched as it sounds, considering that the Millennial Laws mandated: "Floors in dwelling houses, if stained at all, should be of a reddish yellow, and shop floors should be of a yellowish red.")

Since my first introduction to those two Shaker benches, I have looked at a dozen benches in other Shaker museums—Fruitlands in Harvard, Massachusetts, and the Shaker Museum in Old Chatham, New York—as well as a few in private collections. While these represent only a fraction of the total number of Shaker workbenches that must have been made (every Shaker family had a woodworking shop, and the large families, such as the New Lebanon Church Family, had both a joiner's and a cabinetmaker's shop), certain patterns begin to emerge.

I chose to focus my attention on the Shaker workbench at Hancock Shaker Village, shown on p. 32, for several reasons. It

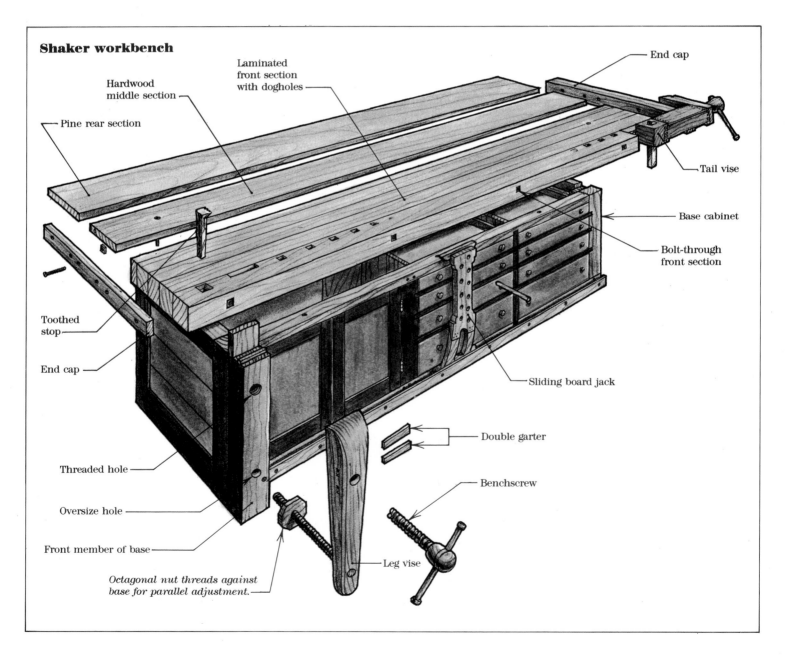

Shaker workbench

Pine rear section

Hardwood middle section

Laminated front section with dogholes

End cap

Tail vise

Base cabinet

Bolt-through front section

Toothed stop

End cap

Sliding board jack

Threaded hole

Double garter

Oversize hole

Benchscrew

Front member of base

Octagonal nut threads against base for parallel adjustment.

Leg vise

is well made and in good condition and does not appear to have been materially altered. In its dimensions and construction, it is as fine an example of a Shaker bench as any I have seen. And it is the only such bench I am aware of that remains in everyday use in a working, Shaker-style cabinet shop, albeit in an interpretive museum. I will describe details of other Shaker benches I have seen as they differ from the Hancock bench or further an understanding of it.

As my first impression suggested, Shaker benches tend to be massive. The Hancock benchtop is 11 ft. 9 in. long and 38 in. wide. The main body of the top is 3¾ in. thick. The smallest Shaker bench I found (at Fruitlands) is *only* 8 ft. 1 in. long. The largest (at Old Chatham) is 16 ft. 7 in. Most of the others are between 12 ft. and 15 ft. long. Indeed, it would seem that a small Shaker bench would be anything under 10 ft. long—several feet longer than what would be considered a large workbench today. (This may not have been unusual at the time, given the 18th-century Dominy workbenches [p. 13] and the French workbenches described by Roubo [p. 21].)

The top of the Hancock bench is comprised of three separate sections (as shown in the drawing on the facing page), built stoutly and purposefully. The front section is 16 in. wide and laminated from four pieces of 3¾-in.-wide maple or birch and a 1-in. strip of pine, glued and bolted together with four hand-forged bolts. (The 3¾-in.-square laminates would have been convenient to work with.) This area houses the dogholes and vises, and functions as the primary worksurface; maple or birch was used on this part of the bench, as it was on all the others I've seen. (Due to the age and patina of the bench, it is often difficult to determine the exact species of wood used; the woods I describe should be considered 'educated guesses.')

The midsection of the top is a single chunk of 9¼-in.-wide chestnut or oak. Although hard and dense, the open-grained wood provides a rougher benchtop texture than that of the front portion, and was presumably acceptable for a secondary worksurface. The 12¾-in.-wide back section of the top is made of knotty, hard pine. Both the middle and back sections are 1¾ in. thick, supported by spacers that rest on the base frame. Both ends are covered by simple, bolt-on end caps with captured nuts fed from the underside of the top. No tongue-and-groove or splined joints were used to attach the end caps. They were merely intended to conceal the end grain on the benchtop and, in the case of the end cap on the right end of the bench, to serve as the nut for the tail-vise benchscrew.

The very size of the enormous top offers some interesting clues to Shaker woodworking. "It's never big enough," according to Joel Seaman, the cabinetmaker who has been making restoration Shaker furniture on the Hancock bench for over ten years. Seaman could lay out all the parts of a cabinet on the top and still have room to use the vises.

The order and cleanliness of the Shakers is legendary, however, and it's unlikely that the benches were built large to accommodate such expansive work habits. (Even the woodshed and tool room of a Shaker brother in Union Village, Ohio, was impeccably organized: "...every stick of wood was exact in its place.... His little work shop exhibited the same care.") In part, bench size may be explained by the institutional nature of the Shaker dwellings and the size of the joinery and furnishings required for them. In every community these buildings are imposing structures, with high ceilings and wide hallways. As shown in the photo below, some of the most remarkable case pieces stand over 8 ft. tall; built-in cupboards, housing dozens of drawers and cabinets, may run floor-to-ceil-

Members of the Shaker Village, Canterbury, New Hampshire, stored their off-season clothing in the more than 80 drawers and seven walk-in closets that line the walls of the 'New Attic' of the Dwelling House. Photo by Linda Butler.

Freegift Wells

The following notes are from the journal of Freegift Wells (Western Reserve Historical Collection, V:B-296), a respected Elder of Watervliet, New York. This excerpt begins in 1857, when Wells was 72 years old.

1857

may 20   It is altogether probable that this will remain my permanent workshop, while I am capable of performing hand labor. I have fixed an accommodation on the front of my bench for holding boards & the like for jointing.

jul 22   Went to Albany, bought a small bench vice & other tools. The vice cost $2.50.

nov 28   Been hewing out a couple of long screws, fixing up the lathe & ruff turning them.

dec 4   Been fileing up...our large screw auger which I made for cutting plates for vice screws perhaps 35 years ago, & I do not believe it has ever been filed up since till today....It cuts for 1¾ inch screws or nearly.

dec 30   I have finished my vice & lathe—So that I now have a vice bench, vice & lathe.

1858

jan 1   ...Today I have made a pair of clams to fit the vice, for the purpose of holding saws to file &c...

jan 9   ...Oiled the new vice & vice bench also the lathe bench which is going to the mill.

mar 11   Dressed out stuff for a drawer to go under my vice bench—dove-tailed it & glued it together...

mar 17   Been making racks, or fixtures in one of the bench draws for keeping small files.

apr 6   Been fixing my small iron vice...to screw on to the side or end of my bench.

apr 13   This morning the Elders gave me little Thomas Almond for an apprentice, & a fine boy he is to (he was born July 25th 1847). Made him a bench to stand on & set him to turning at my little lathe.

jul 29   Altered over some fly nets & did other necessary chours, such as to learn Thomas to make mortises & saw tenons &c. With my instructions he framed 4 sticks together, which looked quite workman like.

1859

june 23   ...Turned & fitted in a screw for Thomas's end of the bench.

jul 27   Repaired my vice at the head of my bench which had been out of order for a considerable time.

1860

dec 17   Began to repair my work bench, plained off a part of the top...

dec 18   Plained off the iron plate & a part of the hook that composes my vice at the head of my bench.

dec 20   Finished dressing up my hook...so that it will pinch a piece of paper all around.

dec 29   Cut my screwplates in the south end of the bench for the purpose of forming a vice on the end to hold boards for marking out dovetails &c.

1861

jan 5   Been making a jaw for a hold fast on my bench.

jan 8   Laid out 9 mortices on the front side of my bench bored them & morticed out 5.

jan 9   Finished my mortices & made a dog, or hook, to go in them...

jan 14   Made a beginning at Thomas's bench vice, plained & bored the hook shaft &c.

jan 15   Worked out the big hole for the screw & bored the 4 holes for bolts & burned them out with a hot iron.

jan 25   Finished my bench to day, got in the screw all complete, so that Thomas has now got a complete head screw to his end of the bench.... He is 13 years & 6 month's old to day.

1865

mar 4   The care of the Saw mill is given up to Thomas Almond...he is a smart fellow, & I hope he will always do well, & honor his privilege by faithfully bearing his cross to the end of his days...my blessing will always remain with him, & he will receive a rich reward for all his labors. What have I been writing! Likely as not he will get a peep at it some time, surely I hope it will not do him any hurt.

On March 1, 1867, Thomas Almond (21 years old) eloped with one Ada Woods and left the Shakers. Wells died on April 15, 1871 in Watervliet, New York, a month before his 86th birthday.

ing and the length of a long hallway. All this work, plus the miles of pegboard circumnavigating the rooms, would have been more easily hand-planed and joined on a long bench. While there was some specialization among Shaker woodworkers, records indicate that a typical woodworker's week would have been spent in a wide variety of pursuits. As the communities stabilized and eventually began to shrink, there would have been less new furniture (apart from chairs for sale) to build. At the same time, fewer craftsmen would have had to perform an even more varied range of tasks.

There is also reason to believe that more than one person worked at the bench at a time. Entries from the journals of Freegift Wells, an Elder and woodworker of considerable stature from Watervliet, New York, depict what was probably a typical relationship between a cabinetmaker and his apprentice. In these notes, excerpted at left, Wells tells us that he installed a vise at the opposite end of his own workbench for his apprentice, Thomas Almond. There are also frequent references in other Shaker letters and journals to projects undertaken by two or more craftsmen working together.

Without exception, all the Shaker benches I've seen have an enclosed base, which contributes substantial mass and storage space, while it restricts any clamping to the ends or the narrow overhang along the front edge of the top. One thing I have never seen on a Shaker bench, but which is common on other benches out in the world, is an open tool tray. This tray, whether built into the top or between the stretchers of the base, collects debris and allows tools to knock about, damaging their edges. To an early Shaker, an open tray would have seemed like an open sewer—seductively convenient, perhaps, but unsanitary and hazardous.

Mother Ann could have been lecturing her woodworking followers when she said: "...take good care of what you have. Provide places for your things, so that you may know where to find them at any time, day or by night...." Just as the walls of the Shakers' dormitories are lined with built-in cupboards, so their workbenches are equipped with substantial cabinets that fully occupy the area between the legs and beneath the top. They are also unique in that the drawers and cabinets are usually built *into* the base framework, a tedious and exacting process. It would have been much easier to support the top with a basic four-leg structure and to install an independent tool-cabinet carcase between them. (Norm Vandal chose this approach when he built the Shaker-style bench shown on p. 43.) In the case of the Shaker workbenches I have seen, the members of the carcase itself—posts, drawer dividers and the frame-and-panel ends—generally function as the legs and stretchers of the workbench. This may have been preferred for aesthetic reasons, or simply to lend continuous support to such a large worksurface.

On the Hancock bench, like most of the others, the base is divided into a succession of drawers that progress in size from the smallest on the top to the largest on the bottom. A portion of the base consists of open shelves, which are reserved for storage of items that won't fit in the drawers (large tools or specially prepared stock, perhaps). These areas are always enclosed by doors. The insides of the door panels on the Hancock bench display remnants of different-color paint, indicating that they were borrowed from some other project and reincarnated in the workbench.

The order and cleanliness provided by the enclosed base cabinet had many practical dividends for the workbench. The problems of racking and sliding, which are inherent in an open-frame base, are automatically resolved by the rigidity of the casework and the sheer weight of the structure. Loaded with tools, as it presumably was, the cabinet anchored the whole bench to the floor and to move it would have taken a small army. Workbench storage would have made it easier to keep track of tools in a large community. "No one should take tools, belonging in charge of others, without obtaining liberty for the same...," the Millennial Laws decreed. "The wicked borrow and never return."

If I had any doubts about the ability of the drawers to carry tools, these were quickly dispelled in examining the Hancock workbench. Joel Seaman keeps the largest of the lower drawers of the bench (32 in. wide by 8¼ in. high) loaded with his collection of 54 wooden molding planes. It is so heavy that it takes two people to lift, yet slides smoothly without sticking on the runners in the carcase. These have become tracked with deep grooves over the more than a hundred years of the

The enclosed base of the Hancock bench, fitted with drawers and cupboards, is one of the hallmarks of a Shaker workbench.

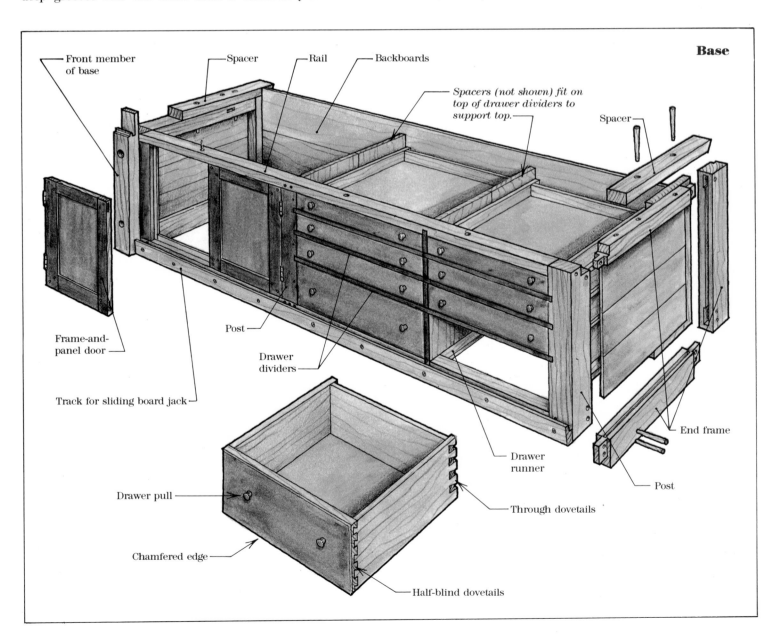

Base

Front member of base — Spacer — Rail — Backboards

Spacers (not shown) fit on top of drawer dividers to support top.

Spacer

Frame-and-panel door

Post

Drawer dividers

Track for sliding board jack

Drawer runner

End frame

Post

Drawer pull

Through dovetails

Chamfered edge

Half-blind dovetails

Despite the grooves worn in the runners and dividers of the carcase, the heavy drawers still slide smoothly.

bench's life. The upper drawers, which are only 3¾ in. high, are perfect for keeping smaller tools safe, well-organized and readily accessible.

Although I have never found any dividers (or signs of there having been any) inside the drawers of a Shaker workbench, these would have been in keeping with the Shakers' strict sense of organization. The few wall-hung tool chests that I have seen are a rabbit warren of compartments and notched racks designed to hold specific tools. Fastidious half-blind dovetails on the drawer fronts (through dovetails on the drawer backs) and mushroom-shaped turned pulls are two typical hallmarks of a Shaker drawer. The Shakers felt so strongly about superfluity that on July 4, 1840, a New Lebanon Ministry sister's journal noted: "David Rowley [master cabinetmaker] has been employed for several days in taking out Brass knobs, and putting in their stead wood knobs or buttons. This is because brass ones are considered superfluous, thro spiritual communication."

While the drawer fronts of many Shaker cabinets overlap the carcase with a rabbeted stop, all of the drawer fronts I have found on these workbenches fit flush within the frame. A flush-fit drawer front must be carefully built into the frame to avoid any unsightly gaps, but it was probably preferred to the rabbeted drawer front because it is much simpler to make. Some of the drawers, such as the ones in the bench at the Fruitlands Museums, shown below, are exceptionally well crafted. Others, such as the ones in the Hancock bench at left, have had the edges of their fronts chamfered slightly, perhaps to allow the maker some leeway in their alignment.

This bench at Fruitlands Museums in Harvard, Massachusetts, is an extraordinary example of Shaker craftsmanship. The drawers are fitted flush with the front of the carcase.

In the absence of a rabbeted drawer front, the Hancock drawers are stopped directly against the backboards of the bench by projecting drawer-side dovetails, as shown at right. Their length would have been easily trimmed when the drawer was installed. (On the Hancock bench, it also allowed the maker to leave the overflow of crusty hide glue on the back of the drawer, without this interfering with its fit.) Otherwise, the drawers are typical of the period. The undersides of the drawer bottoms are planed to fit in grooves routed in the drawer sides and front; a few nails in the back hold the drawer bottoms in position. Over the years several of the grooves have split out, and these have been reinforced with nails.

Tracing the provenance of a Shaker workbench can be every bit as troublesome as tracing a Shaker chest. Benches changed hands frequently, were often made by one maker for another and were sometimes built by several men working together. These factors, in combination with the Shakers' philosophy of shared property, would lead one to believe that the workbench was just another piece of communal chattel. To a certain extent that may have been true. But in their journals, Shaker woodworkers often differentiate between "my workbench" and benches made for other craftsmen. Freegift Wells even had his bench shipped home to Watervliet, New York, when he returned from a temporary posting in Union Village, Ohio.

As might be expected, Shaker furniture was not commonly signed, and the workbench is no exception. The only examples extant are a dated workbench in a private collection and the workbench at Mount Lebanon, which has been signed and dated on the undersides of several drawers: "Moved to the Brick Shop Feb 9 1871 by the Maker. February 1853 Orren N. Haskins Maker." In lieu of a signature, a peculiar method of construction can sometimes provide a clue to the origin of a workbench. The workbench at the Fruitlands Museums, for example, has pine drawer sides tapered in thickness from $\frac{3}{8}$ in. at the top to $\frac{1}{2}$ in. at the bottom, as shown at far right. The drawer itself is squared up, so it fits tightly in the carcase; the inside faces of the drawer sides slope. It's possible this curious detail enabled the maker to lighten the drawers slightly without compromising the strength of the bottom groove, or they might be the result of splitting quartered stock from a billet. Regardless of their purpose, tapered drawer sides are a clue to the benchmaker's identity, as this feature is known to have been employed by several cabinetmakers of the Hancock Bishopric, which also included Tyringham, Massachusetts, and Enfield, Connecticut.

The cabinet of the Fruitlands bench has one other unique and practical feature—a lidded drawer that provides a clean surface on top, and dust-free storage within. Like the rest of the bench, this detail is neatly executed. The upper edges of the drawer sides are beveled, to leave room for one's fingers to lift the hinged cover. I can only surmise that the drawer may have been used as a writing surface and perhaps to store drawings or drafting tools. The only other similar feature I have discovered on a Shaker bench is a simple pull-out board, located directly below the top of the 16-ft. workbench at the Shaker Museum in Old Chatham, New York.

With one exception, the drawers on all the Shaker benches I have seen open from the front only, suggesting that the benches were commonly positioned against a wall. (Orren Haskins' bench at Mount Lebanon has seven drawers on the back side, indicating that it must have been freestanding.) The back of the Hancock bench is sheathed with three $\frac{3}{4}$-in.-thick pine boards. These are shiplapped at their joints and fastened with cut nails to the end frames of the base. Although the bench currently stands in the middle of the workshop floor, the 'unweathered' condition of the backboards implies that it spent most of its life against the wall.

By placing the workbench against a wall, the Shaker woodworker was able to hang large tools, such as saws—or even complete tool cabinets—directly behind the bench within easy reach. The critical work areas of the bench—the vises—could be positioned under windows to improve visibility in an otherwise dark workshop. Furthermore, with the benches against the wall, the shop would have been less crowded, and the coveted Shaker order and cleanliness more easily maintained. (As was the custom in Federal-period homes, furniture lined the perimeter of a room and was brought out only when it was required. The Shakers took this one step further in their dwellings by actually hanging chairs and other furnishings on the wall, or going to great lengths to build in their cabinets.)

The dovetails on the ends of the Hancock drawer sides (above left) extend beyond the drawer back, stopping the drawer against the backboards of the bench. The sides on the Fruitlands drawer (above right) have been tapered—perhaps to lighten them without sacrificing strength in the bottom rabbet.

A hinged lid has been fitted to one of the top drawers of the Fruitlands bench. The lid protects the contents from dust and the drawer is at the right height to provide a convenient writing surface for a craftsman seated on a stool beside it.

The Shakers may have been conservative in their religious and moral practices, but they were quick to adopt new technology where it proved expedient. (Early Shaker craftsmen frequently used 'buzzsaws' and planers in their shops, run off leather-belted line shafts and water-powered turbines.) Accordingly, every Shaker workbench I have seen includes a well-built tail vise, a feature that probably had not been in common use for much more than a couple of decades before 1800.

The tail vise on the Hancock bench is typical. It is neatly dovetailed, houses a single benchdog in the front of the vise, and is operated by a large 2-in.-dia. beech screw with a turned hickory handle. The nut is tapped in the end cap of the bench itself, which extends into the cavity of the vise. A ½-in.-thick maple top cap covers the screw cavity and is pinned into a rabbet at both ends with seven small pegs. One Shaker bench I've seen in a private collection has a removable top cap; its ends are cut at an angle so that it can be slid out from the front of the vise to ease lubrication of the screw. (The top cap on Norm Vandal's Shaker-style bench on the facing page is screwed down.)

The right end cap of the bench extends into the tail vise and is threaded for the benchscrew. A heavy guide bar slides through a notch in the end cap and along the underside of the bench to add stability. The rear jaw is fitted with a garter, which slides up through a tight-fitting mortise to engage a groove turned below the head of the benchscrew. Photo by Richard Starr.

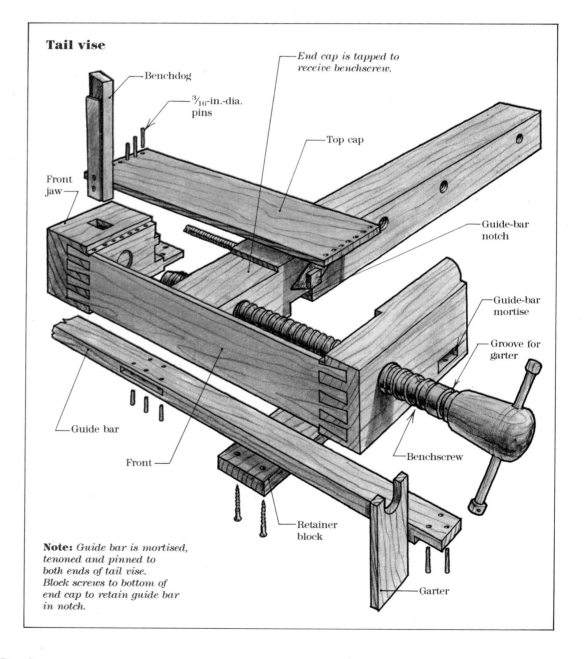

Tail vise

Benchdog

End cap is tapped to receive benchscrew.

³⁄₁₆-in.-dia. pins

Top cap

Front jaw

Guide-bar notch

Guide-bar mortise

Groove for garter

Guide bar

Benchscrew

Front

Retainer block

Garter

Note: *Guide bar is mortised, tenoned and pinned to both ends of tail vise. Block screws to bottom of end cap to retain guide bar in notch.*

The Hancock tail vise is of typical Shaker construction: neat dovetails, a well-turned beech benchscrew and an iron benchdog. Photo by Richard Starr.

A Shaker-inspired bench

"I wanted a bench that had storage space (and would keep dust off the tools) but that also looked decent," Norm Vandal says. "I didn't want something modern." Inspired by the bench at Hancock Shaker Village, Vandal decided to incorporate many of its features into his own workbench, but he amended them to suit his needs. He built an enclosed cabinet below the top, as on the Hancock bench, and he installed a tail vise and leg vise at opposite ends. But he also shrank the bench considerably. At 29 in. wide by 7 ft. 10 in. long, the bench is somewhat of a midget by Shaker standards, but then Norm Vandal is a reproduction furnituremaker in Vermont, not a Shaker. Vandal spread the construction of his Shaker-style workbench over several months, finishing it in the fall of 1985. If he had started from scratch and worked flat out, he figures that it would have taken about 3½ weeks to build.

Norm Vandal adapted the main elements of the Hancock Shaker bench when he built his own scaled-down version.

The Hancock benchtop has 23 dogholes, spaced approximately on 5-in. centers and angled toward the tail vise. These holes are cut to fit large, hand-forged iron benchdogs—1 in. by $\frac{7}{8}$ in. by $11\frac{1}{4}$ in. long, with a $\frac{1}{8}$-in.-thick spring pinned at the bottom. A shallow shelf chopped at the top of each slot receives the larger head of the dog and prevents its dropping through the benchtop.

Like many other woodworkers of the period, the Shakers seem to have preferred metal benchdogs to wood. According to Joel Seaman, metal dogs on the Hancock bench are able to grip a thin piece of wood firmly with only the top $\frac{1}{8}$ in. of the dog protruding. After more than a century of use, the springs on the Hancock dogs continue to work well and the dogs can be positioned at almost any height without slipping. Not surprisingly, Seaman estimates that he uses the dogs nearest the tail vise about five times as often as the rest of them, although the farthest dog (9 ft. away from the tail vise) is indispensable when it comes to working large case pieces or long architectural moldings.

The dogholes were undoubtedly bored first and then chopped square, either by hand or on a mortising machine. (From journal references, we know that the Shakers built and used mortising machines in their production work; these mortises would have required a tilted table to produce the angled doghole slots in the benchtop.) Near the left end of the Hancock bench, another mortise has been cut for the shaft of a toothed stop. This store-bought stop (Taylor patent 1846) is particularly interesting, as it appears to be a transitional development between the single stop used on earlier benches (such as the Roubo bench described in Chapter 2) and modern benchdogs, which are used in conjunction with a tail vise. The teeth grip the end grain of a board sufficiently to hold it for planing without requiring any clamping at the other end. For production planing, the stop would be very quick to use. Seaman objects to the marks left on the end of his work, however, and notes: "I'm a little too compulsive." So he uses it only occasionally where the tooth marks won't matter, as shown in the photo at far right on the facing page.

The cabinet below the bench measures 26 in. by 63 in. and was built independently, before the top and legs were installed. The pine carcase was assembled like a kitchen cabinet—as a separate unit, with a $\frac{3}{4}$-in.-plywood bottom resting on 2x4 braces on the floor. The back and ends of the cabinet have raised panels with a thumbnail molding run around the insides of the 1-in.-thick frames. (Vandal wasn't sure when he built the bench that it would end up against the wall, so he paneled the back.) The legs were erected around the finished case and lag-screwed from the inside, adding weight and rigidity to the structure. The rails of the side and back panels serve as stretchers between the bench legs. In addition, a maple stretcher is mortised into the front legs to carry the sliding board jack. The benchtop was then dropped down and fitted to tenons on the tops of three legs. (The left front leg extends above the top for the vise.) The back legs fit in enlarged tenons to allow for movement in the top.

Vandal chose this approach over the common Shaker practice of using the cabinet alone, without legs, for the base. It seemed an easy way to keep the whole unit square without relying too much on the carcase to carry the top's weight. It was also easier to cut the mortises on the underside of the top to match the tenons on the legs, rather than try to fit the carcase within a preconstructed frame.

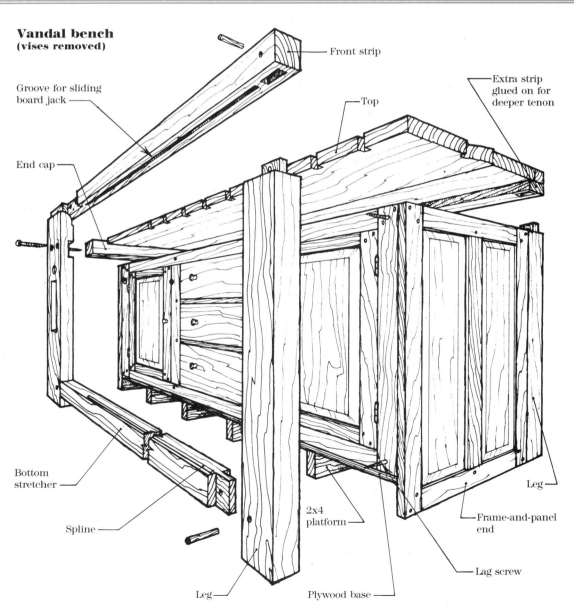

Vandal bench
(vises removed)

Groove for sliding board jack

End cap

Front strip

Extra strip glued on for deeper tenon

Top

Bottom stretcher

Spline

2x4 platform

Leg

Plywood base

Frame-and-panel end

Lag screw

Leg

Leg

The dogholes on the Hancock bench are designed for a heavy, 11¼-in.-long, wrought-iron dog. The dog's cross-hatched face helps grip the work, and a spring is riveted to one side so that the dog will maintain its position at any height.

The toothed stop is used for planing without the tail vise. Work is simply pushed against the stop; pressure against the serrated edge holds it in position.

The workbench resides against the wall in the far corner of Vandal's tidy shop under the warm glow of a bank of south-facing windows. The bulk of the top is made of two maple boards (about 10 in. and 15 in. wide), dressed down to about 1½-in. thickness and fitted with a long, sliding tenon to the end caps. A 2½-in.-wide by 3-in.-thick front strip covers the dogholes cut in the top. All maple parts are coated with a pigmented oil stain (to add about a century to the bench's appearance) and finished with tung oil.

The benchdogs are quite small—only ¾-in.-square oak pegs—spaced 6 in. apart, and the bottoms of the dogholes are blocked by the cabinets below. Vandal tried to align them as closely as possible with the screw in the tail vise and didn't want them to drop through (there are no shoulders in the slots, or springs on the dogs). As a result, he has to blast the holes with compressed air on occasion to keep them from plugging.

The tail vise is constructed like its Shaker predecessor. A screwed-on top panel is easily removed to provide access to the 2-in.-dia. wooden screw, should it need future lubrication or attention. "I'm a WD-40 freak," Norm allows. The front leg vise is not of typical Shaker design, but it would certainly appeal to the Shakers' sense of distilled function. It has a self-paralleling scissors device, which Norm cannibalized from

The height of the vise jaws allows room for shaping (above), and the cast-iron scissors mechanism (left) keeps the jaws parallel.

another old workbench (although it could be fabricated from metal or wood parts). The mechanism, which is shown in the inset photo at left, is made of cast iron and permits the vise to be opened to a maximum of 7½ in., with the jaws held parallel in all positions. The top arms of the scissors are hinged to a pin housed in the bench leg and the matching face of the vise. The bottom arms slide on inlaid metal plates in both faces. After only a few months of operation Vandal has found that there's some slop in the vise action, although he says that it bothers him more than it actually affects the operation of the vise. The vise jaws are 5½ in. wide and extend 6½ in. above the surface of the top to provide plenty of room for three-dimensional shaping.

The screws in both vises are of yellow birch, which is easier to thread than maple. They were turned from 3-in.-square stock and cut with a 'reaming' tap. "The tap works fine," Vandal says, "but it's a bear to use because you need a hardened 'T-handle'" to withstand the torque that's required to rotate it. The sliding board jack runs in a groove located on the underside of the top and on a narrow spline on the bottom base stretcher. Vandal used a spline on the bottom instead of a groove to keep it from clogging with sawdust. The jack can be removed through a gap in the spline on the left end of the stretcher.

One of the distinctive features of the Hancock bench is the vertical fruitwood leg vise, installed near the left front corner. It is neatly shaped, with heavy chamfers on both edges. It's good for clamping small work or, with one end of a board or molding held in the vise and the other supported by the sliding board jack, for all manner of edge-jointing. It's also useful for members requiring three-dimensional carving, such as the leg of a candlestand or the arm of a chair.

As far as Seaman is concerned, the leg vise has only two drawbacks. Being flush with the top of the bench (32¼ in. high), it's a few inches too low for comfort, so he often pulls up a stool and works sitting down. In addition, because the screw is in the middle of the narrow jaw, there's not much room to hold a board vertically for dovetailing. The board must be clamped either to one side of or above the screw, neither of which is very secure. When he's cutting dovetails in a wide board, Seaman prefers to use a homemade wooden vise (see p. 134).

The Hancock leg vise has an unconventional mechanism for keeping the jaws parallel, a constant problem with leg vises. Most leg vises have a horizontal beam at the bottom of the vise that fits a mortise in the front leg of the bench. A pin is inserted in one of several drilled holes in the beam and may be moved to accommodate larger or smaller work. On the Hancock vise, a large octagonal nut turns on a 1½-in.-dia. wooden screw, as shown at right below, to provide an infinite range of adjustment. I suspect that with some practice a craftsman could operate the large nut with his foot while simultaneously opening or closing the main screw with his hand.

The threads on both the upper and lower screws are in remarkable condition. While they have been chipped in several places, the actual threading surfaces are nearly pristine. Seaman explains the damage: "I go by the theory that when benches were abused it wasn't by the maker, but it was a hundred years later." Knowing what we know about the Shakers, I suspect he's right.

One last, typical feature of the Hancock bench is the board jack, which slides in two grooves—one on the underside of the top, the other at the base of the bench. (Not all the surviving Shaker benches I've seen still have the jack, but all have the grooves.) The sliding jack is an almost indispensable method of supporting one end of a board while the other is clamped in the face vise. The curved design of the Hancock jack allows it to slide the length of the bench unhindered by the drawer pulls.

Most of the Shaker benches I've seen have vertical leg vises. A few of them have a face vise, which, in operation, is quite similar to a modern, bolt-on iron vise. This type of vise may be built of either metal or wood—both work the same way. The jaw is attached to a square, hollow beam that slides in a box lag-bolted to the underside of the benchtop. The vise screw passes through the jaw and runs the length of the beam, exiting at the other end through a nut fastened in the box.

I've seen this vise on several Shaker benches that also display the worn holes of a previous leg vise, so it appears that the horizontal face vise was a later 'improvement.' Not requiring parallel adjustment, it is quicker to operate than the leg vise. The screw is also completely encased and protected from clogging with sawdust or accidental damage. Some of the normal stress exerted on a vise screw is absorbed by the beam, which serves the same functions as the parallel guide rods on a modern face vise. The square corners of the beam register in the corners of the box to strengthen the vise and keep the jaw level with the benchtop.

Finally, in true Shaker fashion, the base of the Hancock bench was painted in two colors—blue on the end panels, door panels and drawer fronts, and a darker color on parts of the carcase frame and drawer dividers. Both colors are so faded now as to be almost indistinguishable from patina, but when freshly painted they would have been striking. According to guidelines set forth in the Millennial Laws, the workbench top and vise were probably either oiled or varnished.

The leg vise works well at holding curved chair parts that must be carved in three dimensions. Used in conjunction with the sliding board jack, at right, the leg vise is also good at holding boards for edge-jointing.

Parallel adjustment is achieved by turning the octagonal nut on the threaded shaft at the bottom of the vise. Photo by Richard Starr.

The range of furniture that Joel Seaman builds at the Hancock bench is probably not a whole lot different from that of the original Shakers. Like some of his Shaker predecessors, Seaman has also worked with two or three other people at the bench at once—sometimes his partner in the shop, Seth Reed, or students who have attended the workshops he occasionally conducts. Seaman builds everything from small production runs of pegboards to elaborate sewing desks, trestle tables and chairs of curly maple. All of his benchwork is performed using hand tools, working from sawn and jointed stock. He surfaces the wood with a hand plane, and cuts and chops dovetails with a saw and chisel. If the bench seems low to Seaman (as it would to most other modern woodworkers), it has a certain practicality for intensive hand-tool operations that require pressure from above.

During most of the ten years that Seaman has spent at the bench, he didn't give it much thought. "I guess I just accepted it for the beast that it is," he says. But in that decade, several of its characteristics have insinuated their way into his work habits—the large worksurface, the bank of drawers, the tail vise and dogs. When Seaman gets around to building his own bench—"it might be in the next century," he admits—these features will probably be incorporated. Instead of a massive, one-piece Shaker workstation, though, Seaman thinks in terms of component parts in a knockdown frame, perhaps with a removable tool chest in the base. He has an Emmert patternmaker's vise that would be more flexible than the leg vise. "I'm more interested in some hybrid solution than in being a traditional, 19th-century cabinetmaker," Seaman says.

Seaman likes to point out that the Shakers were not really that different. Despite the mystique that currently surrounds their woodwork, they were essentially pragmatic craftsmen. While the Shakers paid close attention to proportion and dimension in their furniture, Seaman argues that they often used what woods they had available and they were not averse to employing filler strips, blocks or shims where they had to. "Today we're so used to precise measurements," he says, "that some people consider this shoddy workmanship. But it's not, really. The Shakers were just trying to make good, functional furniture by 19th-century standards."

At 5:00 p.m., when the last visitors have left the grounds of the Hancock museum, Seaman sweeps off the old Shaker bench, walks to the back of the shop, swings open a heavy door that leads to another workroom and flips on the power. There, surrounded by tablesaw, jointer, and a leviathan 20-in. thickness planer that would bring an approving smile to the lips of even the most taciturn Shaker woodworker, he prepares the stock he will assemble and finish on the bench the next day. "For the public I do handwork," Seaman says. "What I should do is machine work…this was, after all, an industrial, agricultural community."

When I look back at the features that distinguish a Shaker workbench—size, cabinets, dovetailed tail vise, sliding board jack, etc.—it's clear that any one (or even several together) might be found on a worldly workbench of the same period. Indeed, I have come across a few benches outside of Shaker domain with banks of enclosed drawers and cabinets, or an occasional long top. It is not the individual characteristic, however, that makes a Shaker workbench—any more than a single Shaker comprises a community. As Robert Meader writes in the introduction to his *Illustrated Guide to Shaker Furniture*, "…it is very often not an *individual* feature that will identify a piece as Shaker, but rather the *totality* of features." The piece either " 'feels' right or does not."

Anyone making a workbench in the Shaker style would do well to consider the advice that Brother Thomas Damon offered in his letter to George Wilcox on December 23, 1846, regarding the construction of his desk: "You will please suit yourself as to size and formation, 'For where there is no law there is no transgression.' "

In this wooden face vise, the beam and jaw are cut from a single fruitwood knee. The benchscrew hole is capped with a wooden plate where it exits the jaw.

Shaker face vise

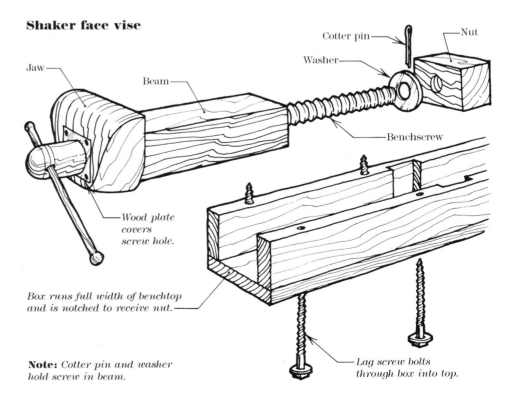

Box runs full width of benchtop and is notched to receive nut.

Note: *Cotter pin and washer hold screw in beam.*

Frank Klausz's new workbench is rooted in a venerable European tradition.

An Old-Fashioned Workhorse

Chapter 4

I met Frank Klausz before I met his workbench. He was seated next to me in the front row during a lecture given by Ian Kirby (Chapter 6). Klausz himself was scheduled to speak about wood finishing the next day. Kirby's talk centered on his workbench, an example of which he'd brought along. Although the bench had evolved out of Kirby's own English tradition, from Klausz's Hungarian perspective it was hardly a cabinetmaker's bench at all. It had no tail vise and the front vise was of the metal, quick-action persuasion. Klausz fidgeted through most of Kirby's talk, and at the first break he sprang from his chair and led me to the bench, where he passionately ennumerated his objections.

To understand the depth of Klausz's convictions, you need to know about his background. Thirty years ago in Hungary, at the age of 14, Frank began his woodworking career in an apprenticeship system that had remained essentially unchanged since the Middle Ages. What was unusual about it, even by European standards, was that Klausz entered into a formal, contractual apprenticeship with his own father. "I paid the highest price for my trade," Klausz explains. "Once I apprenticed, I didn't have a father, I had a master." And a stern master at that. Of the half-dozen workers in his father's cabinet shop, it was Frank who was taken to task if something wasn't quite right. Perhaps wary of his own son's competition, the elder Klausz withheld certain construction tips until the very end of

Frank's apprenticeship. Watching his father work, Frank asked, "How can you do that so fast?" His father replied, "After ten or fifteen years you're gonna be a pretty good beginner yourself."

At the end of four years, Frank became a certified journeyman cabinetmaker, on his way to becoming a master (which required one year of work in each of three different shops). Ten years later, Frank and his wife, Edith, packed their lives in three suitcases and left Hungary. Like the journeymen of old, Frank was on the road—except that his only tools were his hands and head, not chisels and saws in a toolbox strapped to his back. By 1969, the couple was living on Long Island, where Frank ran through a succession of jobs—carpentry, casework, and so on—trying to find his way back to the work he'd been trained to do. It was five more years before he could set up his own shop in a two-car garage in New Jersey. Finally, in 1985, Frank and Edith built the shop they'd been dreaming of.

I went to visit Frank in his Pluckemin, New Jersey, workshop and to meet his workbench in the flesh. My first and most startling impression was of the workshop itself. I had primed myself for an old-world sweatshop, with young apprentices chained to their benches. In Hungary, Frank's father had two small workrooms—one for the benches and another (unheated, even in winter) for the machinery. When lumber had to be cut from a 20-ft. log, the workers fed it through an open window at one end of the machine shop, across the bandsaw and out again through the opposite window over rollers placed at the sill.

Klausz's own shop couldn't be more of the 'new world.' The single-story, cinder-block building sprawls a full 100 ft. in length. Painted off-white inside, it is bright and airy, with windows on all sides and large skylights. If Frank had to mill a mast for the *Constitution*, I doubt that he'd even have to open a window.

Frank takes me on a quick tour of the shop to show me their work. While one of his four employees might be building a set of computer cabinets of walnut-faced plywood, another could be restoring an 18th-century English grandfather clock or stripping an office desk. At the far end of the building, we pause for a moment while Frank sprays the handrails for a casket he has built for an elderly client, whose house he has almost entirely restored. In the old country, Klausz explains, there was a cradle-to-grave relationship between the craftsman and his client. As his last commission for the deceased, the cabinetmaker would appear at the funeral, in his Sunday best, to drive the nails into the lid of the box. Clearly, a workbench in this shop needs to be versatile.

According to another old-world tradition, Frank explains, workbenches were passed on from one generation to another. The woodworker was the custodian, not the owner, of his bench—just as he was the custodian of the knowledge of his trade. The workbench took on a life of its own; it became somehow larger than the sum of the men who had planed upon it.

For Frank that chain had been broken. He had brought no workbench with him from Europe, and had to use commercially made benches for years, never having the time to make his own. But, when he found that there weren't enough benches to go around in the new shop, he decided to build one. "The reason I made one is that you can't buy one good enough," he told me. It seemed to Frank that commercially made workbenches were growing smaller and lighter, even as they got more expensive. I also suspect it was Frank's way of saying he'd come home.

There wasn't any guesswork or design involved when Frank built his bench. He didn't reinvent the wheel. "It's a copy," he says. He had measured two benches at his father's shop, a third in Vienna and another in Belgium. They were all within an inch of each other. "Apart from little touches like the stops

and oil dish, the only difference I found was that some craftsmen treat their benches with loving care and some don't.... Except for the metal vise screws, my bench is the same as my grandfather's.... [The design] is so well worked out—if it hadn't been good, Grandpa would have done something about it."

When a customer enters Frank's shop, he encounters the workbench, which also functions as a desk and business counter. Even if the visitor doesn't comment on the bench, it's a fair bet he's noticed it. If Klausz could fit his workbench in his wallet, he would hand it out like a business card—it is his best foot and he puts it forward.

Klausz begins to explain his workbench by underlining a point too often overlooked—location. As the most important tool in the shop, the bench's placement with respect to work flow (of materials, to and from machines, for finishing and so on) is crucial. Lighting is also important, and ideally should cast no shadows on the benchtop. Hand tools should be readily accessible. Frank's are kept in a wall-mounted cabinet, only 5 ft. from the shoulder vise of the bench.

Of equal importance is the auxiliary set-up table near the bench, shown below. This low table is the right size (40 in. by 60 in. by 27 in. high) for all kinds of gluing, assembly or finishing. Anything that's too messy or large for the workbench can be done on the table, leaving the benchtop free for trimming joints and other last-minute tasks. Rather than cluttering the main bench with drawers, Klausz built open shelves and storage bins in the base of the set-up table to hold hardware, small power tools and accessories.

This 27-in.-high set-up table is a versatile companion to Klausz's bench. It helps organize hardware and portable power tools, and provides a nearby, convenient surface for gluing and finishing. Hardware is stored in 12 plastic bins, and three drawers pull out from beneath the 40-in. by 60-in. particleboard and plywood top.

Klausz, with a glued-up tail vise joined with hand-cut dovetails.

In my travels I'd seen several variations on Klausz's work-bench, variously referred to as Scandinavian, Danish, Swedish and European. These workbenches all have as a common denominator the 'dog-leg' shoulder vise. I thought I had heard most of the arguments for and against this vise; as far as I had been able to discover, the only craftsmen who liked it were those who had trained on it, usually in strict apprenticeships. I posed the same objections I'd heard to Klausz: The vise isn't strong enough to withstand heavy clamping pressure. It's awkward to work around the large corner. The pivoting clamping board often has to be held with one hand to keep it from binding as it's wound in and out. You can't clamp a board anywhere on the bench for crosscutting.

Frank's initial response was a reflex: "If you're a cabinetmaker, if you do casegoods, frames, if you plane, saw or sand wood, if you do dovetails...I can't see anything quicker or better." Later, he explained that the floating clamping board grips well on tapered stock, and one end of a long board (or a door) can be clamped firmly behind the screw while the other end is supported by the portable bench slave, shown at right. But it was only when I watched him dovetail a drawer that I truly began to appreciate the shoulder vise.

Through dovetails are one of the traditional cabinetmaker's preferred joints, and when Frank cuts a dovetailed drawer he puts the bench through its paces. "Good craftsmen," Frank says, "not only do things well, but do them with speed.... If you want to make a good joint, you can do it just about as fast by hand as with a machine, especially if you're doing just one." With the drawer parts milled to length and thickness, Frank uses a mortising gauge to scribe the thickness of the stock across the ends of the boards. Then he slaps the first piece—the drawer front or back—upright in the shoulder vise.

The quick-action feature of a Record vise is nice, Frank admits, but he rarely has to move the screw on his vise more than a single turn. Because there are no guide rods or screws running below the vise, a long board such as a drawer front can be clamped through the opening, not just gripped in the top few inches of the jaw or along one edge. The work won't twist and there's no need to block the other edge of the vise to keep the jaws parallel. (The clamping board pivots on the end of the screw to accept tapered work, and it should move freely without needing to be guided by hand.)

At 33 in. high, Frank's bench is lower than I'm used to, but the shoulder vise helps to compensate by allowing you to clamp work securely at many different heights. If it's too high, it vibrates; if it's too low, it's uncomfortable. Frank holds the top edge of the drawer front about 4 in. or 5 in. off the bench—a comfortable sawing height—and clamps the board tight. He wheels around to grab a backsaw from the tool cabinet and begins cutting pins—without stopping to lay them out. In about as much time as it took him to rip a bagel on the band-saw during a coffee break, he defines all the pins at one end of the board with six sawcuts. Once the pins are cut on both ends of the drawer, front and back, the action moves to the other end of the bench. Frank C-clamps the part to be chopped in

Klausz uses a bench slave (top right) to support one end of a long board clamped in the shoulder vise. When he cuts dovetails (right), he clamps the board vertically in the shoulder vise, centered directly behind the screw. There are no guide rods to interfere with the work. He aligns his arm and body with the direction of the cut and, standing above it, he can keep an eye on both sides of the line.

To chop the pins, Klausz stacks the drawer sides and clamps them to the bench in front of the tail vise.

Klausz reverses the dogs in their holes and opens the tail vise gradually to take a chair apart for repair.

With the vise snugged tight, Klausz sets the benchdogs with a few smart taps on the end of the board and the head of the dog. A piece of scrap protects the board from the face of the metal dog.

front of the tail vise—never on it: The force of the mallet blows is transferred directly through the leg to the floor. Whether he's working on one drawer or six, all the parts are stacked and staggered, one on top of another, so that he has ready access to all the joints. At one point, Frank demonstrates how, during a full day of this work, he would drop to his knees to rest his back (there's a rubber mat in front of the bench to provide a cushion). This places the work at his chest, instead of at his hip, and gives him a closer view as well.

Next, with the parts laid out on the benchtop, he marks the tails from the pins. Then he's back at the shoulder vise to cut them out. During these operations, the tool tray holds the marking gauge, backsaw, pencil, square, chisel and mallet—out of harm's way but easily retrieved. The tray isn't a repository for yesterday's project, and Frank keeps it swept clean. Cutting dovetails by eye requires having your wits about you, or it won't be long before you're cutting pins on one end of a board and tails on the other. A clean, orderly benchtop is essential.

Before the drawer is assembled, Frank planes the machine marks off the inside of each piece. He moves back to the other end of the bench and gently closes the tail vise on a drawer side, using a piece of scrap between the metal dogs and the ends of the board. It doesn't take much pressure. The benchdogs do two jobs: they grab the wood and, because they're angled, pull it down. Their down-clamping action is important, especially on thin stock, which will chatter if suspended in midair. To make the most of this feature, Frank taps down on each end of the board in front of the dog, seating it on the bench. He prefers metal dogs to wood because of their strength; they're more effective at pulling the work down, and he can knock the dogs up an inch or more above the benchtop without having to worry about flexing or breaking them. The dogs can also be reversed and used to pull apart a piece of furniture. For this reason, the dogholes are cut at an 88° angle—any steeper and the dogs might slide out of their slots.

"When I plane, I use my body weight and just push down," Frank explains. "This gives me hours of easy planing without pushing and shoving." The bench has to be the right height for

The tool tray helps keep the benchtop clear by holding the materials not immediately required (above left). To plane the outside of the glued-up drawer, Klausz clamps it in the tail vise, with one end supported by the slave (above).

this. To demonstrate his formula for bench height, Frank stands next to the bench with his arms at his side and his palms turned down—the benchtop grazes his palms. He is 6 ft. tall and his bench is less than 3 ft. high. Frank planes in two motions—a long, cutting, power stroke and a feathered return as he tilts the plane slightly to lift the blade off the wood and to resume position for the next cut. The bench doesn't move under the pressure of his strokes.

Sometimes if a piece is small, Frank glues up right on the bench, spreading a cloth to protect the top. But more often, he turns to the set-up table behind the bench. A quick swipe with a wet rag removes any errant drops.

When the glue is dry, he planes the drawer sides by gripping the frame in the front jaw of the tail vise. The other end of the drawer, which sticks out from the vise, can be supported at any height by the bench slave.

To finish the job, Frank planes the top and bottom edges of the drawer flush. He reclamps the drawer flat on the bench between dogs—using the front doghole in the tail vise when he can and keeping the vise opening small. "You want to have the workpiece on the bench as much as possible, not on the vise," he says. "It puts less stress on the vise itself. The strongest support is up front, over the legs."

Klausz's drawer demonstration answered many of my doubts about the shoulder vise: It doesn't need to be immensely strong, because the screw is always centered behind the workpiece. Whatever inherent awkwardness exists in its design is at least partially offset by this convenient feature, which cannot be found on any other conventional front vise. The clamping board rarely binds in everyday use, because generally it is adjusted in small increments, and it's the only vise I know of that clamps non-square stock as easily as square stock.

I had one final reservation, though. There's no easy way to clamp a board for crosscutting anywhere on the bench. "You don't have to," Klausz says. All that's necessary is a small stop, and he flips up the pivoting bench stop at the right end of the bench and pushes a board against it to demonstrate. Its location on the right end of the bench is also more convenient for a right-handed worker than crosscutting off of the left end.

If my own reservations about the bench were mainly resolved, it was clear that Frank had none at all. "If you're a cabinetmaker, you *should* have a bench like this," he said.

This flip-up padauk stop is all that is needed to hold most boards for crosscutting.

An Old-Fashioned Workhorse 53

Making the bench

In its construction, Frank Klausz's bench is traditional. The basic top is comprised of two wide boards (the body) and two front strips, one slotted for dogholes. Two end caps, a tool tray and the two vises complete the top, as shown in the drawing on the facing page. Massive dovetails are used for all major joints, both for their strength and, Frank acknowledges, for their visual impact. The base members are heavy, too—their joinery a scaled-down version of that used in the traditional timber-frame house. The vises use wood-on-wood guides, just as they did in the 18th century. Only the benchscrews and nuts and a few lag screws and machine bolts are metal. Frank says he would have replaced the benchscrews with 2-in.-dia. wooden screws if he could have found them. (He does admit, however, to a "crazy idea that's still in my head," to hook

up a foot-operated motor to the tail-vise screw. "It has to go faster than by hand.")

For a utilitarian object like the workbench, Frank uses inexpensive wood wherever he can, preferring to use the good stuff where it will count most. He had been collecting wood for years and built the top out of a melange of hardwoods—quartersawn maple for the body, beech end caps, a bird's-eye tool-tray backboard and spalted-maple front strip. The front arm of the shoulder vise is also beech and the support block is spalted (to match the front strip in the top). The tail vise is made of locust and maple. All the accessories are padauk.

If you, too, combine different woods, be sure they are of like density and moisture content. As Frank advised a friend who is planning to build a new bench out of 175-year-old white oak from a dismantled pole barn: "Get it inside, make sure it warms up, rough it down and let it move. Then it's ready to use."

The details of construction.

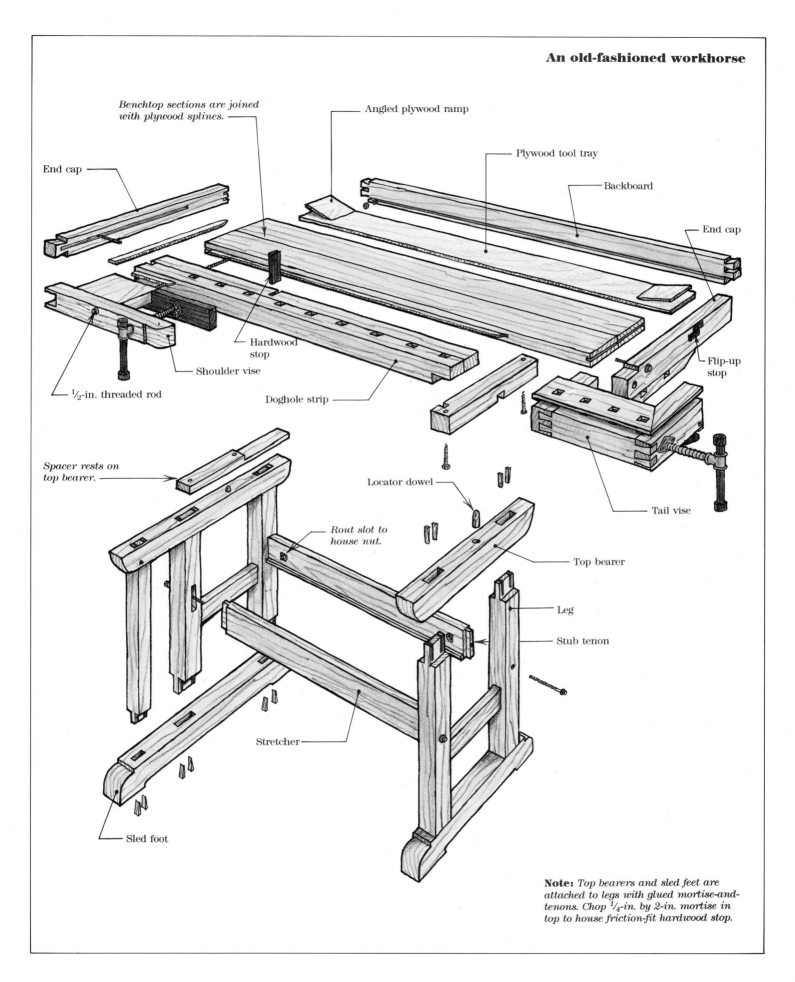

Benchtop sections are joined with plywood splines.

Angled plywood ramp

Plywood tool tray

Backboard

End cap

End cap

Hardwood stop

Shoulder vise

Flip-up stop

½-in. threaded rod

Doghole strip

Tail vise

Spacer rests on top bearer.

Locator dowel

Rout slot to house nut.

Top bearer

Leg

Stub tenon

Stretcher

Sled foot

Note: *Top bearers and sled feet are attached to legs with glued mortise-and-tenons. Chop ¼-in. by 2-in. mortise in top to house friction-fit hardwood stop.*

The base is built of 'utility-grade' oak, beech and white ash—knots and all. Frank builds the base first—it's an ideal platform for assembling the top. Through-wedged mortise-and-tenons connect the legs to the top bearer rails of the trestles; the sled feet may be either blind- or through-mortised. Adjust the height if necessary by lengthening or shortening the leg between the stretcher and the foot.

The stub-tenoned stretchers fit in blind mortises in the legs, and are attached with hex-head machine bolts and captured nuts. Bore the bolt holes in the legs outward through the center of the mortise. Then assemble the base and drill back through the holes into the ends of the stretchers. Remove the stretchers and lengthen the holes as required. The washers and nuts are housed in cavities routed on the insides of the stretchers. On Frank's bench, the bolt heads are not countersunk, but are spray-painted green to be unobtrusive. He doesn't bolt the top to the trestles; bullet-shaped dowels glued in the tops of the trestles position the top, and its weight keeps it in place.

The base is built of utility-grade hardwood. End trestles are of glued, mortise-and-tenon construction, bolted to the stretchers. The top sits on bullet-shaped dowels. Top photo by Dick Burrows.

To make the top, first prepare the body. The shoulder vise is reinforced with a $\frac{1}{2}$-in.-dia. threaded rod that runs through the top, so remember to bore a hole through each of the top boards before assembly. Frank glues the boards together using $\frac{1}{2}$-in. by $1\frac{1}{2}$-in. plywood splines to align the surfaces; you can insert the rod to align the holes.

Frank uses a dado blade on the radial-arm saw to mill the angled doghole slots in the front strip. After the first cut, he removes the waste in $\frac{1}{4}$-in. bites until the dog fits, with about $\frac{1}{16}$-in. play left in the slot for the shaft. Rather than using a jig, Klausz marks the slots individually and tests the dog's fit in each one. After they're all cut, he chisels the notch for the dog's head by hand.

Next, lay out the position of the tail vise and glue on the two front strips. (Forming the tail-vise notch accurately when gluing up avoids awkward sawing and cleaning up later.) Make sure that all spline grooves and the hole for the threaded rod in the shoulder vise have been cut and aligned properly. After the top is glued up, Frank cuts the end-cap spline grooves by standing the top on end and running it over the dado blade on the tablesaw. If that makes you shudder, as it does me, use a router or follow Michael Fortune's method (see p. 74). Always reference the rip fence or tool against the top surfaces of the benchtop and end caps so the grooves will line up.

The end caps are an integral part of both the shoulder and tail vises, so the caps can't be fully installed until the other vise members have been fitted to them. Bolts and captured nuts fix the end caps to the top; bore these bolt holes and chop or rout the slots in the underside of the top for the captured nuts now. Klausz also leaves the bolt heads exposed on the end caps and uses them as mini-anvils to flatten hardware or to blunt nails so they won't split wood when driven.

With the top and base complete, you're ready to tackle the vises. As always, tailor the vises to fit your hardware. The shoulder-vise construction is straightforward, as shown in the drawing on the facing page. The thread length on the screw Klausz used for the last bench he built was about 12 in. long, or a few inches longer than he figured was necessary for the shoulder vise. Since too large an opening will weaken the vise, he shortened the length of the threads to 9 in. (a 13-in. overall screw length). He removed the cast handle fitting and, with one end of the screw clamped in a 3-jaw chuck and the other end supported by a tail center, he turned off about 3 in. of steel. It took about an hour, using a 4-speed Rockwell wood lathe and a cutter reground from a star drill. He found that the trickiest part was cutting through the metal threads. This is a difficult task and you might want to look around for a friend with a metal lathe. (If your screw is only slightly longer, you could get away with beefing up the thickness of the arm, or simply allowing the head of the screw to protrude.)

To counter the considerable force developed by the vise screw, Klausz uses a single, massive dovetail at the corner joint, reinforced by the $\frac{1}{2}$-in. threaded rod that runs through the vise and benchtop. I wondered about the wisdom of running the rod across more than 18 in. of maple—what happens when the wood expands and contracts with moisture changes? "It puzzled me quite a lot," Frank admits. "I would have hesitated to mention it if I hadn't tried it [on another bench] for ten years. The only difference is that, as the seasons change, the washer sinks in slightly or the head gets loose.... It moves only as much as ⅛ in., I'm sure." This detail is one of the few departures Frank made from his father's bench. That original bench (and several others I've seen) had a bolt running through the shoulder vise and a few inches into the benchtop, where it was fastened from below with a captured nut. It resolved the problem of movement in the benchtop, but was a weak spot in the bench construction. Frank recalls that the vise was in frequent need of repair. "The threaded rod really holds that bench together," Frank says.

Cut the dovetail carefully and, using the drill press for accuracy, bore the arm for the vise screw. Klausz first bores a large hole on the back face to house the vise-screw nut, then uses the same center for the smaller screw hole. Also bore the hole for the rod through the support block, the arm and the splines, and cut the grooves for the splines. Clean out the pre-drilled holes in the top by running a bit on a 12-in. extension from both ends of the hole, then bolt the end cap and shoulder vise to the bench. The dovetail joint and the support block contribute greatly to the strength of the vise, so they must fit tightly. The clamping board should be loose enough to slide easily without binding. Frank recommends making the neck about ⅛ in. thinner than the opening between the bottom of the support block and the top bearer of the base. The neck can then be shimmed with a laminate or a dense hardwood to get the best fit. He also mounts the swivel fixture at the end of the screw to the clamping board so that the opening faces out, as shown at right. This means that the clamping board can be taken off only by removing the two screws, but Frank finds that it holds the block in a better position—perpendicular to the bench-screw and, therefore, less likely to bind.

The clamping board slides in the space between the top base bearer and the vise support block (above). With the swivel fixture mounted to the clamping board so that its opening faces out, the vise tracks more easily without binding (left).

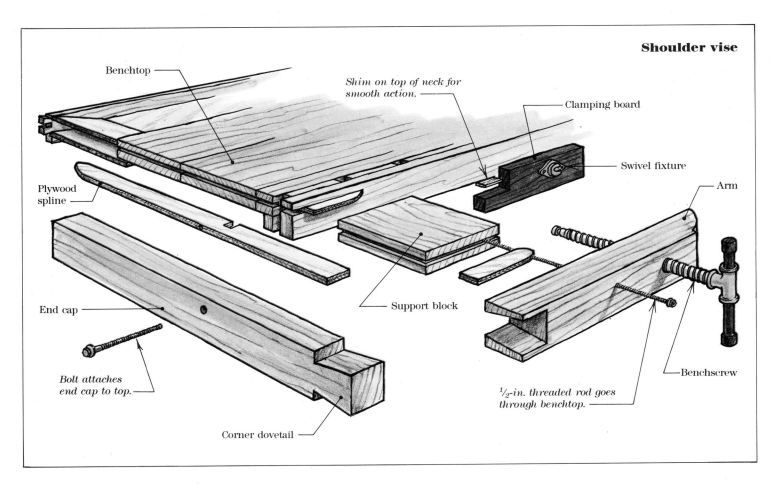

Shoulder vise

Benchtop

Shim on top of neck for smooth action.

Clamping board

Swivel fixture

Plywood spline

Arm

Support block

End cap

Benchscrew

Bolt attaches end cap to top.

½-in. threaded rod goes through benchtop.

Corner dovetail

The tail vise is more complicated, so follow the drawing on the facing page carefully and refer to the measured drawings on pp. 230-233. For a longer or shorter screw, you'll have to lengthen or shorten the jaw assembly and the opening in the benchtop. (The vise Frank built for the photos here was made to fit a longer screw than the vise shown in the drawings. Except for the length and the number of dogholes, the process for making both tail vises is identical.)

The tail vise has two parts—a sliding jaw assembly and guide blocks bolted to the underside of the benchtop. The jaw assembly consists of a heavy front and rear jaw dovetailed to a face piece and a back runner. The rear jaw houses the screw, the front jaw provides a clamping surface and the face piece is slotted for benchdogs. This solid structure is further strengthened by two mitered top caps, which cover the screw and are planed flush with the benchtop after the vise is installed. The guide block lag-screwed beneath the end cap is notched for the jaw-assembly runners, and a bench runner joining the two guide blocks rides in a slot in the underside of the front jaw. The vise-screw nut is housed in the end cap.

To build the tail vise, mill the four main members of the sliding jaw assembly and lay them out as they will be positioned in the vise. The front piece is the same thickness as the front strip of the benchtop so that the dogholes will be in line.

I've found that old benches frequently have only two dogholes, one near the front and one near the back of the vise. This limits your clamping options and encourages working on top of the tail vise, which is not a good idea. In addition to the dogholes in the front and back jaws, Frank cuts two or three evenly spaced dogholes in the front piece.

Scribe the dovetail shoulder lines around the ends of the jaws and on the front piece, then lay out the dogholes. Those in the jaws fall on the scribed shoulder lines. Cut the dogholes in the front piece like those in the benchtop, but angle the slots in the opposite direction. The slots in the front and rear jaws must be tapered to create the angles. Frank does this on the radial-arm saw. Reset the saw to crosscut at 90°; set a bevel gauge to 88°, wedge the jaw until it matches the gauge, then cut the slot. (Notches to house the dog heads will be created when you glue on the top caps.)

A lot of the action takes place at the right end of the bench. The tail vise is used mainly for planing, and the flip-up stop holds work for crosscutting. Photo by Dick Burrows.

Tail vise

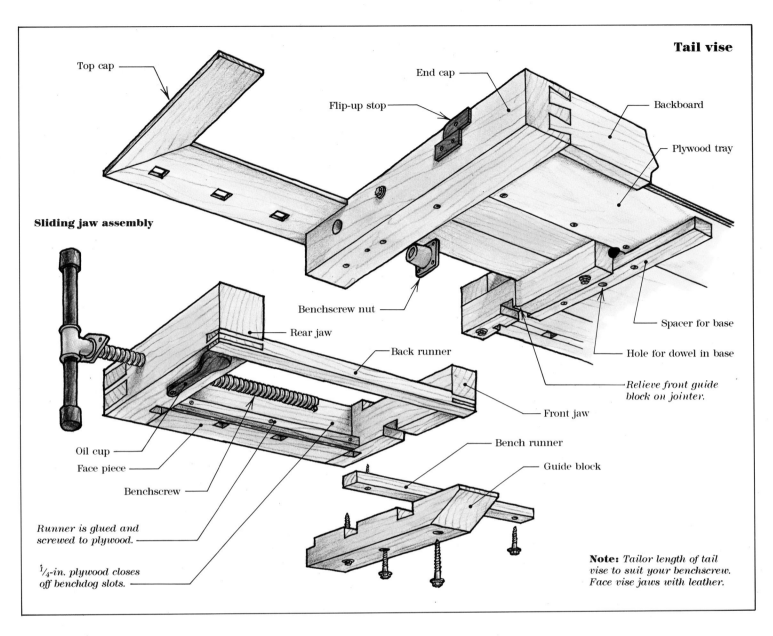

Top cap

End cap

Flip-up stop

Backboard

Plywood tray

Sliding jaw assembly

Benchscrew nut

Rear jaw

Back runner

Spacer for base

Hole for dowel in base

Relieve front guide block on jointer.

Front jaw

Bench runner

Oil cup

Guide block

Face piece

Benchscrew

Runner is glued and screwed to plywood.

¹/₄-in. plywood closes off benchdog slots.

Note: *Tailor length of tail vise to suit your benchscrew. Face vise jaws with leather.*

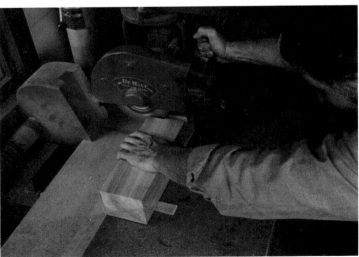

After the dovetail shoulders have been marked on the jaws and the front piece, Klausz lays out the dogholes (left). To cut the angled dogholes in the front and rear jaws, he sets the radial-arm saw at 90° and angles the jaw with a small wedge (above).

Through-dovetail pins are cut on the bandsaw to a 10° angle (left), the maximum tilt on Klausz's saw table. Half-blind pins are cut with a frame saw (above), then routed and chiseled clean. Klausz uses the tail vise to hold the work at an angle while sawing the half-blind pins. Tails on the front piece (right) are cut on the bandsaw after they have been trimmed to thickness.

The back runner slides on the underside of the end cap. Above, Klausz measures to include the thickness of the top cap as he locates the runner. At right, the tail vise is assembled dry before gluing to make sure everything fits.

Cover the dogholes with plywood and glue and screw a maple runner to the bottom edge (left). Klausz also relieves the guide block so that only the first 6 in. contacts the front jaw (above).

Next, lay out the through-dovetail pins on the rear jaw and the half-blind pins on the front jaw. Klausz cuts the three pins in the rear jaw on his 14-in. bandsaw. The angle isn't critical—for Klausz, the 10° maximum tilt of his bandsaw table determines the angle of his cut. He clears the waste with a narrow-bladed frame saw, then chisels the bottoms of the sockets flat and to the shoulder lines. It's common practice to very slightly undercut dovetails for an easy fit, but for his bench-vise construction, Frank prefers to fit them tight, since much of their strength comes from good contact.

Frank saws the half-blind pins in the front jaw by hand, using a wide blade in the frame saw. He removes the waste neatly and quickly by routing the end grain, then cleans the corners with a chisel. Klausz prefers this hand-cut method, although you could rip the front jaw, cut through-dovetail pins and re-glue the jaw. Next, mark the tails on the front piece from the pins. To avoid interfering with the dogholes cut in the jaws, Frank notches the front piece, thus reducing the thickness of the tails by the width of the doghole. He then bandsaws the tails and trims their outside shoulders using the radial-arm saw or the tablesaw.

Klausz joins the back runner to the jaws with decorative through dovetails, but finger joints would work as well and would be easier to cut. The back runner slides on the underside of the end cap, so mark the thickness of the end cap on the rear jaw, making sure to include the 1-in.-thick top cap in your measurement (see photo, facing page, center far left). Transfer that mark to the front jaw and lay out the same joint at both ends, leaving at least 1/4 in. of wood on the runner below the jaws. Frank bandsaws the through dovetails.

After the joints are cut, check their fit by assembling the vise dry. Then glue the vise together, clamping across the joints, and make sure it's square. Glue a 1/4-in. plywood cover over the insides of the dogholes in the front piece, then glue and screw the maple runner to the plywood, flush with the bottom edge of the front piece.

Turn the benchtop upside down to mount the tail vise. Clamp the vise, end cap and front guide block in place, as shown on the facing page, bottom right. (Remember that the doghole strip on the benchtop is thicker than the rest of the benchtop, so it must be notched for the guide block.) It is critical that the front of the vise be in perfect alignment with the front edge of the bench. Adjust the vise position by moving the end cap forward or back. Frank relieves the front guide block on the jointer so that it is in contact with the front jaw for only 6 in.

The bench runner should be located near the middle of the tail vise; its precise location is not important, but it must be parallel to the two runners on the vise. Lay out and cut the notches for the bench runner on the guide blocks and front jaw and for the two vise runners on the guide block attached to the end cap. Bore a relief hole on the inside face of the end cap to house the benchscrew nut, then bore a smaller hole through the end cap for the benchscrew centered in the relief hole. Mark the center of the benchscrew hole on the rear jaw of the tail vise. Transfer the center mark to the outside face of the jaw and bore another hole. (Frank's benchscrew is 1 1/4 in. in diameter; a 1 1/2-in. hole works fine in both the end cap and the tail vise.)

You can now finish fitting the end caps at both ends of the bench. Trim them to length, install the benchscrew nut for the tail vise, bolt the end caps to the bench and dovetail them to the back rail. Glue the dovetails well at the back of both end caps and at the corner of the shoulder vise. (Frank also glues his end caps to the benchtop. While this practice generally is not advisable because of movement in the top, Frank's bench has never split. Well-seasoned wood, a stable shop humidity, periodic oiling and the 1/2-in. rod through the benchtop probably all have something to do with this.) Clamp the vise, guide blocks and bench runner in position and pull the tail vise back and forth by hand to make sure it's working smoothly. If further adjustment is required, you can enlarge the slots for the runners, insert shims where needed or move the guide blocks. Then bore pilot holes and install the lag screws that attach the parts to the underside of the bench. Insert the benchscrew, center it in the hole in the rear jaw and screw on the flange. To complete the vise assembly, cut slots for the doghole heads in the top caps and glue the caps on. If the tail vise has lost some height in the process, you can correct it now by making the top caps thicker. When the glue has dried, the caps should be planed flush with the surface of the benchtop.

To obtain the best possible mating of the front jaw and the bench jaws, close the vise and cut between them with a sharp, fine-point saw—don't let the saw wander. For a final touch, Frank glues top-grain cowhide to each face to protect both the work and the jaws. The leather can be put on with white glue and peeled off with water and replaced when it is worn.

To finish the bench, level the top with a sharp jointer plane and sand it smooth. Klausz applies two coats of Waterlox (a polymerizing tung-oil-base sealer) to every wood surface and several more coats to the top. He rubs on paste wax for a beautiful and protective shine.

When you use the bench, Klausz says, you have two main responsibilities: protecting the work and protecting the bench. Put a blanket on the bench or use an auxiliary table for painting, glue-up and assembly. At the end of each working day, every bench in Frank's shop is dusted off and the tools are put away. Any glue that was spilled is removed that same day. Every few months the benchtops are refurbished. They are wiped clean with warm water to loosen residual glue, then scrubbed with steel wool, re-oiled and waxed. This kind of treatment pays off in the long run; Frank's bench is used hard every day and yet it resembles an altar. As Frank points out, "It's not my design—it's a thousand years old. The only credit we can take for this is that we made it and it's nicely done."

With the vise and guide block clamped to the bench, mark the location of the runners in both guide blocks.

Michael Fortune and bench in his Toronto workshop. Photo by Jack Ramsdale.

A Modern Hybrid

When Michael Fortune arrived as a neophyte design student at Sheridan College outside of Toronto in 1970, the workbenches in the woodshop resembled the flight deck of an aircraft carrier. They were large, flat work stations of laminated maple, about 30 in. by 60 in., with a small Record vise at each end, perched atop battleship-gray steel cabinets. Fine for drawing or building models out of all kinds of materials, they presented a stiff surface to pound on and were easy to keep clean. They would have worked equally well for tap dancing or kneading bread.

At the time, these general-purpose benches served Sheridan's needs. Furniture-design students, like Fortune, were as likely to be vacuum-forming thermoset plastics as steambending solid wood. Rather than learning how to sharpen tools, how to square up stock with a hand plane or how to cut neat dovetails, students learned how to draw, how to fashion molds and jigs, and how to work with a variety of materials and methods. In the process, they acquired the broad vocabulary of industrial design.

When he graduated from Sheridan, however, Fortune felt incomplete. He had learned more than he realized about design, but didn't feel that he had been given enough ammunition to earn a living *making* the furniture he'd been trained to design. To fill the gaps, he embarked upon an ad hoc apprenticeship that took him to England and Sweden.

His first and most important stop was a two-month apprenticeship with Alan Peters in Devon, England. For Michael, Peters' workshop was more than a continent away—it was an immersion in the foreign world of the venerable English hand-tool tradition. Peters worked almost entirely in solid wood and used hand tools with a facility that Michael had never before witnessed.

It was in Peters' shop that Michael first encountered a traditional English workbench. As appropriate as the Sheridan benches may have been for the multimedia design program at the college, Michael's body told him that the tail-vise and benchdog system on Peters' English benches had some obvious advantages for hand woodworking. For starters, he found that he could hold work in place with a quick turn on a single screw—no messing about with C-clamps and blocks. He also discovered that the 14-in.-wide worksurface of the English bench—less than half the width of the Sheridan benches—made the bench more versatile. It was easy, for example, to clamp a piece of furniture to the top and still have free access to it from either side. What's more, Peters' benches, which were commercially made, had been propped up high for comfortable, backache-free planing.

Two months hardly constitute an apprenticeship. But in that short time, Fortune absorbed lessons he wasn't even aware of. Years later, he credits Alan Peters for having taught him that high-tech equipment is not necessarily the salvation of the small, flexible cabinet studio. When he returned to Canada, he found himself doing more handwork than he'd ever done before. The big, flat worksurface he'd made just before graduating from Sheridan became an encumbrance. Modeled on the Sheridan benches, it was 32 in. by 72 in., a couple of inches thick, with a metal vise at one end. To hand-plane something, he had to clamp it to the top, invariably situating the clamps right where the plane wanted to go. If he clamped stops to the bench, he begrudged the time it took to adjust the stops and clamps to change the position or length of the work. And he discovered that after he had been planing for a couple of hours on the 32-in.-high benchtop, his back would give him serious trouble.

Obviously, he needed a new bench. Borrowing heavily from the design of Alan Peters' benches, Michael made the new bench narrow (14 in. wide with a 4-in. tool tray) and a full 39 in. high. "It had honest-to-goodness benchdogs," Michael says, and two vises. He mounted a Record #52½D quick-action vise on the front and used commercial front-vise hardware to build another vise with wooden jaws across the right end of the bench.

The improvement was amazing, but it still wasn't quite right. He'd chosen the hardware for the end vise to avoid the complicated joinery of the traditional all-wood, L-shaped tail vise and because it suited his minimal shop technology. A drill press was all that was necessary to bore the holes for the screw and two guide rods. He installed a single benchdog in the vise and a row of dogholes down the middle of the bench, to try to keep the clamping pressure in line with the center screw. Having the dogs 7 in. away from the edge of the bench was awkward and it didn't take long before the jaw began to sag anyway, and become skewed out of square. "I began to see traditional cabinetmaker's benches as having quite an advantage."

Despite its shortcomings, Fortune worked on this bench for almost a decade. As his skill developed, he understood more all the time about what he wanted in the next one. The first priority was a row of dogholes near the front edge of the bench and a real wooden tail vise that would place the screw in line with the dogs, thereby eliminating racking. As a full-time woodworker and part-time teacher at Sheridan, however, Fortune could never find time to build a new bench. In 1981, Sheridan decided the woodshop needed new benches, and Fortune was put in charge of the project. He took the opportunity to put his bench ideas into practice. The result was the prototype of the bench described in this chapter.

About two years later, I approached Michael with a plan to design and build a workbench for sale. I had stumbled blindly through the construction of my own first bench by following an old Scandinavian pattern. I had seen Michael's bench and was impressed with his ability to adapt a traditional design for modern tools and hardware. We agreed that the best commercially available benches had shortcomings and decided to re-examine every element of the top, base and vises. Though the design we ended up with incorporated many of the features of Michael's bench, we learned a lot about benches during the design process. The following summary of our investigations should be useful to anyone wishing to design or alter a workbench.

The first consideration is basic dimensions: height, width and length. Instructions for determining the 'ideal' bench height often sound more appropriate for fitting a new suit. "When you stand straight, with your hands at your sides, the workbench top should reach your second knuckle." Or is it the third? Or am I supposed to turn my palms down? Like most formulas, these guidelines are fine for average people and average use, but what and who is average? Your height, your work and your personal preference should all determine the design of your own bench.

Basically, the lower the bench the more pressure you'll be able to exert on the work, and the less likely the bench is to wobble. This is useful if you're removing a lot of wood with a hand plane or boring through lignum vitae with a brace and bit. Leaning over the work allows greater tool control than working upright at arm's length, but it can also guarantee backache. If you're half as myopic as I am, cutting fine dovetails or carving detail is easier if you don't have to bend over to see what you're doing. More and more woodworkers use their bench in conjunction with hand-held electric routers and drills, which require control and visibility more than pressure, and therefore call for a higher bench.

Although I was skeptical at first, never having worked on a bench over 33 in. high, I quickly became converted with each opportunity to use Michael's 39-in.-high workbench. Michael, who is 5 ft. 10 in. tall, spoke persuasively on behalf of the higher bench. He was using it comfortably for everything from hand-planing and chopping dovetails to sanding and, increasingly, for holding production jigs while routing.

Spacers installed below the feet raise the bench height to a comfortable level for detailed work. End-cap bolt holes are tight at the front and slotted at the back to allow the top to move away from the tail vise.

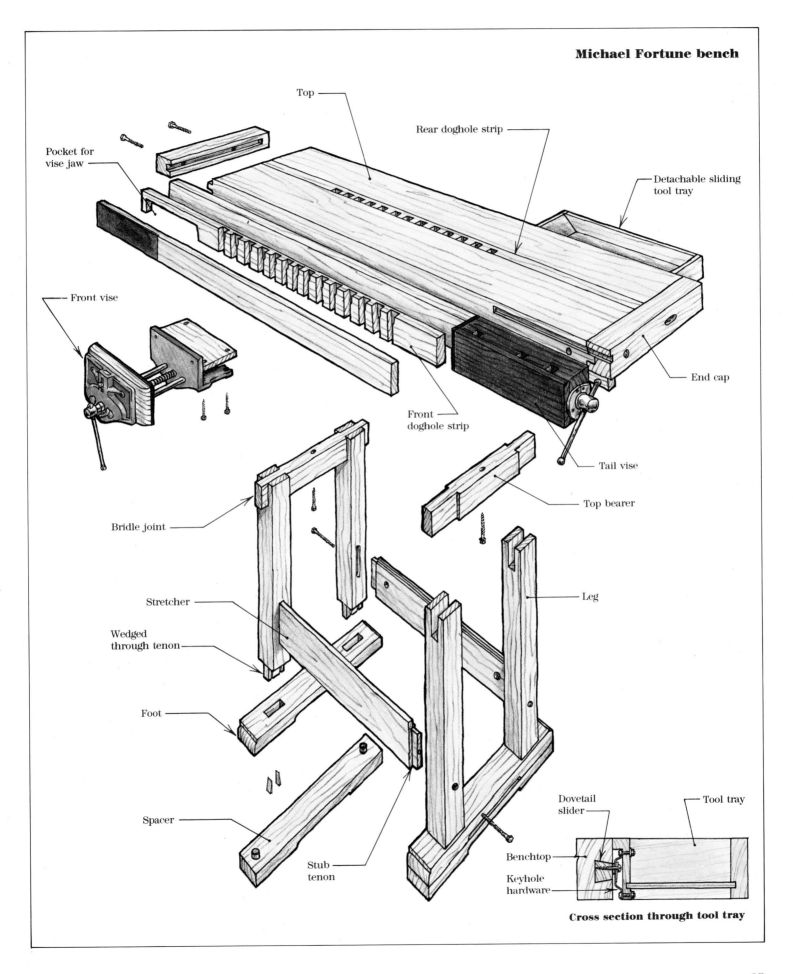

Top

Rear doghole strip

Detachable sliding tool tray

Pocket for vise jaw

Front vise

Front doghole strip

End cap

Tail vise

Top bearer

Bridle joint

Leg

Stretcher

Wedged through tenon

Foot

Dovetail slider

Tool tray

Spacer

Stub tenon

Benchtop

Keyhole hardware

Cross section through tool tray

Fortune's narrow bench provides easy access to all parts of a chair. Photo by Jack Ramsdale.

But because people come in different sizes, do different kinds of work and have different height preferences, we resolved to be as flexible as possible. People doing a variety of work on one bench may find it useful to be able to alter the bench height to suit the job at hand. We made our basic bench about 35½ in. high, and provided 2½-in.-high spacers that could be slipped beneath the feet to raise the bench. In my travels for this book, I've also seen hinged blocks attached to the tops of the trestles, as shown at left in the drawing below, which serve the same end.

The spacers are fine for making gross adjustments in the height of the bench, but they don't help at all when it comes to the dips and sags found in most concrete or wooden shop floors. Once a bench is positioned, it often requires shimming to stabilize it and tape on the floor to mark the location of the feet in case the bench gets moved. It's a general nuisance, and recently Michael devised the levelers shown at right in the drawing below. "I got so frustrated sweeping up the shims," he says, that he installed four levelers on his own bench—one on each end of the trestle feet. For the pad, Michael uses a heavy-duty nylon leveler with a ³⁄₈-in. post. This is threaded into a heavy-duty T-nut fitted to the underside of the foot. The top of the post is drilled and tapped to accept a ¼ x #20 Allen-head machine screw, which is adjusted from the top of the foot. The head of the Allen screw remains slightly proud of the wood, making it easily accessible. Michael's levelers provide sturdy adjustment of up to ½ in.—enough to keep the bench from rocking no matter where it travels in the shop.

The 14-in.-wide top on Michael's bench is particularly well suited to the willowy furniture he makes. With work clamped to his freestanding, narrow bench he can dance around it, working from virtually any position and angle. A wider top suits larger work, such as wide panels or heavy cabinets that might overhang a narrow bench or make it top-heavy. Bench stability is at least partly determined by the correct proportion of height to width; if the bench is too high for its width, it might walk across the floor or even tip during planing. Still, we couldn't discount the fact that a lot of people buy work-benches like they buy art—by the square foot.

Hinged height adjusters

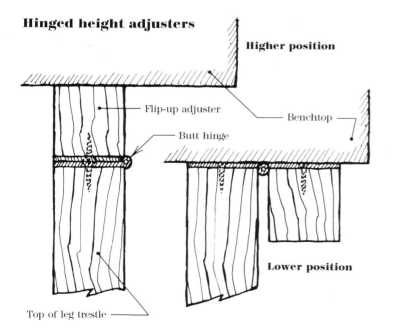

Higher position

Flip-up adjuster

Benchtop

Butt hinge

Top of leg trestle

Lower position

Adjustable levelers

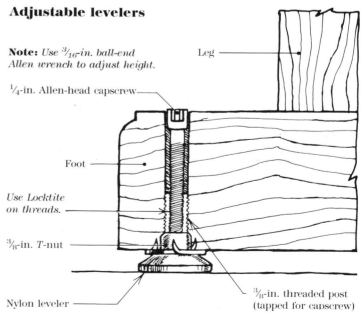

Note: *Use ³⁄₁₆-in. ball-end Allen wrench to adjust height.*

Leg

¼-in. Allen-head capscrew

Foot

Use Locktite on threads.

³⁄₈-in. T-nut

Nylon leveler

³⁄₈-in. threaded post (tapped for capscrew)

We compromised with a 21-in.-wide by 75-in.-long top—quite a bit wider than Michael's bench, but still several inches narrower than most of the other benches on the market. The length was determined by our desire to be able to trap long work between dogs. (When the tail vise is fully extended the capacity is 66 in. between dogs; an extra stop clamped on the left end handles work up to 78 in.) By using $71\frac{1}{2}$-in.-long pieces in the main body of the top, we minimized our waste by cutting our stock from 12-ft.-long boards, carefully selected to avoid end checks and other defects.

A massive benchtop helps absorb the blows of a mallet and is less likely to wander. Most workbench tops are thicker (often as thick as 4 in. or 5 in.) on the front edge than in the center (sometimes as thin as 1 in. or 2 in.), presumably to match the tail-vise thickness and to beef up the dogholes. A thick edge also creates the illusion of mass. We opted to achieve the desired mass by making the top a uniform $2\frac{3}{4}$ in. thick. The full thickness across the bench's width makes it easy to clamp work to the top or hang drawers without the nuisance of having to block out an irregular underside. It also simplifies the lumber order—all the stock can be milled out of 12/4 material.

We covered the end grain of the top with 'breadboard' end caps bolted over a continuous 1-in.-thick tenon. These end caps are attractive and help to restrain untoward warping of the top. If the end cap were run right across the full width at the left end of the bench, however, it would interfere with the operation of the vise. Every time an object was clamped in the left corner, the vise would push the end cap back, making it impossible to get a good grip. We solved the dilemma by running the front pieces of the top assembly by the end cap, as shown in the photo at left below.

Like politics and religion, the workbench tool tray elicits passionate opinion. Those who like it appreciate the extra storage capacity. Those who don't, see it as a hiding place for shavings and lost tools. Michael and I are of the latter school, so we didn't build one into the benchtop, but planned to offer the optional, removable, sliding tray shown on p. 65.

At the same time that you establish the bench's essential dimensions, it's important to consider materials. There's no point in planing 12/4 stock down to 10/4 because that's what the plans call for. Likewise, if you're sold on a 4-in.-thick top and can find only 12/4, you're going to have to laminate, cut your stock differently or change your plan. The choice of wood species depends on personal preference, price and availability. Any reasonably hard, reasonably stable wood will do, though softwoods are perfectly satisfactory for many types of work. Unstable wood, regardless of its other qualities, is more trouble than it's worth—any bench must stay flat to be of use.

Europeans find the right combination of quality, price and availability in beech; North Americans favor hard maple, a dense, close-grained and stable wood. But you'll note that the benches throughout this book are made of everything from plywood to Hawaiian eucalyptus. Michael's own bench, for example, is made of 3-in.-square lengths of Indonesian ramin, a dense, relatively stable but open-grained wood. We chose hard maple for all parts of the bench except the padauk tail vise, and we specified that the maple for the top be kiln-dried to about 8% to 10% moisture content.

The dimensions of the base should be commensurate with the thickness of the top and the overall size of the bench. We decided to follow the design for Michael's original base, which was a standard trestle construction with a sled foot and top bearer. This trestle design allowed us to inset the legs a couple of inches from the edges of the benchtop to provide clamping room around the front of the bench without reducing the width or stability of the footprint. The two other requirements were that the trestles be positioned between both vises, but as close to the ends of the bench as possible, and that the top bearers not obstruct the dogholes. (There are many other ways to build a base, each with its own advantages, and these are discussed in greater detail on pp. 99-107.)

The joinery must be sturdy and well-braced to transfer blows from the top to the floor and to resist racking. We used a conventional knockdown construction to bolt the stretchers to the glued-up trestles. The drawing on p. 65 shows the sturdy mortise-and-tenon construction. The relief on the bottom of each foot provides a four-point stance for stability, which is a virtual requirement, given the washboard floors most of us work on.

The front few laminates are run the length of the top for full vise clamping support.

Each end trestle is glued at the top in a strong bridle joint (left). Stretchers join the trestles with a stub tenon, pulled tight with a machine bolt and a tab-weld nut (right).

The underside of the Record face vise. A maple spacer is bolted to the vise casting, and four lag screws attach the assembly to the bottom of the benchtop. The rear jaw fits in a pocket, bandsawn in the doghole strip before the top is glued on.

The front vise is mounted flush with the end of the bench to grip the wood securely near the point of cut.

For vises, we went with a basic front-vise/tail-vise configuration. Neither of us felt that the wooden, European shoulder vise (see p. 57) was worth the trouble it took to build. Instead, we chose a Record #52½D. Its smooth, quick-action operation, the sliding metal dog, 9-in. wide jaw and 13-in. clamping capacity seemed appropriate. The Record #53, with a 10-in. width and 15-in. capacity, although it cost only a few dollars more, seemed unnecessarily large and out of proportion with the rest of the bench, which we were trying to streamline. The front vise is mounted at the left corner of the top, so that with the vise's wooden 'shoe' in place, a piece being trimmed can be clamped securely right next to the cut. The stationary jaw of the vise is housed in a pocket on the underside of the top, so that the front edge of the top functions as a wooden cheek. This makes it easy to clamp a long board to the front edge of the bench.

We wanted a tail vise that would be reliable, not wholly comprised of wooden parts that stick in summer and rattle in winter, and it had to be as adjustable as possible. Michael had established that a tail vise was much stronger and less likely to work loose than simpler arrangements, such as the center-screw and guide-rod vise he built across the end of his first serious bench. The principle of the tail vise is to place the pressure exerted by the screw as closely in line with the dogholes as possible. We chose a Record benchscrew and had the movable parts machined from iron flat stock. In designing the wooden components of the vise (shown in the drawing on the facing page), we avoided the *L*-shaped tail-vise construction common to the more traditional European benches (see p. 59). The bottom leg of the *L*, which accommodates wooden guides, is unnecessary with the metal hardware. (Because this custom-made hardware is difficult to make, the vise shown in the drawings on pp. 236-237 and described at the end of the chapter is based on one made by Tom Nelson of Bellevue, Washington, which uses commonly available hardware.)

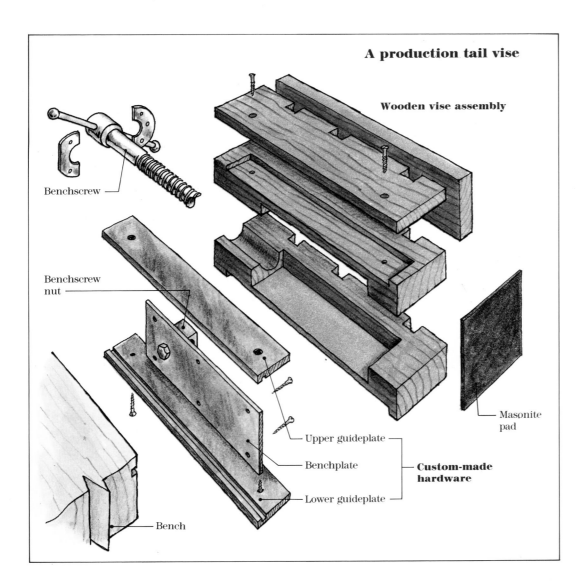

A production tail vise

Wooden vise assembly

Benchscrew

Benchscrew nut

Upper guideplate
Benchplate — **Custom-made hardware**
Lower guideplate

Bench

Masonite pad

In addition to the row of dogholes along the front of the top, another row in the middle makes it possible to set up a triangulated clamping pattern on frames or large projects. The holes are angled slightly, which causes the tip of the dog, rather than its entire face, to bite into the work so the dog need not stick up from the top very much to get a good grip—even on an irregular surface. The staggered spacing of the holes in the tail vise allows clamping pieces of different lengths with a minimum of tail-vise adjustment. For the vise, we selected African padauk, a bright-red wood that darkens like blood with exposure to the air. At about $6.00 per board foot, it might have been an extravagance if its only benefit was its stunning contrast with the maple. But at least as important is padauk's stability. The tail vise is the critical moving part of the top and the workbench is either successful or not depending upon its consistently smooth action. The $15 it tacked onto the cost no longer seemed extravagant.

The five prototype benches we built in 1983 constituted our total production. There were a host of reasons for the project's demise. I like to think it had nothing to do with the quality of the bench, unless it was that we built more of it in than we were able to charge for. Michael moved on to other projects, and I moved on to write about workbenches instead of writing receipts for them.

Two rows of dogholes in the top, and one in the vise, provide a firm three-point grip on wide or irregularly shaped boards.

Hardware: Finding the right stuff

Hardware can be the bane of a woodworker's existence. We usually design first and then try to bolt and screw our dreams together with off-the-shelf components. If we're lucky it works, but just as often either the hardware or our dreams have to be cut down to size. For some specialty purposes the hardware can also be prohibitively expensive.

Such was the case when Michael Fortune and I and another partner set out to design our tail vise. Instead of paying full price for off-the-shelf hardware, we decided to have our own custom-made. It would require a much bigger up-front investment, but we felt that it would pay off over the long haul. We would get just what we wanted, and we would be able to guarantee the quality.

With entreprenurial confidence, I embarked on a walking tour of Toronto machine shops. I located a source for 18-in., Acme-threaded Record benchscrews and needed only the three metal plates, cut to length, drilled for screws and machined with grooves, and a nut for the Record screw. A simple project. Or so I thought. Instead I was turned away from more shops than I care to remember. The odyssey taught me more about the economics of production in two weeks than I would have learned in a year of business school.

In the sprawling suburban outskirts of town I finally found a willing, if not enthusiastic, machine shop—for a minimum order of a hundred units. Our initial run was for five prototype benches, but we were optimistic. We persuaded Atlas Machinery Supply Ltd. (see Sources of Supply at the back of the book), a local tool store, to join us in the order. The milling machines began to roll. We took our five vise assemblies back to the shop and the other 95 were delivered to Atlas, where I feel certain most of them can still be purchased. We were ready to begin making benches.

Planning for production, Fortune designed the tail vise shown in the photo below and in the drawing on p. 69, which we built out of African padauk. The four-piece construction required four router jigs to cut the recesses for the vise hardware. A core-box router bit was used to cut the half-round channels for the benchscrew. Two more router jigs were needed to prepare the mating face of the workbench for the mounting hardware.

The beauty of having hardware manufactured to your own specifications and using production jigs is consistency. The jigs result in a neat, predictable tail vise, and the hardware fits perfectly. The fifth vise will work as well as the first. But, unless you plan to build the other 95 benches, it's a lot more efficient to buy standard components and make the vise without jigs. Some dreams are best trimmed down to fit reality.

Machining the grooves in the guideplates for a hundred vises.

The parts for one padauk tail vise ready for assembly. The custom-made guideplates, benchplate and nut are fitted to an 18-in. Record benchscrew. The vise block is laminated out of four separate pieces, prepared with router jigs.

Building the bench

I went back to Toronto when I began this book to make one more bench with Michael—this time for the record. He orchestrated the project, while I helped mill stock and took notes and photos. We made the top first, from roughly 30 bd. ft. of 12/4 maple and 5 bd. ft. feet of 5/4. You might want to build the base first, so you'll have something to set the benchtop on when you make it. Honeycomb is common in heavy stock that has been pushed through the kiln too fast, so it's worth taking the time to look carefully at the end grain for telltale checking, although it has a way of appearing only after you've taken the plank home and cut it open.

The basic top consists of eight pieces, as shown in the drawing on p. 65. Two 8-in.-wide pieces sandwich the rear doghole strip, two narrower pieces flank the front doghole strip, and an end cap abuts each end. In our first night, Michael and I machined all the wood in the spacious Sheridan shop. As big as your shop may be, you'll be surprised at how it shrinks when you're building a workbench.

The first step is to trim the lumber roughly to length and mill it to size. Although we had a 20-in. jointer and thickness planer, the top's widest components are only 8 in., and you can glue them up from narrower strips. Mill all pieces slightly thick, then run them through the planer at the same setting for a uniform final thickness. Hand-planing is admirable, but I recommend saving your strength—take the rough stock to a millwork shop to have it dimensioned if you don't have the equipment or the space to do it yourself.

From Sheridan, we moved production to Michael's own shop. Located behind his house just off one of Toronto's main downtown streets, the shop is a small, red-clapboard building. Adjacent to an outdoor wood storage shed, the shop squats beneath a Byzantine church structure to the south and a high-rise apartment complex to the west. Across the street, trucks spend the day rolling candy bars away from a Neilson factory; the pungent odors of chocolate and diesel comingle on Michael's side of the street with the fragrances of maple and padauk. A narrow drive that you might easily miss leads back to the shop and the house where Michael lives with his wife, Janice, and daughter, Jennifer. The house and shop could be transported to any country location, and I was startled to find myself in downtown Toronto every time I turned out of the drive.

The details of construction.

Doghole sled

Note: *Angle of runners determines angle of dogholes. Distance between indexing peg and sawcut determines spacing between dogholes.*

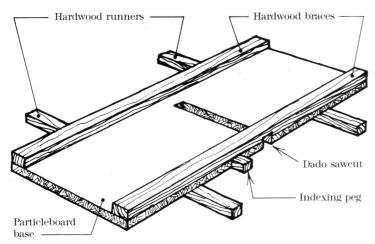

Hardwood runners

Hardwood braces

Dado sawcut

Indexing peg

Particleboard base

Note: *Indexing peg fits tightly in dogholes. It is screwed from below into a notch in jig.*

1. *Remove peg to cut first slot.*
2. *Install peg and slip first slot over it to cut next slot.*
3. *Continue until all slots are cut.*

Doghole 'blip' jig

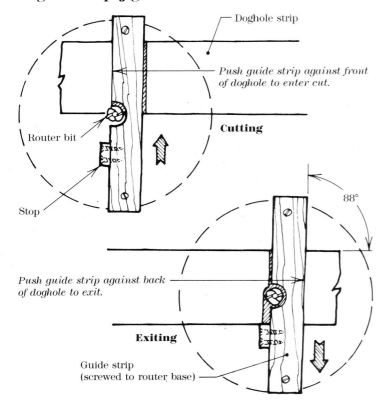

Doghole strip

Push guide strip against front of doghole to enter cut.

Router bit

Cutting

Stop

88°

Push guide strip against back of doghole to exit.

Exiting

Guide strip (screwed to router base)

Fortune cuts the dogholes on the tablesaw (above right), using a dado blade and a particleboard sled. Waxed hardwood runners on the bottom of the sled maintain the proper angle of the slots. A small block of wood screwed to the front of the sled (shown in the top drawing above) maintains the spacing between them. The router cuts the 'blip' as it is pushed into the slot. The long guide strip on the jig should be slightly narrower than the width of the slot to allow you to move the bit out of the cut as it is withdrawn (right).

When we resumed construction, Michael cut the slots in the two doghole strips, using a dado blade on the tablesaw and a sliding particleboard sled. The sled holds the strips at the correct 2° angle and has an indexing peg that maintains the proper interval between slots, as shown in the drawing at left. The benchtop requires one sled and the vise another, angled in the opposite direction. It's remarkably easy to cut the dogholes in the wrong direction, so take a moment to check the sled angle and to make sure that the correct edge of the strip faces the blade. You can also cut the slots without jigs on the tablesaw or the radial-arm saw or with a router. (To make this bench, we used jigs developed for producing batches of workbenches. For one bench, jigs are often unnecessary, so I'll mention alternative ways of doing things where appropriate.)

We cut a shallow shoulder, or 'blip,' at the front of each slot to house the head of the benchdog and keep it from slipping through the top. This can be chopped with a chisel after the top is glued up, but it's easier to cut it at this stage using a small router jig, shown in the drawing. The guide strip slides in the doghole slot with just enough play to be pushed away from the cut as the bit is withdrawn. A small stop is screwed to the guide strip on the jig to stop the cut at the correct depth.

Rather than setting the inner jaw of the face vise in a routed inset on the front edge of the bench, Michael prefers to insert the jaw in a pocket behind the front edge, as shown in the top photo on p. 68. This makes for quick installation and provides a solid rear cheek for the vise. (Together, the pocket and the wooden 'shoe' on the front jaw reduce the vise's clamping capacity by about $1\frac{1}{2}$ in.) Bandsaw the cavity in the slotted doghole strip. Be sure to leave at least $\frac{1}{2}$ in. of wood above the cavity—this piece may appear flimsy, but it's immobilized once the top is assembled.

Michael faces the vise cheeks with $\frac{3}{16}$-in.-thick tempered Masonite, held in place with double-sided tape. These facings protect the wooden cheeks and the work, and can be easily popped off and replaced. If you want to do the same, bandsaw the relief for the Masonite on the face strip.

Now the top is ready to be glued up. First glue the two 8-in.-wide pieces to the long doghole strip. Gluing up chunks of maple this size requires speed, lots of stout clamps and all the hands you can muster. Because our pieces were cut to length first (to avoid running large rafts of wood over the tablesaw), we had to be careful to avoid slippage during glue-up. We used an Elu biscuit cutter to register the pieces at both ends; dowels or splines would work as well. In our previous benchtop-gluing episodes we used pipe clamps—almost every one in the shop, as I recall. We pressed Jorgenson bar clamps into service this time around. A maple workbench top is a big chunk of wood, so if you go with pipe clamps, be sure that the pipe is heavy-gauge and that you have plenty of them available.

Because it allows longer open-joint time, we used white glue instead of yellow glue and spread it on both mating surfaces. Most of us overglue our work in a futile attempt to compensate for less-than-perfect joinery. Better too much glue than too little, I suppose, but in this case you'll pay for your excess with a lot of cleaning up—especially in the dogholes. And, as Michael points out, "I've never seen an object delaminate from the middle." (You can minimize glue in the dogholes by spreading glue for that joint only on the slotted strip.)

Set the first clamps in the middle of the bench and work out in both directions, alternating adjacent clamps on opposite sides of the top. Before cinching the clamps tight, check that the ends and surfaces of the pieces are all aligned. To aid surface alignment, we clamped several pairs of battens above and below the top, bolting them through the center doghole strip wherever possible, as shown in the photo below.

When the glue has cured to a rubbery consistency, take off the clamps and remove the glue squeeze-out on top and bottom with a scraper. It's easier to clean the excess glue out of the dogholes now with a chisel, before it hardens. (You can also mop up glue in the dogholes by running a damp rag through each slot while the glue is still wet.) Next, follow the same procedure to laminate the three narrow strips that comprise the front section of the benchtop.

Alternate clamps above and below the top for even pressure. Battens bolted through the dogholes help keep the top flat.

Before you can glue the front section to the main body of the benchtop, you have to rout the groove for the tail-vise top guideplate and cut the tenons for the breadboard end caps. Consult the tail-vise description on pp. 76-79 and make sure you understand how the hardware works before routing the guideplate groove. (Remember to make whatever adjustments are necessary for the hardware you plan to use.) Michael used a router jig mounted on the edge of the large bench section to align the groove; a router-mounted fence would work as well.

Michael cut the 1-in. by 1-in. tenon on both ends of the large benchtop section with a circular saw, chisel and router. He made a straightedge jig to guide the circular saw and the router; its front lip clamps square to the front edge of the top and it extends beyond the back edge to provide bearing surface for the tools at the start of the cut. A simple straightedge fence clamped to the benchtop section will work fine. Make multiple passes with the circular saw to remove the bulk of the wood, then chop the chips away with a chisel. Clean the tenon to its finished dimensions with a router.

At last, the two sections of the top are ready to be joined. It is very difficult to rectify an error made in the alignment of these two sections, so take particular care that exactly the right amount of room is left for the tail vise and that the dogholes in both strips are in alignment across the top. Also make sure that the surfaces of the sections are flush. If they're not, the tail-vise jaws will not be square to each other, and it will take a lot of hand-planing to flush the top surface. Once again, we used Elu biscuits to keep the two sections from slipping lengthwise and clamped battens above and below to keep their surfaces flush.

Fortune cuts the tenons on the ends of the top using a circular saw and a router. After making a series of quick cuts with the saw, he knocks out the chips with a chisel and then cleans the tenon with a router. A special plywood base screwed to the base of the saw enables him to use the same jig for both the saw and the router.

An Elu biscuit at both ends of the front section of the top will help keep it aligned during gluing.

Alternate end caps

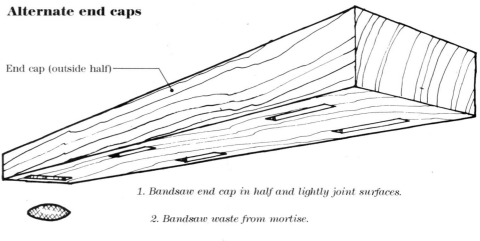

End cap (outside half)

1. Bandsaw end cap in half and lightly joint surfaces.

2. Bandsaw waste from mortise.

3. Prep both halves for alignment biscuits and reglue.

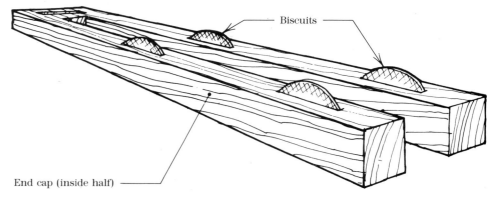

Biscuits

End cap (inside half)

To complete the top, make and install the end caps. On our original benches we cut out the bulk of the stopped mortise using a dado blade on the tablesaw. The remaining corner was drilled out on the drill press and cleaned up with a chisel. If you don't have a dado head, you could rout the mortise. Michael also devised another method requiring a bandsaw and a biscuit joiner, as shown in the drawing on the facing page. He ripped the end caps in half on the bandsaw, rejointed the ripped surfaces lightly, then bandsawed the waste from the mortise. The two pieces were prepped for alignment with an Elu biscuit cutter and then reglued.

Fit the end caps to the tenons and extend the tail-vise guide-plate groove through the end of the right end cap. The end caps are attached with two countersunk $\frac{3}{8}$-in. machine bolts that thread into captured nuts installed from the underside of the top; this avoids screwing lag bolts into end grain. The front bolt fits snugly in a $\frac{13}{32}$-in. hole, while the shaft of the rear bolt floats in a $\frac{5}{8}$-in. hole bored through the end cap, allowing the top to expand and contract away from the vises. An elongated slot must also be made for the rear bolt in each end cap to leave room for movement around its countersunk head and washer. Michael used a router and a simple jig for this, but you could do it on a drill press by boring a series of connected holes, then trimming with a chisel. Bore the holes in the end cap, then clamp the end cap in position and bore into the top. Flip the top upside down and bore the holes for the captured nuts, being careful not to go too deep. If all the pieces in the top have been aligned carefully during glue-up, the joints need only be scraped or planed very lightly and the whole assembly is ready for the tail vise.

I suggest you build the base before tackling the tail vise—it's easier to work on the vise at bench height. We milled the trestles from 12/4 maple, making the feet $\frac{3}{8}$ in. wider than the legs to avoid having to flush surfaces. There is nothing unusual about the trestle joinery. The legs attach to the sled foot with through tenons, double-wedged from the bottom. A bridle joint fixes the legs to the top bearer. Be careful to clamp the trestles square when you glue them up, or your bench will go through life with a permanent wobble.

The bottoms of the trestles are relieved, which provides stable, four-point contact with the floor and makes it easier to keep the bench level. Michael cut this relief on the jointer, as shown in the photo at right, lowering the outfeed table to the same depth as the infeed table and setting stops for both ends. Beginning with a light cut, he lowered the foot down on the exposed cutterhead and pushed it between the stops. Both tables were lowered slightly for each successive cut to the desired depth. (You can achieve the same result by running the feet over a dado blade on the tablesaw in several passes with two stops clamped to the fence.)

The stretchers are $1\frac{1}{4}$ in. thick and $5\frac{3}{4}$ in. wide to make the most out of the 6/4 by 6-in.-wide stock we had, but as with most of the bench parts, there is nothing sacred about those dimensions. The stretchers are positioned on the legs by stub tenons, and attached with $\frac{3}{8}$-in. by 4-in. machine bolts and tab-weld nuts, which seat nicely in the drilled hole in the stretcher and provide a convenient protruding tab for easy installation. They are shown in the drawing on p. 234. These holes should be bored in the same sequence as those for the end-cap bolts and nuts.

The elongated countersunk slot in the end cap can be cut with a router and a simple jig.

Fortune relieves the trestle feet on the jointer by lowering both tables to the same depth and pushing the foot between stops.

The tail vise

Michael Fortune built his tail vise using custom-made hardware and six router jigs to speed their production in small batches. Since most people make only one vise, I adapted a similar but more straightforward vise made by Tom Nelson, of Bellevue, Washington, for this bench. Nelson, who has taught several workbench-making classes in Bellevue, designed his vise to be made with readily available hardware and no jigs. He's supervised the building of about 35 benches to date, and figures he's worked out most of the bugs.

Like Fortune, Nelson dispensed with the bottom of the *L*, found on the traditional vise. For stability, it is necessary to build the all-wood vise in the shape of an *L* to support the wooden slides that run below the bench. The sliding metal plates make the *L* shape redundant, and remove the temptation to clamp wood between the leg of the *L* and the end of the bench. Such a practice strains the vise and throws it out of alignment by placing pressure too far away from the benchscrew. "The temptation is too great to use what I call the 'tonsil,'" Nelson explains. "So I give all my benches a tonsillectomy."

The guts of Nelson's vise are comprised of a steel benchplate screwed to the edge of the bench and two grooved steel guideplates that are bolted to the vise and slide on the benchplate. No other guides are required to make the vise function properly. Tom uses Hirsch hardware made by Gebruder Busch, a German firm, available through several suppliers listed in the Sources at the back of the book. Other brands and sizes of similar hardware will work as well, but may require adaptation of the dimensions of the vise and related bench parts. This shouldn't be a problem once you understand how the vise works.

The Hirsch hardware is readily available in two different sizes—with a nominal benchscrew length of 14 in. or $17\frac{1}{2}$ in. Tom Nelson uses the larger hardware on his bench, which is much larger than Michael Fortune's. For the purpose of this discussion, however, I have followed Nelson's method to install the smaller vise on Fortune's bench. It more closely approximates the hardware Fortune and I used on our benches and is less disruptive of the bench's proportions and features. You will note that the photos of Nelson's bench all show the larger vise. The measured drawings on pp. 234-237, however, are based on the smaller hardware.

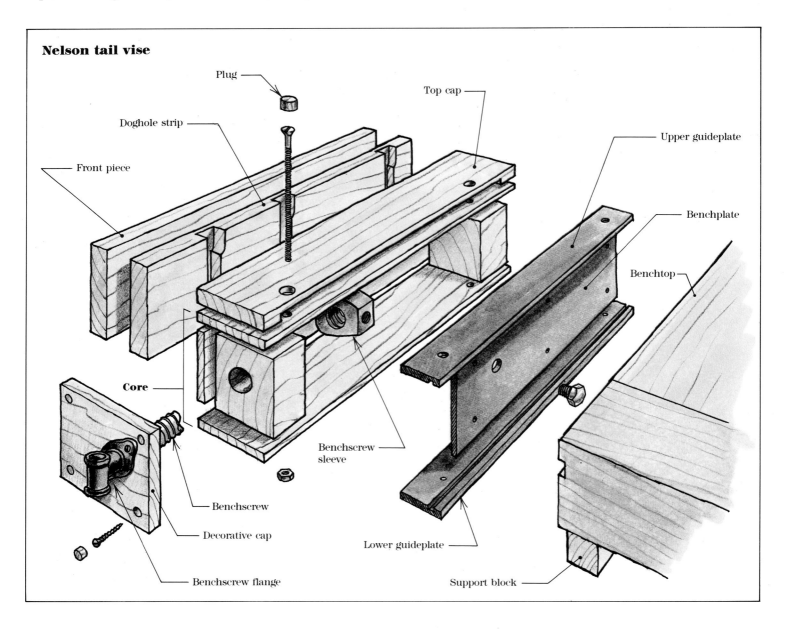

Nelson tail vise

Plug

Doghole strip

Top cap

Upper guideplate

Front piece

Benchplate

Benchtop

Core

Benchscrew
sleeve

Benchscrew

Decorative cap

Lower guideplate

Support block

Benchscrew flange

Tom Nelson's tail vise is built with conventional store-bought hardware.

The instructions that sometimes accompany the vise hardware suggest excavating a single, solid chunk of wood to accommodate the metal parts—a tedious, and potentially dangerous process, as Tom's first bench class discovered. For his next class, Nelson created what he calls a 'tail-vise core' of five separate pieces glued together, as shown in the drawing on the facing page. To this core, he glues a separate top cap and a front piece, which includes the doghole strip. The internal cavity is simply created by leaving a void, and the benchscrew hole can be bored in one end of the core on the drill press before the core is assembled and glued. After making several vises, Tom realized that not all of the benchscrew's length was being used, so he added almost an inch in the cavity. "It's like fine-tuning a car to make it a hot rod," Tom says. "You may only use it once a year, but when you do, you say 'all right!' "

It can't be said too often, especially of the tail vise: get all your hardware before you build your bench. And never assume the hardware to be as specified in the catalog. Measure it carefully and be prepared to adapt the wood parts to fit. From bent bolts to coarse screw threads and irregular plates, I can think of at least half a dozen places where the hardware may not be accurate. Assemble your benchplate and the upper and lower guideplates and check their fit *before* you build your vise.

The height of the assembled hardware in relation to the height of the wooden core is critical. When the grooves in both guideplates are seated on the benchplate, the guideplates should be about $3\frac{1}{4}$ in. apart, or about $\frac{1}{32}$ in. more than the height of the tail-vise core, as shown in the drawing at right. If the core is too large, the guideplates will not seat squarely on the edges of the benchplate and the vise will wobble. If the core is undersize, the guideplates can be shimmed square from below and the shims can be easily adjusted as the bench parts swell and shrink. Shimming used to seem like cheating to Nelson. That was before he dismantled a Gibson Mastertone banjo (the Cadillac of production banjos) and found a matchbook shim between the heel and rim.

Fitting the vise core

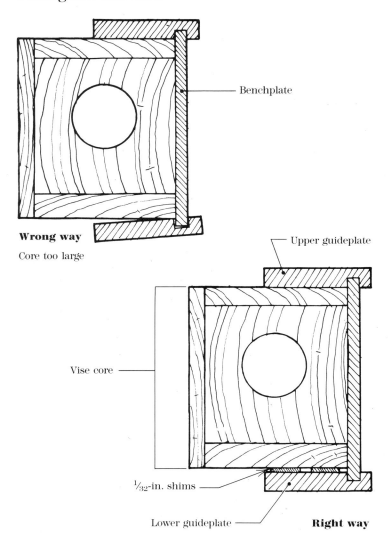

Benchplate

Wrong way
Core too large

Upper guideplate

Vise core

$\frac{1}{32}$-in. shims

Lower guideplate

Right way

When you've checked the hardware and altered the vise dimensions if necessary, mill the wooden vise-core parts. Because there's so much glued surface area, no other joints are necessary. Before gluing the core together, bore the benchscrew hole in the appropriate end. Nelson uses a spur bit (a $1\frac{1}{8}$-in. bit will work for the small hardware), which cuts cleanly if you work slowly and clean the chips out so the bit doesn't overheat. I've also used a spade bit, reground with a smaller point. Take care to glue the core together squarely; dowels or small finish nails will help align the parts. It's easiest to make the parts precisely and glue the core together carefully—that way, you can scrape off the glue and you're ready to go. If the core is slightly oversize, or if parts are not precisely aligned, it will be necessary to plane it to exact dimensions.

If you haven't routed the $\frac{3}{4}$-in. by $\frac{1}{4}$-in.-deep upper-guideplate groove in the edge of the benchtop, do so now. This is much more easily done before the front doghole strip is glued to the top, but you can still rout most of it afterward, completing the groove with a chisel if necessary. For any benchtop up to $4\frac{1}{4}$ in. thick, the lower guideplate clears the underside of the top and does not require a groove. If your top is thicker, rout another groove or a recess in the underside (depending on thickness) to allow the lower guideplate to slide freely. If your top is thinner than $4\frac{1}{4}$ in., glue or screw a block to the underside of the bench to support the bottom of the benchplate.

The next step, mounting the benchplate on the front edge of the benchtop, may be the most critical step in the whole workbench-building process. "Everything rests on where that plate is located," Tom emphasizes, "so it's nice to get it right on the money." It takes a little time, but careful work now will save hours of grief and a pile of plugged screw holes.

First, align the top edge of the plate parallel to, and $\frac{15}{16}$ in. below, the top of the bench, as shown in the photo at left below. (Tom's vise has a thicker top cap, so the plate is farther down in the photo.) Mark all the holes in the plate and bore pilot holes for #12 flat-head wood screws in the middle of the upper-left and the lower-middle screw holes. Bore a $1\frac{1}{8}$-in.-dia. hole about $\frac{7}{16}$ in. deep to house the nut that attaches the threaded benchscrew sleeve to the benchplate.

With the sleeve bolted to the benchplate, screw the plate to the top, using just two screws. Position the core and two guideplates on the benchplate and clamp the guideplates to the core, 1 in. from the front end of the core. Shim the lower guideplate if necessary so that it's square to the benchplate. Check that the core fits snugly against the benchplate. If you've done a neat job, the outer face of the vise core should line up with the doghole-strip glueline in the benchtop. When you're satisfied with the fit, check that the guideplates are still 1 in. from the front end of the core and mark the location of the bolt holes that attach the plates to the core. (You can use screws to attach the guideplates, but bolts are easier to adjust.) Bore the bolt holes on a drill press, and attach the plates to the core.

Loosen the guideplate bolts enough to allow you to mount the core on the benchplate, tighten the bolts down and wind the benchscrew into the sleeve. Center the benchscrew in the hole and screw the mounting flange into the end of the core. (Nelson installs a decorative cap on this end later, to which the mounting plate is refixed.) You now have the skeleton of a working tail vise. Tom has his students use the vise in this condition to make final adjustments. It's at this point that they finally realize that the vise is actually going to work.

Open and close the vise to be sure it slides smoothly and fits well. It should be snug, with very little play when fully extended. When closed, the jaws formed by the core and bench should fit tightly and square to one another. The vise core and top should be parallel; check by measuring with the vise closed and open, and correct by moving the benchplate up or down at one end as necessary. If the plate has to move up on the right, for example, remove the vise core and bore a third pilot hole in the upper-middle hole in the benchplate. Back off the bottom right screw and install the new one. If that levels the plate, leave the new screw in place, enlarge the pilot holes and replace the original two screws with #14 screws. Then replace the upper-right screw with a #14 and install #14 wood screws in the remaining holes.

The object, as Tom explains, is to "get that rascal level!" Don't assemble the rest of the vise until you are completely satisfied that the benchplate and tail-vise core are level.

Nelson makes sure that the top of the benchplate is parallel to the benchtop.

Clamp the guideplates on the benchplate and to the core, shimming below the core, if necessary, to get the guideplates square.

The top cap and front piece (including the doghole strip) finish the tail-vise assembly. Rout the cavity that houses the upper guideplate in the top cap. Glue the top cap first, making sure it fits pefectly flush with the front face of the core. (If you work carefully, you can glue the top cap and front piece right on the mounted vise core, which helps align all the parts.) In Tom's vise, the top cap permanently traps the bolts; you can still remove the vise, by loosening the lower guideplate, but the bolt heads are covered. To make them accessible, simply drill holes in the cap and fit them with removable plugs.

When the top cap is attached, glue the front piece to it and the core. Both top and front pieces should be slightly proud of the surface of the bench so they can be planed flush with the front edge and top of the bench, which is a lot easier than planing the top to match the vise. The vise-jaw end should require only a light touch with a scraper or block plane to clean the gluelines. If the jaw does not mate tightly with the bench, you'll have to trim it to fit on a large radial-arm saw or a tablesaw. You can also line both faces of the tail-vise jaws with pads of $\frac{3}{16}$-in. tempered Masonite, held in place with double-sided tape.

As a finishing touch, Tom covers the right end of the vise with a decorative cap, screwed not glued in place. The cap is essentially cosmetic—it covers the end of the upper guideplate and the end grain of the tail vise—but it also provides a better anchor than the end-grain core for the mounting screws that attach the benchscrew flange. "Make it pretty if you want," Tom suggests. He made it out of rosewood for his first bench, to match the rosewood details all over the bench.

You know the vise will work well, because you've already tested it. The nuts that hold the bottom plate need not be cinched tight. It may be necessary as the bench seasons with time, use and climate, to re-tune the tail vise. Simply loosen the two bottom nuts and add or remove shims between the guideplate and the core. When the vise is completed, you can keep it running smoothly by rubbing paraffin on the edges of the benchplate and in the guideplate grooves and making sure the screw and sleeve are free of sawdust.

While Tom prefers wooden benchdogs, he made the dogholes in his bench large enough to accommodate the commercially made metal variety. Michael Fortune and I made the slightly smaller wooden dogs shown in the photo at right. Their shafts are beveled to accommodate a thin spring of white ash, which is glued and screwed at the bottom. We included three with each bench—two of padauk to contrast with the top and one of maple to contrast with the padauk tail vise. Make them easy to find, we figured, on a late night in the shop or buried beneath drifts of sawdust.

Like any other piece of woodwork, the workbench should be coated with a protective finish. A good finish extends the life of the wood by replacing some of the moisture given up during the drying process. The wood becomes less brittle and slower to move with changes in humidity. Finishing also toughens the surface of the bench against knocks and scratches as well as spilled glue or cups of tea. Fortune finishes his benches with a sealing furniture oil, such as Watco. Nelson prefers Watco cut with turpentine and a touch of Varathane varnish to raise the sheen. There are many other recipes—some using linseed or tung oil—that will work as well. Whatever you choose, an oil-based finish is preferable to a straight varnish coating. The oil penetrates deep into the wood, polymerizes to seal it and is easily renewed without sanding.

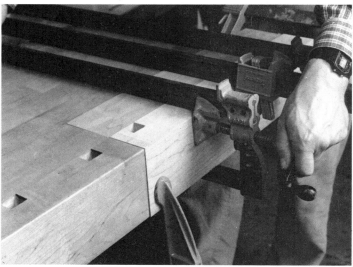

You can glue the front piece on while the core is mounted on the bench. Note that the gluelines in the vise and the benchtop are aligned for a neat job. The front and top piece of the vise have been left slightly oversize so they can be planed flush with the benchtop.

Hardwood dogs in contrasting colors with white-ash springs are a useful and attractive finishing touch.

The essential workbench, according to Ian Kirby, is simplicity itself. Photo by Scott Jansen.

A Basic Bench

lose your eyes for a moment and picture a cabinetmaker's workbench. There's a top and four legs, a vise on the front, a tail vise on the right-hand corner and a row of dogholes marching down the length of the bench. That describes most of the workbenches I've seen across the country, and many woodworkers, like Michael Fortune and Frank Klausz, wouldn't have it any other way. But there's no denying it—building this type of traditional workbench is a lot of work.

Before you leap headlong into lamination, dogholes and an intricate tail vise, you might want to examine the way you work and consider just what you expect from a bench—particularly if it's your first. As I explained earlier, workbench design is inextricably bound to working methods—the nature of the work you do, the tools you use and the way you use them. You wouldn't use a shaving horse to join jewelry boxes or a Japanese planing beam to carve chair parts. Beginning woodworkers, uncertain of their skills or the direction their interests will lead them, may not want to put a lot of time, effort and money into a complicated cabinetmaker's bench right at the start. A basic bench, easy to build and maintain, can help beginners gain the woodworking experience necessary to decide whether they want or need a more complex bench.

Before I began researching this book, it seemed to me that the traditional workbench with tail vise was the cabinet-

maker's ideal, the end of the evolutionary trail. But in my travels I discovered skilled cabinetmakers who prefer simpler benches. Ian Kirby, a transplanted Englishman who now runs a school and workshop outside of Atlanta, Georgia, has been using a workbench without a tail vise all his life, and speaks passionately on their behalf. Kirby builds two different benches for his shop and for sale: a straight-ahead wooden bench, which is rooted in his English background, and a combination workbench/veneer press made of tempered Masonite and medium-density fiberboard in a torsion-box construction (see the sidebar on p. 88). As workbenches, both models function in the same way.

The workbench, according to Kirby, is extraordinarily simple. "It's a flat surface with square edges," he says. "As far as you can get absolutes in wood, it's a geometrically regular piece of wood. That's all." Boiled down, Kirby's workbench is a top with four legs. Kirby is a furniture designer and maker, and like most contemporary woodworkers, he uses machines to dimension wood and to cut basic joints. Hand tools are used for fine joinery and detail work, and to deliver quality of surface—cutting dovetails or removing machine marks with a razor-sharp plane, for example.

More than anything else, Kirby's bench is designed with the hand plane in mind. When Kirby planes a piece of wood at the bench he simply pushes it against a single stop. No dogs, no tail vise, no benchscrew. It's simplicity itself. "Sure, you could put the work in a tail vise," Kirby says, "but you would pay a price. Once you start holding wood you immediately divorce

Ian Kirby relaxes on his workbench/veneer press.

The benchroom at Kirby Studios is a mix of the old and the new. The students get a chance to try out Kirby's recently patented workbench/veneer press (in the foreground), as well as the traditional, solid-wood models.

yourself from a large amount of 'data feedback.'... The wood can't respond—it's held.... You won't be able to sense the incorrect use of the tool. It's like trying to listen to good music with earmuffs on."

Moreover, Kirby points out that when you trap wood in compression between dogs, you run the risk of bending it. Kirby's method of pushing the wood against a stop allows the wood to maintain its undistorted shape while it is being worked. To accommodate work of different dimensions and to enlarge the range of operations that he can perform at the bench, Kirby employs several other aids, such as panel stops, shooting boards and bench hooks. Because of the utter simplicity of the workbench, these items are really much more than accessories—they're an integral part of his working method.

To see Kirby's methods and his bench first-hand, I visited him in his temporary workshop and studio in Cumming, Georgia, where he recently moved from Vermont. The shop takes up somewhat more than half of a pre-fab metal building. The main working area is the huge benchroom, outfitted with several rows of workbenches—about 20 benches in all. The older, maple-top benches are in use alongside Kirby's new, veneer-press models. With a low-pitched roof, exposed I-beams and cement floor, the shop has all the old-world charm of an airplane hangar.

To demonstrate the utility of his workbench, Kirby grabs a Record #07 jointer plane and puts a bench through its paces. The bench stop, a simple wooden device about 3 in. wide and 1 in. thick, slides up through a mortise in the benchtop to the left of the front vise. Below the top, the stop is slotted for a bolt and wing nut, which attach the stop to the adjacent leg of the bench at the appropriate height. Kirby's bench has two stops, as shown on the drawing on p. 86, which can be used singly or together.

Moving quickly and efficiently in his white smock, with his body in a crouch, Kirby looks like a judo master. Butting one end of a 1x2 against the stop, he strips off a shaving, keeping his weight centered over the plane and work to maintain consistent pressure. In an instant, he can flip the piece to plane another surface or turn it end-for-end to plane with the grain. When planing an edge, if the plane isn't square to the board's face, the wood will tell him—it will fall over or he'll feel it. In the worst situation, Kirby explains, when you've got a very irregular edge to plane, or a board so thin that it won't stand on the bench, you can clamp it in the front vise to true the edge.

Next, Kirby demonstrates chiseling. He works directly above the piece, his hands holding the work in position. If pressure is exerted straight down, the work won't move and vises or clamps become redundant. Kirby teaches his students to scrupulously protect the benchtop by inserting another board between it and the work when chiseling.

Kirby places great importance on stance and body movement while working, so bench height is critical. He figures his 34-in.-high bench is just about right. (Kirby is 5 ft. 9 in. tall.) When planing, you need to be able to get on top of the work to exert pressure and maintain a long, even stroke. "A six-foot guy would be doing a lot of bending over," Kirby notes, however, and he suggests adjusting the height up or down for comfort.

The thickness of the top is also a matter of personal preference—a solid-wood top should be thick enough to resist deflection and to allow periodic flattening as it moves. Kirby's bench may be simple, but it is far from crude. To be useful as a planing surface, the benchtop must be absolutely flat and without twist. If the benchtop sags or is crowned, it will be impossible to flatten stock. Kirby planes his top to within +/− 0.002 in. of dead flat, removing between 0.005 in. and 0.008 in. each time. The underframes on the bench are spaced on 51-in. centers. The top is about 3 in. thick. "That leaves you a few chances to flatten it," explains Kirby.

The 66-in. by 30-in. top is shorter and wider than most of the workbenches I've seen, but Kirby finds it sufficient for most of the work he has to do. While you can't work around it as easily as you can a narrow bench, the extra width is useful for planing wide panels. Kirby also uses it as a 'surface plate' for keeping work square during assembly and glue-up. "It could go longer," he says, "but it gets more costly."

What about a tool tray? In England, Kirby worked at times at double-sided benches. One man worked on each side, with a tool well down the middle. It was a convenient way to separate workers, but Kirby concluded that the tool tray, or trough, is more bother than it's worth. "It's just a dirt trap," he says. "Tools get jumbled around in there and planes stick up above the work surface.... I don't believe that anything should be on the bench. The process that is under way at the time is all that should be happening. It should not be a collecting place for tools. It should not be a garbage heap."

Consistent with his simplification of the rest of the bench, Kirby dispensed with end caps, or battens. He was convinced that the standard end cap could not resist a top's tendency to cup. "I finally determined how I would install a breadboard end batten, and it became so long-winded that I took it off," he says. "Why not just use decent materials and keep it flat?" If the bench shrinks, the breadboard end caps stick out. If it swells, the caps become recessed. You can inhibit moisture movement by oiling the top or waxing the end grain.

By pushing the work against a single stop, Kirby relies on a keen blade and good balance to plane the edge of a narrow board. Without a vise to hold it, the wood will fall over if he gets careless.

Kirby eschews clamps when he is chopping dovetails. By keeping his weight centered directly over the chisel, the job gets done quickly and effectively.

Kirby uses panel stops to secure both ends of a wide board for planing. He clamps the front one in the vise (with its blade against the bench stops) and the rear one to the bench.

The bench hook is one of Kirby's most frequently used accessories. Here, Kirby pushes the stock against the top of the bench hook with one hand and saws with the other. He uses another board the same thickness as the bench hook to support the other end.

Kirby relies on several devices to augment his bench, as shown in the drawings below. The panel stop is a little-known piece of equipment that resembles a *T*-square, with its head fixed at a right angle to the blade. When the head is clamped in the front vise, the blade goes right across the bench and is supported by the bench stops. The $\frac{1}{4}$-in. Masonite blade of the panel stop, as its name suggests, provides even support along the entire edge of a thin panel while it is being planed.

Kirby demonstrates another way of deploying the panel stop. He installs one panel stop as described above and pushes the work against it, then clamps another along the opposite end of the panel. Trapping the wood securely between two stops allows him to plane across its width. This strikes me as clumsy—there's a lot of fiddling with clamps and, in the end, it's simpler to put the work in a tail vise. But Kirby explains that, although the panel is held firmly, there's no force of compression to induce twist or warp. "You've got it held at the north and south," he says. "That's all you need."

Like many woodworkers, Kirby uses a bench hook for crosscutting. To demonstrate, he locates one that has been used by several generations of summer-school students. The bench hook should not be all hacked up with wayward sawmarks, Kirby explains, peering at me over his eyeglasses. A regular, right-angle pattern of sawcuts provides a ready guide to the correct saw position. First, he places the stock on the benchtop and lays out the cut with a sharp knife and a square. With one hand holding the piece of wood firmly against the stop on the top of the bench hook, Kirby takes several firm, rapid strokes with a backsaw, stopping just short of cutting through. As he lifts the piece up, the waste portion folds over on the hinge of wood that's left and Kirby flicks his wrist to snap it off. Kirby supports the end of a long piece of stock with another board the same thickness as the bench hook.

Shooting boards are another useful accessory in any woodworking shop. Kirby has several—one for jointing the edges of veneer, another for planing ends square in solid wood, and a mitered shooting board, as shown in the photo at the top of the facing page, for planing the end of a piece of wood at a guaranteed 45° angle. A small lip on the bottom of each board can be clamped in the jaw of the front vise or simply hooked over the edge of the bench.

Panel stop

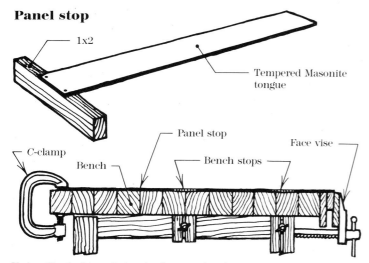

Note: *Head of panel stop is clamped in vise. Tongue is supported by bench stops and C-clamped to bench.*

Bench hook

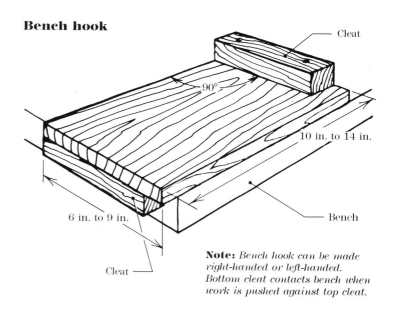

Note: *Bench hook can be made right-handed or left-handed. Bottom cleat contacts bench when work is pushed against top cleat.*

Though the bench may be ideal for working flat stock, it occurred to me that it would be difficult to carve irregular objects, such as a cabriole leg, without clamping the work from both ends. Kirby responds that, although his students have a solid grounding in traditional woodwork, he prefers that they not rely on conventional elements like the cabriole leg. Modern machinery and materials such as fiberboards and plastic resin glues, Kirby argues, give today's cabinetmaker access to a whole new world of design opportunities.

But it happens that one of his students is carving a set of cabriole legs for a small table, and the other students direct me to him with a certain rebellious pleasure. Giving in good-naturedly, Kirby demonstrates how to clamp the leg in a Record bar clamp, which in turn is fixed in the front vise. The clamp is set at a slight angle to allow the leg to rest on the wooden cheeks of the vise for support. From there the blank can be worked with spokeshaves and rotated as necessary. "You can hold anything on earth this way," Kirby says. One of Kirby's students tells me that his ideal setup would be to have two benches—a Kirby veneer-press bench for clamping and planing, and a second, traditional bench with a tail vise and dogholes for carving.

Kirby uses his front vise for all manner of workholding that cannot be accomplished using accessories. Of the many different kinds of metal woodworking vises (see Chapter 10), Kirby prefers the Record #53E, Record's top-of-the-line model. "It's big enough for everything you do," he says. "It's an excellent piece of engineering." The back jaw of the vise fits at a right angle to the benchtop and parallel to the front jaw. The jaws close at the top first and then, as the vise is tightened, at the bottom.

This type of vise has one major drawback. When you clamp a piece of wood vertically, as in sawing dovetails, the guide rods get in the way, making it impossible to center the work behind the screw, as you can with the traditional shoulder vise (see Chapter 4). "The vise goes into convulsions if you overtighten it," Kirby admits. There's nothing you can do about it, except put a matching piece of wood in the other side of the vise. Kirby compensates for this deficiency with supplementary clamping. If he wants to secure a wide piece of wood vertically, he puts one edge of the board in the vise and fastens it across the bench with a bar clamp on its other edge (see p. 143).

This mitered shooting board is one of several different kinds that Kirby uses at the bench. Kirby holds the work firmly against one side of the miter while planing its end to an accurate 45° bevel.

To handle a curved object, like this cabriole leg, Kirby puts it in a Record bar clamp, which is then gripped in the vise. He angles the bar clamp so the leg contacts the vise cheeks for support at both ends.

Shooting boards

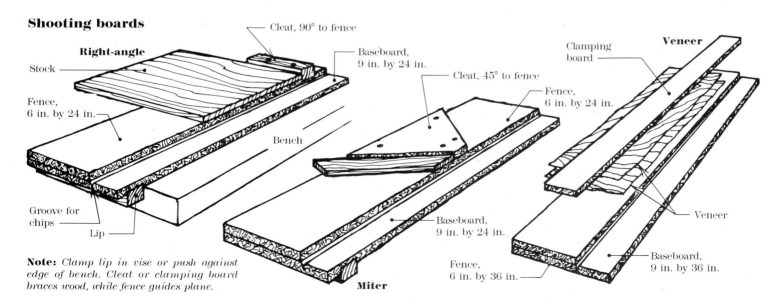

Note: *Clamp lip in vise or push against edge of bench. Cleat or clamping board braces wood, while fence guides plane.*

Right-angle

Stock

Fence, 6 in. by 24 in.

Groove for chips

Lip

Cleat, 90° to fence

Baseboard, 9 in. by 24 in.

Bench

Cleat, 45° to fence

Fence, 6 in. by 24 in.

Baseboard, 9 in. by 24 in.

Fence, 6 in. by 36 in.

Miter

Clamping board

Veneer

Fence, 6 in. by 24 in.

Veneer

Baseboard, 9 in. by 36 in.

A basic bench

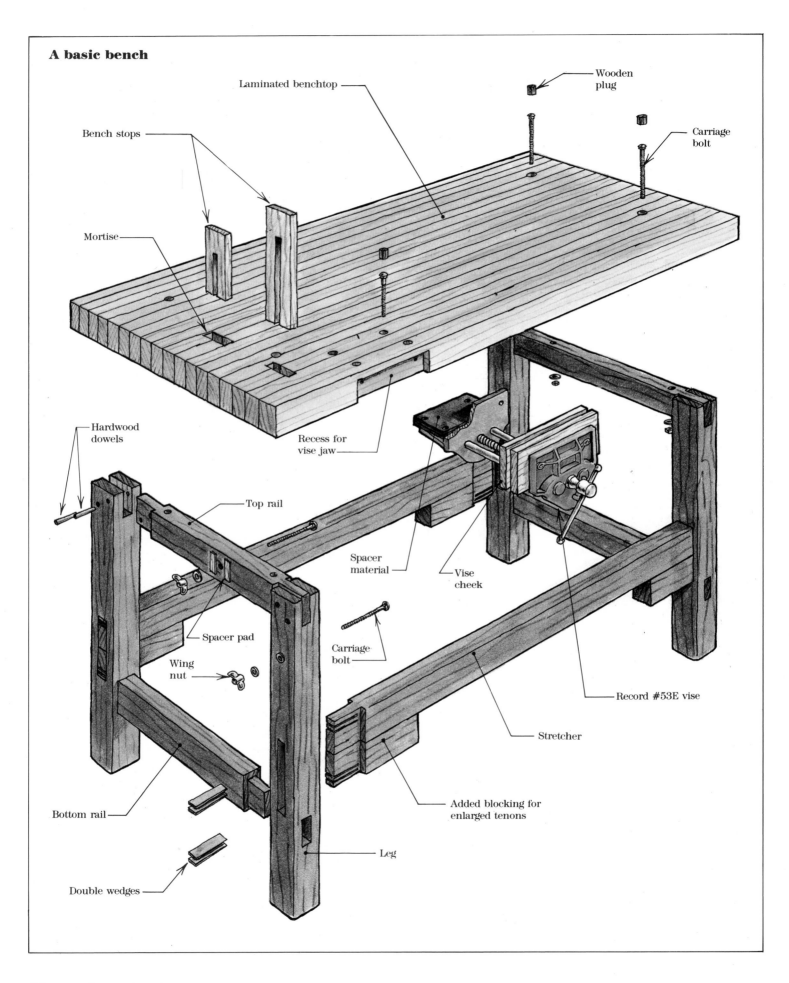

Laminated benchtop

Wooden plug

Bench stops

Carriage bolt

Mortise

Hardwood dowels

Top rail

Recess for vise jaw

Spacer material

Vise cheek

Spacer pad

Wing nut

Carriage bolt

Record #53E vise

Stretcher

Added blocking for enlarged tenons

Bottom rail

Leg

Double wedges

Making the bench

There are three major steps in making Kirby's basic bench: building the top, building the underframe, and installing the vise. The top requires manipulating a large mass of wood—preparing the stock, gluing it up and planing it flat. The base is built of large members connected by mortise-and-tenon joints.

The benchtops Kirby worked on in England were made of quartersawn beech slabs. Beech is the 'donkey wood' in Europe, as maple is here. It's hard to find wide boards that are quartersawn and therefore more stable than flatsawn boards, so Kirby has his benchtops laminated of strips of hard maple. The thinner the laminations (especially if they are quartersawn), the more stable the top.

Some benchtops I've seen that are laminated from ¾-in.-thick strips of maple are exceptionally stable. Because it's difficult, however, to spread glue quickly over so many surfaces and then get them aligned and evenly clamped, many woodworkers prefer to laminate just a few wide, thick boards. And most shops don't have a jointer or thickness planer wide enough to handle the full width of a benchtop. Alternatively, you could build the top in several sections, laminating and thicknessing each one separately before assembling the whole thing. The glue-ups are a lot less hectic, and you have only a few joints to keep straight in the final assembly. (To help keep the sections aligned, you can use dowels or Elu biscuits, as described on p. 74.)

However you do it, making a solid-wood benchtop amounts to a lot of work. "It's really wood engineering," Kirby says. "You can't be too surprised when you start off with gusto and great fervor and you wind up either not completing it or realizing halfway through that you've set yourself one hell of a task and you wish you hadn't done it." Kirby argues that there's no virtue in making a top if you can buy a good one at a reasonable price. Instead, he uses maple tops made by a company that specializes in laminated work. Because the makers use radio-frequency gluing, they're able to achieve a stronger bond than Kirby can with conventional cold-curing glues. "[Such a top] will rarely come adrift."

Kirby counsels against having a commercially made top thickness-sanded. When you eventually have to true the surface with a plane, the imbedded sanding grit will dull the blade in short order. (For tips on leveling a top, see p. 95.) For the large bench stops, Kirby uses a dense, straight-grained, hardwood like maple cut about 3 in. wide, 1 in. thick and 15 in. long—long enough to extend several inches above the top. He routs the slots through the finished top so the stops are loose enough to move freely, but not sloppily.

The underframe of the bench has only one major requirement: it must provide stable, solid support for the top. It is not a storage unit. On Kirby's bench there are no cupboards and no shelves. The top rails run across the bench, rather than lengthwise, so he can pass clamps right under the top for unhindered clamping.

There are many different simple base structures, a few of which are described in Chapter 7. Kirby's base consists of four legs, connected by two long stretchers and four short rails. The legs run right to the floor, rather than to sled feet, creating a four-point stance that transfers the load directly to the concrete and is less likely to slide than a sled-foot base. Legs and

stretchers are built of 4x4 Douglas fir, fastened with wedged through mortise-and-tenons (photo, below left). There are a few knockdown bases around the shop that are bolted together, shown in the drawing below. Rails are made of the same fir stock as the rest of the base. Bottom rails are joined to the legs in a simple through tenon, top rails in a pinned bridle joint.

The top rails protrude about ¼ in. above the tops of the legs (photo, below right), so that if the rails shrink slightly the benchtop won't be forced out of shape by a protruding leg end, and it will always be supported evenly across its width. If the rails themselves aren't parallel, they will twist the top. So, before attaching the underframe to the top, Kirby sights across the top rails to make sure they're parallel.

For maximum strength, Kirby bolts the top rails to the benchtop, boring oversize (⅝ in. for a ⅜-in.-dia. bolt) holes in the rails to accommodate movement of the top. An oversize washer spreads the load, and the countersunk holes in the benchtop are plugged for decoration.

Kirby doubles the effective length of the tenon by gluing an extra block of wood to the bottom of the stretcher. The longer the tenon shoulder, the less likely the joint is to rack. The stretcher tenons are double-wedged, top and bottom, while the bottom rail is joined to the leg with a simple through tenon (above left). The top rail supports the benchtop ¼ in. above the leg's top to provide even support across the width of the top regardless of any seasonal movement (above right).

Knockdown variation

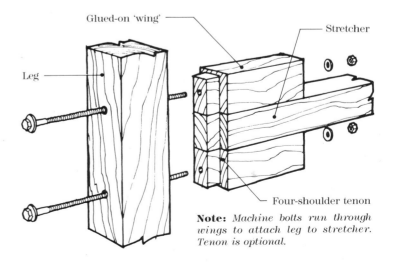

Glued-on 'wing'

Leg

Stretcher

Four-shoulder tenon

Note: *Machine bolts run through wings to attach leg to stretcher. Tenon is optional.*

Kirby mounts the vise on the front of the bench, about a foot away from the end, just inside the left leg, which carries the large bench stop. This arrangement allows him to clamp the panel stop in the vise and support it with the bench stop. Kirby doesn't just bolt the vise onto the bench—he routs a recess in the front edge of the bench so that the rear jaw is flush with the edge and at a right angle to the top. Then he covers the rear jaw with a 1-in.-thick wooden cheek. That way, if the wood he's clamping has any lumps on it, they will be held away from the front of the bench. "There are many times when you want that space," Kirby says. "And when you don't, it's easy enough to clamp a piece of wood the thickness of the cheek to your bench to hold the floating end." (For a more detailed description of the Record vise and how to mount it, see Chapter 10.)

The bench has been turned upside down to show the position of the spacers and the two vise cheeks. Photo by Scott Jansen.

Workbench/veneer press

By breeding his basic bench with a veneer press, Ian Kirby hit upon an unusually versatile workshop accessory. The combination of man-made materials in a torsion-box construction results in exceptional strength, accuracy and stability. Here, Kirby discusses some of the uses for his bench press.

I wanted to design a veneer press because the woodworker of the 1980s and 1990s should be capable of veneering. The quality of veneers you can buy is superb. They're not expensive. The particleboard and fiberboard substrates being spun off by industry are excellent. Anybody who's going to get serious about making furniture with veneers has got to have a press. But presses come damned expensive.

Couldn't an accurate workbench serve as a veneer press? The answer is no. Most people find it hard to believe, but solid wood bends like crazy, especially across its width, compared to a torsion box. That was why I didn't think there was any sense in using traditional materials to make a veneer press.

A veneer press must have an extremely flat bed and some sort of system for selectively applying pressure to a panel. The press I've designed and patented uses two torsion boxes: one acts as the bed of the press and the other is a movable caul that goes on top of the veneer assembly. For economy and strength, I use torsion boxes for the legs. They also provide convenient support for the pipe platforms,

which are used to hold the cauls and battens.

The torsion boxes are made by gluing high-density-fiberboard skins over core strips of medium-density fiberboard on $7\frac{1}{2}$-in. centers; the voids between the strips are filled with Verticel, a resin-coated paper honeycomb (available from the Verticel Corporation, Englewood, Colorado).

The core stock is set out in a grid and the strips are stapled together to hold the butt joints in position. A bead of white or yellow glue is then rolled out on one side of the grid and the skin is applied. (Unless you are using a veneer press to build the box, don't

attempt to glue both skins at once.) Glue the paper honeycomb into the core and then glue on the other skin. A torsion box is very flat, stable and strong. Pressure is applied to the veneered panel with a series of clamps and cambered battens, which are laminated of alternating layers of high- and medium-density fiberboard, as shown on the facing page.

Having made the press accurately, you must also use the thing correctly. With any veneer press you can make a mess, even with the most sophisticated industrial press. After applying glue and positioning the veneer, place the panels in the center of the

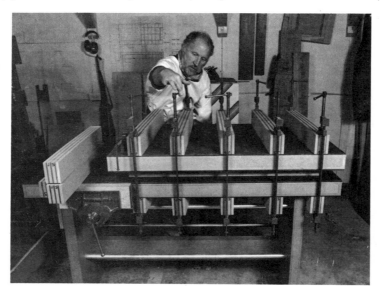

By mating his workbench with a veneer press, Kirby came up with this particleboard and Masonite workstation. Here, with it in its veneer-press mode, Kirby demonstrates the proper arrangement of battens, clamps and caul to obtain even pressure across the width and length of a panel.

benchtop on a platen, a $\frac{1}{8}$-in.-thick piece of high-density fiberboard the same size as the caul. (Treat the platen with wax so that residual glue can be removed.) The veneer and substrate assembly must be in the middle of the platen and sandwiched between it and a second platen the same size as the first. Now add the top caul and battens.

The cambered side of each batten is oriented toward the panel. Place one batten on top of the caul, another under the benchtop, directly below the top batten. It's easy to assess how close to put the battens: clamping pressure is diffused in a fan of about 90° from the clamp head. Use enough battens to ensure that the pressure fans overlap—for the largest pressings, use ten battens, five on the top and five more on the bottom.

Once the battens are in place, tighten the clamps enough to put a little pressure on the battens. Because the battens are cambered, they transfer the pressure from the center to the outside edges as the clamps are tightened in unison. By looking at the gaps between the ends of a pair of battens and the caul, you can make sure you're applying pressure equally. Continue to tighten the clamps on each side until you see the battens flatten out over the area being pressed. You can sense the same amount of pressure coming through the clamp bars.

Don't overtighten the clamps, especially if the panel you're veneering is narrow. If overtightened, the caul will

To protect all surfaces of the bench, Kirby recommends either a penetrating-oil finish, such as Watco, or a simple blend of beeswax and pure turpentine. To make the latter, Kirby shaves a block of beeswax into chips and adds about half its volume in turpentine. The turpentine dissolves the wax to the consistency of soft butter, at which point it may be applied in several light coats with either a rag or a brush.

Watching Ian Kirby work, I was struck by the similarity between his approach and that of the Japanese-trained craftsmen I'd seen. Both exhibit an uncommon degree of physical interaction with their work. No other Western woodworker I'd observed, except for a few craftsmen who work green wood, pays as much attention to the 'feel' of the tool and the wood as Kirby. For although Kirby pushes his plane and saw whereas the Japanese pull, and Kirby stands whereas the Japanese squat or sit, they both work against a single, fixed stop. The wood is never clamped under tension between dogs. That very fact requires an appealing presence of mind on the part of the craftsman.

After dinner on my last evening in Georgia, while we were seated at Kirby's round, torsion-box table, one of his students asked innocently, "You mean there are people out there who actually hold the work clamped from both ends?" I had to confess that there are—lots of them. And I'm one. Curiously, though, by watching Kirby work, I had come away with a new understanding of my own method of work. While I was not ready to take a chainsaw to the tail vise on my bench, I had learned that there's a time and a place for the basic bench and that the tail vise is no substitute for developing a feel for the interaction of the tool and the wood.

Workbench/veneer press

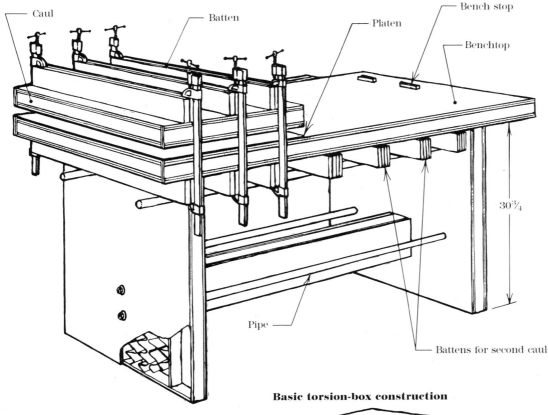

Caul — Batten — Platen — Bench stop — Benchtop — $30\frac{3}{4}$ — Battens for second caul — Pipe

Note: *Each of two cauls provides a $23\frac{1}{2}$-in. by 30-in. pressing area.*
Benchtop ($3\frac{1}{4}$ in. by 30 in. by 65 in.); legs and cauls are fiberboard torsion boxes with core strips on $7\frac{1}{2}$-in. centers.
Pipes make a platform for storing cauls and platens.
Battens (4 in. by $2\frac{3}{4}$ in. by 30 in.) are laminated from four layers of $\frac{1}{2}$-in. medium-density fiberboard, separated by $\frac{1}{4}$-in. high-density fiberboard. They have a $\frac{1}{8}$-in. crown in middle, which flattens when pressure is applied.

PATENTED

Basic torsion-box construction

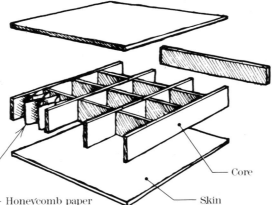

Core — Skin — Honeycomb paper

bend around the edges of the panel and leave an area of low pressure or no pressure in the panel center. The glue will migrate to the low-pressure area and the veneer will ripple as it dries. To avoid this washboard center, use a straightedge to check the top of the caul to make sure it doesn't become convex as you tighten the clamps. Also, when pressing narrow panels, place dry spacers the same thickness as the veneered panel on each side of the panel to help prevent the caul from bending.

Although it is made of completely different materials from the basic solid-wood bench, the veneer press has the same characteristics of accuracy and simplicity. It doesn't have to be a veneer press—it can also be used as a traditional bench.

The weight is more than enough, but this press is much more solid than a wooden bench because of the nature of its construction. The surface, which is roughened on a wide-belt sander to make it less slippery, takes a lot more abuse than solid wood because it's of a higher density. I haven't run into the problem, but it would be easy to repair the top by routing it out and gluing another piece in.

The only trade-off I've found in using the veneer press as a workbench is a minor one. When setting the depth of a plane blade, it's more difficult to see the blade against the reflected dark surface of the fiberboard benchtop than it would be against a light-colored solid-wood top. I was brought up on a solid-wood workbench, but if I had to make a choice I would take this one.

Workbench variations and accessories are limited only by time, money and the imagination. Seattle cabinetmaker Harold Foss, above, built his dream bench with two tail vises, a shoulder vise and a fully enclosed base cabinet. (For more on Foss' bench, see p. 92.)

A Workbench Sampler

What makes a workbench? More to the point, what makes a good workbench? I have to confess that I began this project with a fairly conventional opinion on this subject. My image of a workbench was neatly circumscribed by the traditional forms I've described so far. At some point it became obvious to me that our richest workbench tradition is variation—and that one person's tradition is another's innovation. The kind of workbench you make says much about where you learned your trade, the nature of the work you do, and the tools, materials and technology available to you. It also says a lot about who you are.

Despite this diversity, almost every workbench I've seen shares three common elements: a top, a base and a vise. Because of their importance and complexity, I have devoted two chapters to vises (see Chapters 9 and 10). In addition to the traditional forms of top and base discussed in previous chapters, I came across many ingenious alternatives in my travels, and I'll discuss a few of these here. I'll conclude the chapter with a look at some of the accessories people use to complement and extend the range of their workbenches.

Top

The workbench is most readily defined by its top. Fundamentally, the benchtop is a platform, a flat surface upon which work is set. It must support the work without flexing too much under pressure of planing, sawing or chiseling. Its dimensions must be appropriate to the size of the work done upon it as well as to the method of work, the tools and the work habits of its maker. It should be made of a relatively stable, durable material that is easily cleaned and repaired. It must contribute its share of mass to keep the bench from vibrating while the user is sanding, or from skidding across the floor while planing is being done. The top usually provides a point of attachment for one or more vises and possibly a workholding system of stops or dogs that function in coordination with the vise. It may or may not include a tool tray for temporary tool storage. There are two final considerations—cost and appearance. One is practical, the other aesthetic, but both are matters of personal preference.

Not all of us have the luxury of building the workbench of our dreams from scratch. You may find yourself faced with an inherited bench; a commercially made, general-purpose bench; or a rig picked up on the cheap. Or you may have simply outgrown your old bench. A few years back, the furniture-design staff at Sheridan College of Arts and Design in Mississauga, Ontario, realized their program had developed beyond the capabilities of the benches in their shop. Rather than buy or build new benches, however, they decided they could upgrade the old benches sufficiently by rebuilding the tops.

Almost 20 years before, the college had installed 12 standard, institutional workbenches in the furniture studio. The laminated-maple slab tops, roughly 1¾ in. by 30 in. by 60 in., were bolted to gunmetal-gray cabinets and paired end-to-end, creating 10-ft.-long worksurfaces with two vises mounted on each end. This arrangement had been perfectly adequate for model-making and for working with the variety of materials that were the original focus of the design program. But in the 1970s, under the direction of Donald Lloyd McKinley, the orientation shifted dramatically to include more traditional wooden furniture—and the cabinetmaker's skills required to build it.

After considering the problem, McKinley, Michael Fortune (the other instructor in the studio) and Stefan Smeja (the wood technician) decided that the two main requirements for the new benchtops were dogholes and thick, flat worksurfaces, which would provide more weight and a better anchor for the vises. In addition, they felt that the tops should be at a variety of heights to accommodate a range of students, and that they should be attached to the bases with room left for clamping. They decided to resurface the old tops and sandwich them between two ½-in.-thick maple skins to make a fresh surface. Notched strips were added to provide dogholes. New maple slab tops would have cost about $150 each; considering the labor involved, they probably didn't save any money, but the glued-up sandwich reinforces the glue joints in the core.

Before renovation, the 20-year-old benchtops at Sheridan College display the signs of hard wear. They are beat up, delaminating in places and no longer flat. In addition, they lack dogholes and space underneath for attaching clamps.

One man's obsession

In 18 years of cabinetmaking, Harold Foss, of Seattle, Washington, has built only three pieces of furniture that he considers handmade: a grandfather clock, a table and his workbench. "The handwork is on its way out," he reflects sadly. "I really like it, but you can't get paid for it. Twenty-five percent of our business is veneering panels. It's no fun, but at least you make money."

Behind Harold's three-year-old bench, a queue of oak-veneered desks marches off into a thicket of machines, many of which he has stripped down and built up again from their bolts and bearings. He harbors a missionary zeal toward his machines, taking them in when they've hit bottom and transforming them into good citizens—rather like a Salvation Army captain. But it's as though his own army of born-again machinery has gotten the better of him. The workbench forms a tranquil corner, a refuge, from the grease cups, pinch rollers and rpm that surround him.

Harold's workbench was a labor of love. It took about two months to make, working evenings and weekends. "I enjoyed hand-planing the thing. Sure, you could take it

to some mill and run it through a giant abrasive planer, but if your plane blade's sharp you get a real satisfaction." Harold actually hand-planed the top twice. After the first time, he laid the top on a heavy steel table (to keep it dead-flat, he thought) while he built the rest of the bench. Condensation on the steel caused the top to warp like a potato chip. When he tried to flatten it on a press, it cracked.

Harold invested in his bench the care and the sometimes excessive, but delightful, detail that doesn't pay in his work. The top is 3 in. thick (4 in. on the edges), and has two tail vises on the right end and two matching rows of dogholes. The left end has a dogleg shoulder vise, accented with rosewood molding. The molding circumnavigates the underside of the top and tail vises, and is used on the raised-panel ends of the base, the legs and various other parts of the cabinet. (Harold considered installing a fourth vise in the only free corner, but resisted the temptation.) When he laminated the top, he used a double row of Lamello biscuits in every edge joint for alignment and strength. To protect the doghole slots from

wear, Harold lined the front of each one with 1/32-in. brass, hammered into shape on a form and then screwed in place from below. There is an adjustable board support below the front tail vise and a small drawer on the end that pulls out below the two tail vises.

Harold enjoys talking about his shop machines, but his eyes light up when he talks about his workbench, as they do when he mentions his solid-

rosewood writing desk. An original Sheraton, it awaits restoration under a pile of sawdust on top of an old refrigerator. It's precisely where it was when I visited Harold six months earlier, only with six months more dust on it. In fact, he's had the desk for 13 years—the date was written on the wall when they brought the piece in. "That's fun," he says. But, like the workbench, that one's for Harold—not for sale.

The actual construction steps don't differ greatly from those Michael Fortune employed to make his top (see Chapter 5). A standard face vise mounted on one end of each bench and fitted with special wooden jaws serves as a tail vise, an adaptation that greatly simplifies construction. If you undertake your own workbench-renewal project, it's a good idea to check the top with a metal detector, if possible, before doing any machining. Stefan failed to do this on one of the first tops, which resulted in a nicked planer blade. Rip the top along any faulty gluelines, and reglue into two cores, divided according to the position of the inner doghole strip, as shown in the drawing at right. Add extra strips, if necessary, to obtain the required width. Flatten one face of each core by hand or machine, lightly joint one edge, then thickness-plane to the largest dimension possible. The Sheridan cores finished a 'fat' 1½ in.

Stefan resawed the lumber for the top and bottom skins out of 6/4 maple and dressed these down to $^9/_{16}$ in. The skins must be either quartersawn or laminated, or they will split apart when glued to a laminated core. After gluing them up in panels slightly wider than the cores, he thickness-planed the skins to ½ in. and trimmed them a few inches longer than the cores. Blocks hot-glued to the skins' edges kept them from slipping out of alignment with the cores during gluing. Stefan glued the skins to the cores in Sheridan's veneer press, but you could

Renewing an old top

1. *Glue skins to laminated cores.*

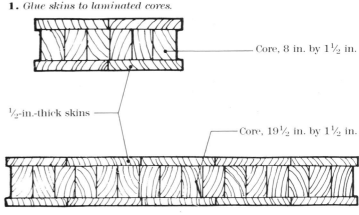

Core, 8 in. by 1½ in.

½-in.-thick skins

Core, 19½ in. by 1½ in.

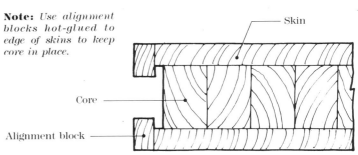

Note: *Use alignment blocks hot-glued to edge of skins to keep core in place.*

Skin

Core

Alignment block

2. *Joint edges and glue top.*

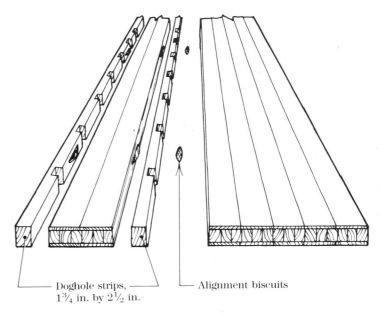

Doghole strips, 1¾ in. by 2½ in.

Alignment biscuits

Harold Foss tried to think of everything when he built his bench. The twin tail vises are great for clamping round tabletops, and a drawer between them keeps frequently used tools handy (facing page). The adjustable board jack (above) is mounted on the right leg of the bench.

Each top is carried by three 29-in.-long spacers, laminated out of three pieces of ³⁄₄-in. plywood. These are made in three heights (3 in., 5¼ in. and 6¼ in.) to accommodate different students. The notch in the top of the spacers holds a stick, which can be pulled out to support the end of the bench when the Emmert vise is in use.

The wooden shoe on the vise jaw is fitted with several dogholes (above). A single doghole in the middle would work fine, and you could span the two dogs in the benchtop with a separate piece of wood to clamp narrow stock. After the tops have been renewed, woodworking can proceed with a lot more flexibility (right).

use cauls, battens and clamps. When the glue has cured, joint and trim the panels to the finished width.

Prepare two doghole strips and dress the panels to their final thickness at the same time. Stefan cut doghole slots in the strips using a dado blade on the tablesaw, as described on p. 72. He didn't rout a shoulder in the dogholes, so that the dogs can be slid out from the underside of the bench when work is clamped on top; bullet catches in the sides of the dogs hold them in place. It's a good idea to do any special cutting required for mounting your vise before gluing up the top, while the pieces are still light enough to flip over easily. Stefan biscuit-joined the doghole strips and matching core panels and glued them together, being careful to keep the dogholes in both strips aligned. After trimming the ends, he sanded the surfaces and oiled the tops.

Three laminated plywood spacers elevate the top to the desired height. These are lag-screwed to the top and screwed to a sheet of ³⁄₄-in. particleboard, which in turn is lag-screwed to the metal cabinets. The spacers create room to attach clamps below the worksurface. The movable jaw of the Record end vise is faced with a maple cap, which houses three dogholes. After seeing the benches in use, Stefan realized that the front doghole gets much more action than the others. Because this will eventually throw the vise out of alignment, he suggests putting only one doghole in the center of the vise jaw. The two rows of dogholes in the top would still be useful for holding wide stock and they can be bridged with a piece of wood to hold narrow material.

Whatever your workbench, if the top is solid wood, it's got to be made flat—and kept that way. You can't expect to hand-plane pieces flat and straight on a warped benchtop. Not everyone has access to a wide planer or thickness sander, and these are unreliable at best for removing cup or twist. It's not difficult to true a top with a hand plane. Usually this needs to be done once when the bench is built, and periodically as its condition dictates. If you build the top out of good, dry wood, keep it well oiled and avoid cutting into it, it may need only light planing once a year to keep it true.

Tage Frid's workbench is perhaps the best known and most copied in North America. When Frid completes a new bench, he hand-planes the top flat with a jointer plane, using winding sticks to ensure that it's out of twist. In the photo below, Frid demonstrates the use of winding sticks by sighting along their top edges. Two sticks are placed at opposite ends of the benchtop. If the edges of the sticks are parallel along their entire length, the bench is flat. If the edges don't line up, mark the high corners and plane the benchtop in these areas until the sticks are parallel.

Frid's hardly a hand-tool purist. He's developed a router jig that can save a lot of elbow grease when flattening a top, and can also be used instead of a thickness planer. As shown in the sketch at right, it consists of a sled and two tracks mounted on a plywood base. The router slides across the sled; the sled slides across the tracks. By using heavy, straight, hardwood stock for the sled, you can span a wide surface and flatten a board of almost any size. It's time-consuming, but it works.

After flattening a top, Frid belt-sands it smooth with a worn 80-grit belt, being careful not to gouge the top. All corners and edges are chamfered slightly, and wherever wood runs against wood, Frid coats the parts with melted paraffin thinned slightly with turpentine (about one tablespoon per 2-oz. block of paraffin). Several hearty coats of raw linseed oil on the worksurface and a few on the rest of the bench finish the job. "Now your bench is completed, and it looks so beautiful that you hate to use it," Frid says. "If you take good care of it, working on it but not into it, it should stay beautiful for years."

Router-jig thickness planer

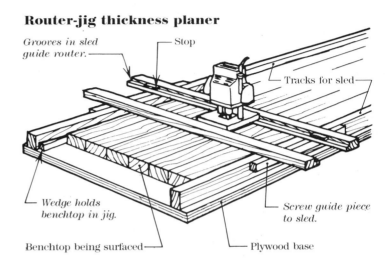

Grooves in sled guide router. — Stop

Tracks for sled

Wedge holds benchtop in jig.

Screw guide piece to sled.

Benchtop being surfaced — Plywood base

Frid sights along the length of the winding sticks to make sure the top isn't twisted. Any high spots are planed down until the sticks are parallel.

The Danish tradition

Tage Frid, who lives in North Kingstown, Rhode Island, was raised in Denmark and learned to woodwork under a strict apprenticeship system. Frid tells several stories that reflect the respect accorded the workbench in Denmark, and how he has incorporated that tradition in his teaching in North America. In the following excerpt from his book, Tage Frid Teaches Woodworking, Book 3: Furnituremaking, *Frid recalls his early career.*

When I came to this country in 1948, I was given a tour of the school where I was to teach. I was guided to a large room and introduced to the teacher with whom I was to work. He did most of the talking because my vocabulary didn't go much beyond "yes" and "no," but using my arms and legs I finally conveyed to him that I wanted to see the woodshop. When I was told that I was standing in it, I just about passed out. In this room were a huge thickness planer, which I think Columbus's father must have brought over, and a few small power tools. I was really flabbergasted when I saw the students' workbenches. These were large, two-person tables with a vise at each end. Most

of the time the students had to hold the work with one hand and use the tool with the other, which is a good way to get hurt. Some students had taken a lot of time to make special contraptions to hold the work so they could use both hands—which I'm sure was the Lord's intention when He designed us with two. (Of course, the Japanese use their feet to secure the work, which also leaves both hands free.)

After being at the school for several months, I realized that the bench I wanted was not available in this country, so I designed my first workbench. It was quite similar to the one I had been taught on.

Eventually we made a bench for each graduate student. Since then, we have been making these workbenches every two years so the students each have their own. This gives them the proper tool for holding their work. In addition, the process of building the benches is a good exercise in learning how to set up machines for mass production and how to work together as a production team. The last time, we made a run of 15 benches and it took us three days from rough lumber to having all the parts ready to fit or assemble, with the benchtops glued together.

A solid-hardwood top is ideal for many kinds of woodwork, but as Ian Kirby's veneer-press bench illustrates (see p. 88), it is possible to make a top of man-made material that will work as well. In addition, modern materials offer some advantages over solid wood. About three years ago, Jim Mattson, of Silver Springs, Maryland, built a workbench with a top made out of plywood and tempered Masonite. In its material and design, the bench is perfectly suited to the laminate, particleboard and plywood work that he does for a living.

At a previous cabinet shop, Mattson had inherited his boss's old bench. A typical cabinet-shop workbench, it had a 4-ft. by 8-ft. sheet of particleboard for a top, a metal vise bolted to one corner and a hastily constructed lumber frame below. Whenever he glued a carcase, Jim found himself struggling to compensate for the top's $\frac{5}{8}$-in. midriff sag. For the sake of his work and his sanity, he decided to build his own bench. He knew that he wanted something radically different from what he was working on, and the bench shown on the facing page began to take shape.

Mattson's most immediate concern was for a large, perfectly flat work surface: "If you're putting two cabinets together and they sit flat on the bench, they should fit pretty well on the site." In a magazine he'd seen a workbench with a double top separated by a tool tray. Jim figured that with the tool tray removed, the space between the two halves of the top would provide advantages in clamping and gluing large cabinets. At the same time, he had a workbench at home made from an old solid-core door that provided the idea for a laminated, rather than a solid-wood, construction. "Most of the work in the shop was plywood," he says, "so anything I built tended to reflect that."

Mattson's most significant departure was to make the back half of the top free-floating. Only the front section is secured to the base. One of the workbench's most important requirements is that it suit the work done upon it, and Mattson's free-floating top is a real breakthrough. Of all the Western workbenches I've seen, only the Black & Decker Workmate (Chapter 16) represents such a radical interpretation of the function of the benchtop. The conventional benchtop supports the work in one or two positions—either clamped flat to the top or against the front edge. Mattson's two tops work together as a clamping device in numerous configurations, some of which are shown in the drawing on the facing page.

To function with such flexibility, the tops had to be heavy, flat, stable and level. All of these conditions are difficult to meet in solid wood, which warps and moves and requires periodic flattening. Mattson's shop had a continual supply of $\frac{5}{8}$-in., 5-ply plywood offcuts in all different lengths. Laminated together, these were heavy and stable enough to meet most of his demands. He ripped the plywood into $2\frac{3}{4}$-in.-wide strips and glued these into two core panels, about 20 in. by 92 in. Such a laminate is unlikely to move, but Mattson further secured it with $\frac{3}{8}$-in. threaded rods every 6 in. to 8 in. and countersunk the nuts and washers on the outsides of the panels. The glued and bolted panels were then run through the shop's thickness sander (the most critical tool for this workbench's construction). It took Mattson a few hours of repeated light passes on a small 10-hp sander to get the tops perfectly flat and parallel on both sides, bringing them down to about $2\frac{3}{8}$-in. thickness. You could speed things up by using a larger sander or trimming the plywood laminates first on a big thickness planer (outfitted with dull knives).

Tage Frid flattens the top of his bench with a jointer plane once when it's built, and every now and then as the surface demands it.

Mattson benchtop variations

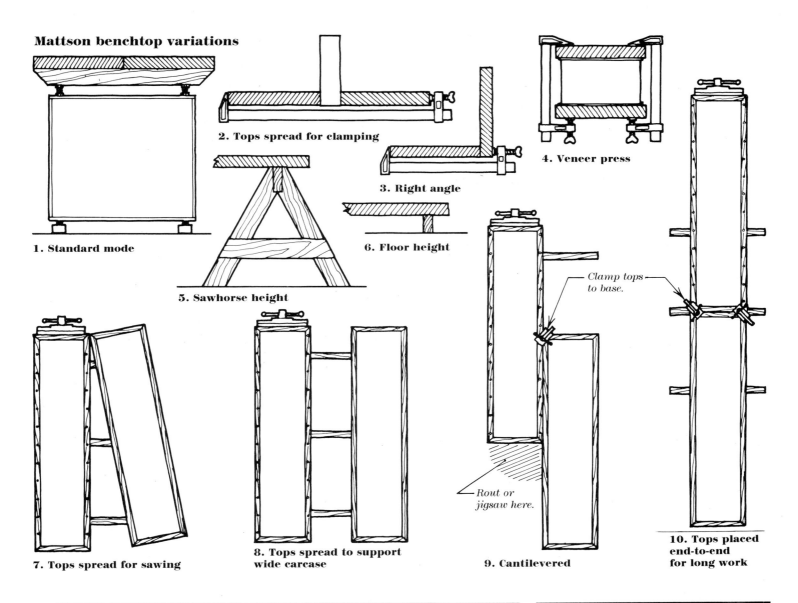

1. Standard mode

2. Tops spread for clamping

3. Right angle

4. Veneer press

5. Sawhorse height

6. Floor height

Clamp tops to base.

Rout or jigsaw here.

7. Tops spread for sawing

8. Tops spread to support wide carcase

9. Cantilevered

10. Tops placed end-to-end for long work

Cabinetmaker Jim Mattson built his double-top workbench out of plywood and Masonite and trimmed it with maple. The tops are stable and massive, and can be moved around to handle all kinds of clamping situations.

Mattson benchtop

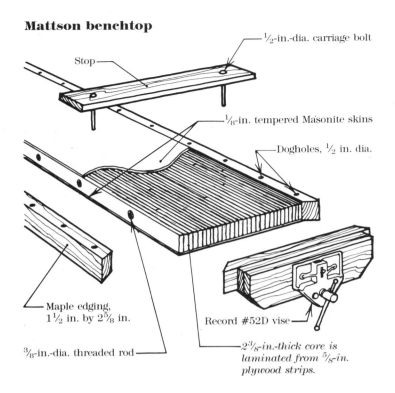

½-in.-dia. carriage bolt

Stop

⅛-in. tempered Masonite skins

Dogholes, ½ in. dia.

Maple edging,
1½ in. by 2⅝ in.

⅜-in.-dia. threaded rod

Record #52D vise

2⅜-in.-thick core is
laminated from ⅝-in.
plywood strips.

Next, he glued the ⅛-in.-thick tempered Masonite skins to both sides of each core, clamping the stacked cores together to form a makeshift press. After trimming the Masonite, Mattson glued 1½-in.-thick hard-maple edging on both panels. The edging protects the Masonite and houses parallel rows of ½-in.-dia. round dogholes on the fixed top. As evidence of the top's stability, he points to the corners where the edging is mitered, still tight after three years. The edged tops were run through the thickness sander one last time to bring the edging flush with the Masonite. Each top weighs about 150 lb. and gives the bench the stability that makes it work. "It's real nice having a heavy bench," Mattson says. "One that doesn't wobble or move on you. That's what I didn't like about those European benches."

The whole project took Mattson approximately three months of Saturdays to complete. "At the time," he says, "you wonder whether all the effort is worth it. I think it makes me about ten percent more efficient."

Mattson has since discovered that the Masonite top is much harder than the maple edging, and he'd consider making the edging thinner on a future bench. If the surface of the top were somehow damaged (and Mattson figures this would take extreme abuse), it could always be sanded down and another Masonite surface glued on. A small wound could be excised easily with a router and patched with a Masonite inlay. The sanded Masonite surface is no more slippery than a smooth maple top, and when it is well oiled (Mattson saturated his with Watco furniture oil), it resists glue nicely.

On one end of the fixed top, Mattson mounted a Record #52D, quick-action vise. On wide stock, the single metal dog in the vise makes a three-point connection with the two rows of dogholes in the top. More often, Mattson uses a narrow maple stop (about 1 in. by 3 in.) to span the width of the top. This registers in opposite dogholes with two ½-in. carriage bolts, allowing him to clamp narrow stock between the metal dog in the vise and the board.

The tool tray is an optional piece of equipment that can either enhance your use of the top or get in the way, depending on the way you work. Some people swear by them, while others swear at them. When he built his bench, Jim Mattson eliminated the tool tray from the design. Frank Klausz (Chapter 4) and Tage Frid always include it, while Ian Kirby would never have one. I've never had a tray on my bench, but I sometimes wish I had one to hold the tools I am using on a project, which could free the top for glue-up or assembly.

In the volley between tool-tray proponents and detractors, I've heard two main objections: a tray is hard to keep clean and it's difficult to clamp around. To make it easier to clean, most people add angled blocks to the ends. A couple of woodworkers I've met have come up with other innovative solutions to these problems. Jonathan Cohen, of Seattle, Washington, left an 18-in. space in the plywood bottom of his tool tray at one end. It's easy to sweep out sawdust, and it offers some clamping advantages over the tray-less benchtop, as shown below. (Cohen's first workbench was an old door on sawhorses. The door had three lights, without glass, and Jon got used to sweeping the bench off through the holes.)

Michael Fortune designed the sliding tool tray on p. 65 to preserve the flexibility of his benchtop. It has angled cleanout blocks at both ends and travels along the length of the bench on a dovetail slider. Because Fortune's tool tray is only 2 ft. long, it can be moved close to the action for convenient tool storage, or to the opposite end of the bench for uninhibited clamping. In the long run, having a tool tray won't make you any more or less organized, and you'll have to decide for yourself whether it's worth the extra time it takes to make.

A gap at one end of Jonathan Cohen's tool tray is handy for sweeping out shavings and for placing a clamp in the middle of the bench.

Base

To be useful, the top must rest at a comfortable working height. This is the job of the base, which must also support a vertical load, distribute weight and contribute mass. Its structure must resist racking forces along the length of the bench and across its width.

The benches described in Chapters 4-6 represent two distinct types of open-frame base: the sled-foot trestle (Klausz and Fortune) and the post leg (Kirby). While each has certain advantages, their differences are less important than what they share.

Both types evolved from a long furnituremaking tradition. Kirby's four-post leg structure is the less sophisticated, tracing its origins directly to the earliest precursor of the workbench— the three- or four-legged stool. One could speculate that stretchers were installed between pairs of legs for added racking resistance. As the stretchers moved closer to the floor, the structure became more rigid. Ultimately, when they reached the floor they became part of a trestle, with all the parts effectively immobilized in a single leg with a wide footprint.

The four-leg structure would have been easy to build and easily adapted to uneven floors. The legs also provide somewhat greater resistance to overall bench movement because of their relatively small surface area. The smaller the foot, the greater its tendency to dig into the floor (especially on a dirt or softwood floor) or to skitter when pushed. The sled-foot trestle, as its name suggests, is more likely to slide. Like wide tires on a car, which do not necessarily supply good traction, the large trestle foot spreads the weight of the bench over a greater surface area.

But I would guess that the trestle structure had two main historical advantages over the post leg. First, the post's end grain, which will dig into a dirt floor, is also a convenient wick for moisture, causing the bottoms of the legs to rot. The trestle foot puts side grain on the floor. Second, the trestle construction locks the legs into a frame that prohibits racking across the width of the bench. Rails connecting the trestles add racking resistance along the length of the bench.

The structural integrity of both trestle and post bases relies entirely upon a single type of joint: the mortise-and-tenon. The workbench has always been subjected to extremes of abuse— the shock of heavy pounding from a mallet and the twisting and pushing that occur during planing (and more recently, the vibration caused by the electric router and orbital sander). The mortise-and-tenon, the essential joint of the timber framer and furnituremaker, has long proven its ability to withstand stress and strain in framed construction.

There are countless variations of the mortise-and-tenon joint used in the base trestle, a few of which are shown below. The most common type has two or four shoulders and may be wedged or drawbored, and pinned to enhance its strength. A bridle joint is easier to cut, and is really a modified mortise-and-tenon. Horizontal mortising machines have made loose-spline joinery a perfectly acceptable alternative to the conventional mortise-and-tenon joint.

Mortise-and-tenon variations

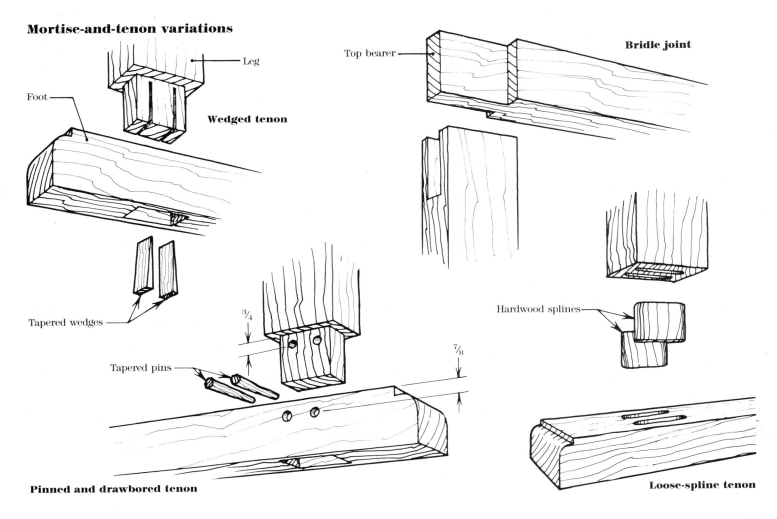

Leg

Foot

Wedged tenon

Top bearer

Bridle joint

Tapered wedges

Tapered pins

3/4

7/8

Hardwood splines

Pinned and drawbored tenon

Loose-spline tenon

Knockdown connections

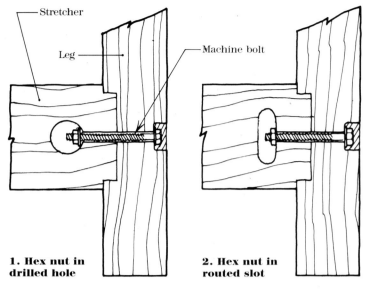

1. Hex nut in drilled hole

Stretcher

Leg

Machine bolt

2. Hex nut in routed slot

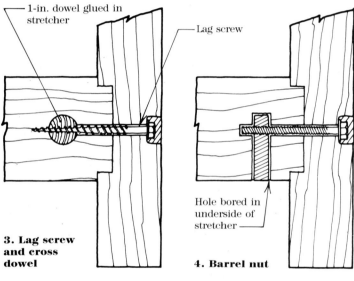

1-in. dowel glued in stretcher

Lag screw

3. Lag screw and cross dowel

4. Barrel nut

Hole bored in underside of stretcher

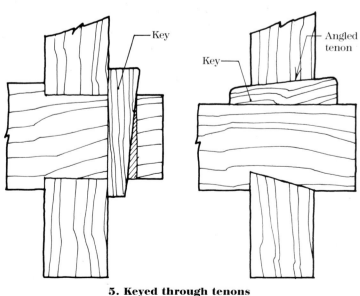

Key

Key

Angled tenon

5. Keyed through tenons

The trestles on most bench bases are usually joined to the horizontal stretchers with knockdown mortise-and-tenons. Bolted or wedged in place, these joints allow disassembly for storage or shipping. Although four shoulders take more time to cut and fit, they do a better job of resisting racking than two. Wider stretchers also resist racking—a 1-in. by 6-in. stretcher, for example, is much more rigid than one 2 in. by 3 in. It's also important that the joint mate snugly—an ill-fitting mortise-and-tenon is not much better than none at all.

I've seen many effective knockdown connections, and a selection of the most useful of these is shown in the drawing at left. (A quicker but less effective variation is the dowel joint, often used in 19th-century commercial benches.) While bolted or wedged knockdown joints work fine to attach the stretchers, relying on mechanical fasteners for end-trestle construction is probably not a good idea. It presents many more opportunities for the base joints to work loose, which eventually weakens the structure.

Sam Bush made the mortise-and-tenons for his base trestles by laminating rather than sawing and chopping, as shown below. Bush, who works in Portland, Oregon, leaves voids in the laminated feet and top rail for the mortises, and runs the center laminate long on both ends of the legs to form the tenons. The two outside laminates on each leg are made slightly longer than necessary and their ends are trimmed square on the tablesaw. (All pieces are made oversize in width, with their edges jointed flush after glue-up.) Note that the inside ends of the mortise are angled slightly to allow for the wedged-tenon construction. Be sure to run all the center laminations (for both horizontal and vertical members) through the thickness planer at the same setting and glue the laminates on a very flat surface. Dowels, splines, or Lamello or Elu biscuits will keep the laminates from wandering.

Laminated mortise-and-tenon

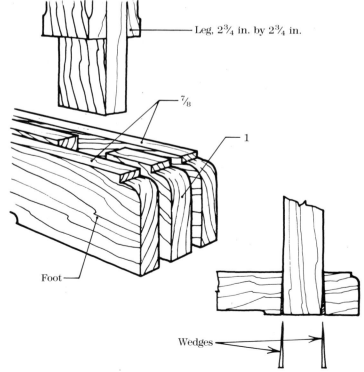

Leg, 2¾ in. by 2¾ in.

⁷⁄₈

1

Foot

Wedges

Some benchmakers have eliminated joinery altogether. John Economaki, also of Portland, Oregon, makes a rigid, sturdy bench base from extruded aluminum angle. What is accomplished in the traditional base by the ubiquitous mortise-and-tenon is achieved with ¼-in. bolts in the aluminum model. As a result, these bases are easily knocked down and almost as easily built (Economaki made three in one evening).

"For over two years this was the only workbench I had," John says. The 2½-in. by 2½-in. by ¼-in.-thick aluminum angle was cut using a triple-chip carbide blade on the tablesaw (John wore a face mask for protection). After cutting the pieces to length, Economaki C-clamped them together in position and bored the bolt holes. In addition to being quick to construct, the aluminum-angle erector-set design is eminently flexible—it would be easy to bolt in supports for shelves at any height.

All the weight of the bench is focused on the area of the four aluminum-angle feet, rather than spread out over the larger surface of the 4-in.-square wooden post feet on other benches in his shop. They simply don't pull across the wooden floor.

Aside from stability, the greatest asset these benches provide, according to Economaki, is that the tops can be easily restored. Economaki's company, Bridge City Tools, makes beautifully detailed measuring and marking tools, so he and his assistants are more often riveting, gluing and pounding than chiseling or planing on the benches. When a benchtop really gets abused, all Economaki has to do is unbolt it, rip it in half and run it through a thickness sander. He reglues the halves and, presto, a new top.

A bolt-together base

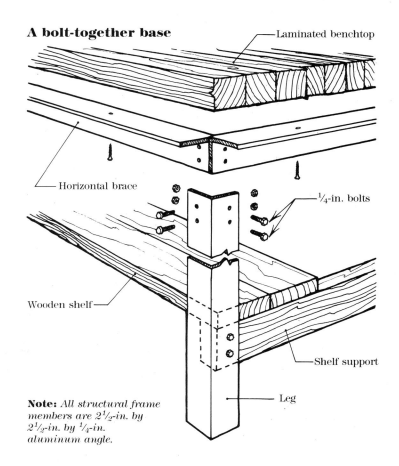

Note: All structural frame members are 2½-in. by 2½-in. by ¼-in. aluminum angle.

John Economaki's bolt-together workbench has a 38-in. by 81-in. laminated maple top.

Bench with knockdown 2x4 base

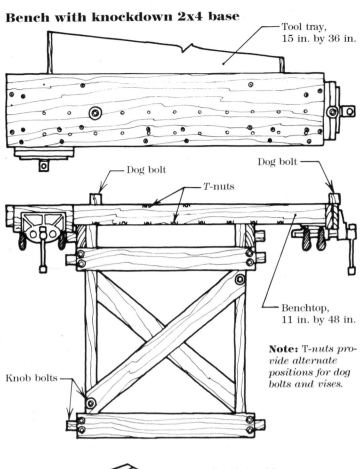

Tool tray, 15 in. by 36 in.

Dog bolt

Dog bolt

T-nuts

Benchtop, 11 in. by 48 in.

Note: T-*nuts provide alternate positions for dog bolts and vises.*

Knob bolts

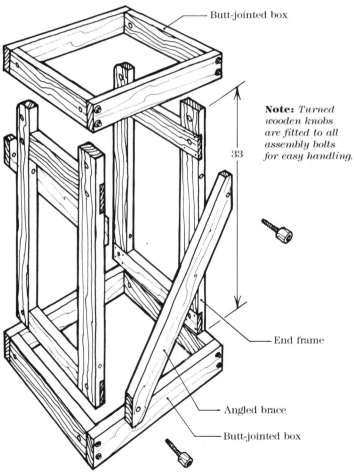

Butt-jointed box

Note: *Turned wooden knobs are fitted to all assembly bolts for easy handling.*

33

End frame

Angled brace

Butt-jointed box

Along the same lines, Clifford Metting, of St. Clair Shores, Michigan, designed the knockdown workbench shown at left. Metting's bench is much smaller and less formidable than Economaki's, but it's a handy hobbyist alternative. Metting's base is made from salvaged 2x4s, and assembles in a matter of a few minutes with a handful of store-bought bolts and *T*-nuts. The end frames, which are built with glued and screwed half-lap joints, are bolted at the top and bottom within butt-jointed boxes. Two angled braces lend rigidity to the whole structure.

When Metting built the bench ten years ago, his ideal was a compact, portable worktable with a lot of flexibility. He was living in a small apartment, doing his woodwork and metal sculpture on the fire escape. He wanted the bench to collapse for home storage and fit easily in his car for transport to local art fairs. To save space, he designed the bench with a removable top that could be easily replaced with either a homemade circular saw/router table or a drafting board. (These auxiliary tops hang on the wall when not in use.)

The top is riddled with *T*-nuts, which makes it easy to move the vises or mount a jig. But because Metting has so little time and money invested, he has no qualms about drilling a new hole whenever he needs to mount another jig—a claim not many woodworkers can make. "It's simplicity itself," Metting says, and he assures me that almost anybody could build this bench in a day.

Most people opt for a knockdown bench for one of two reasons: simplicity of construction or ease of transportation and storage. John Economaki was most interested in the former, Clifford Metting in the latter. But several makers I met went even further, relying on the knockdown elements to pull all the parts of the base together and strengthen it.

About ten years ago, when he was teaching in Hobart, Tasmania, Don McKinley built the bench shown in the photo on the facing page, which carries the knockdown form to an extreme. His main objectives were that the bench be easily disassembled and stored (for shipping home to Toronto, Canada) using a minimum of tools and without compromising its rigidity and strength. By using an ingenious self-locking, three-way joint (shown in the drawing on the facing page), McKinley was able to make a base with no glued frames and no mechanical fasteners—it breaks down into an easily packed collection of sticks. The interlocking joint is complicated, but it works extremely well; in fact, the joints are tightened with increased load on the bench. "I like the principle of the thing—that it 'collects' itself together," McKinley says.

The structure is symmetrical—the joints at the top and bottom of the legs are identical. To assemble the base, the tenons on the front and back stretchers are slipped into the corresponding mortises in the legs and then locked in place with the top and bottom end rails. The joints are roughed out on a radial-arm saw and bevels are cut with chisels guided by angled blocks. In order to fit locating sockets in the underside of the benchtop (a basic slab, $1\frac{7}{8}$ in. thick by 3 ft. by 6 ft., made of Tasmanian oak), the front legs are cut $\frac{3}{8}$ in. longer than the rear legs.

Other features of the design reflect McKinley's reaction to the Australian workbenches he saw, as well as his own background. The typical Australian bench had a massive top, supported by "a perimeter of 2x12 'joists' and 4x4s stuck in there somewhere for legs." Besides providing lots of support, the 2x12 apron offered a clamping surface for a vertical board anywhere around the bench. But it was correspondingly difficult to clamp pieces flat on the top for mortising. So McKinley inset his base by several inches from the edge of the top and mounted a Record vise on the left front corner. (He had been raised on the Emmert patternmaker's vise, but there was no hope of finding one in Australia, so he modified the Record to make an 'imitation Emmert,' as described on p. 141.) He had no desire to make a European-style bench: "I find the classic European bench depressingly narrow and something I just haven't grown accustomed to. I've never made a great deal of use of an end vise." The adjustable bench jack, shown in the inset photo below, which hooks over and slides on the top stretcher, supports work clamped in the vise. The face of the jack is the same distance from the edge of the benchtop as the fixed jaw of the vise. A quarter-turn on the support block disengages the two lag screws in the back, allowing it to be adjusted up or down or removed from the jack entirely. The screws are set off-center by one-half the distance between the notches in the jack, to double the number of height positions by inverting the block.

Self-locking joint

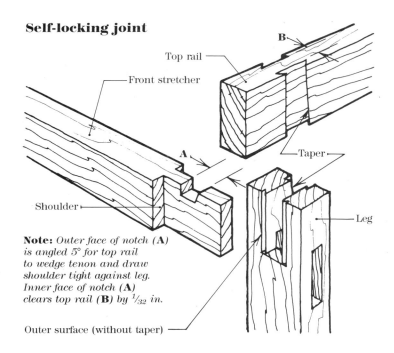

Note: *Outer face of notch (A) is angled 5° for top rail to wedge tenon and draw shoulder tight against leg. Inner face of notch (A) clears top rail (B) by $\frac{1}{32}$ in.*

Don McKinley's accessories (above, right to left): bench hook, board jack, hooked wedge and rails, and aluminum bench stop. The jack hooks over the top stretcher (right) and slides along the front of the bench to support a long board. Photo at right by Donald Lloyd McKinley.

Michael Podmaniczky's bench does a good job with only one hefty stretcher in the base. The legs are recessed where they overlap and are tied together by a keyed through tenon.

Single-stretcher base

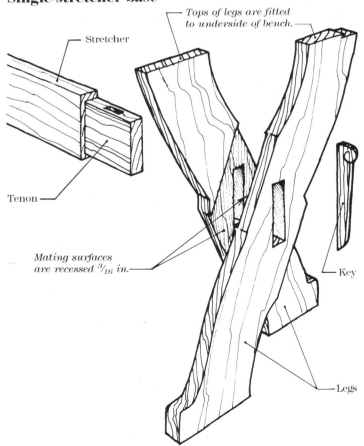

Stretcher

Tenon

Tops of legs are fitted to underside of bench.

Mating surfaces are recessed 3/16 in.

Key

Legs

One of the most unusual knockdown bases I've seen was built by Michael Podmaniczky, of Wilmington, Delaware. The design is based on the workbench of an old Swedish boat-builder, Seth Persson, with whom Podmaniczky apprenticed more than 12 years ago. The crossed legs, which are built out of 2x8 maple, are each recessed about $\frac{3}{16}$ in. where they overlap, creating a very strong, shouldered joint. Michael cut the recess using a miter gauge and a dado blade on the tablesaw and then cut the mortise for the single stretcher. The heavy, keyed tenon creates a firm, locked joint, which is easily knocked apart for storage or transportation.

On three benches Michael recently built for the Williamstown Regional Art Conservation Laboratory in Williamstown, Massachusetts (where he trained as a conservator), he glued the legs together where they overlap. This is structurally unnecessary. The tops of the legs are fitted to the uneven underside of the benchtop. (The top is $2\frac{1}{4}$ in. thick in the middle, and 4 in. thick on the front edge, at the dogholes. The particleboard tool tray is screwed through a $\frac{5}{8}$-in. spacer into the underside of the top.) Once the bench is assembled, the first few shots of a mallet on a mortising chisel, for example, will drive the top onto the legs. This spreads the legs and locks the long shoulders of the lap joint.

The wider the middle stretcher, the better the bench resists racking lengthwise. The stretcher on Michael's bench is about 2 in. by 8 in. at the joints, and tapers slightly in width at the middle. It's also important that the crossed legs be as far apart as possible to lend rigidity to the whole structure. This is not a problem on a large bench, such as Michael's own 7-footer. But the new benches are smaller (24 in. by 66 in.) and the legs are positioned just inside the end caps on the top for maximum spread. This means that the tenons in the base stick out slightly on both ends of the bench and might scrape a shin or two.

The area below the vises is unobstructed by legs, so it's easy to work on objects that hook under the bench. The bench is remarkably sturdy, resistant to movement and much easier to build than the typical end-trestle construction—there are only four large joints to cut. Three benches were built over a couple of weeks at Williamstown, but Michael estimates that he could probably build a single bench in one week.

There are two fundamental schools of thought about bases among workbench builders. So far in this section I've introduced only one. This group asserts that the base should be unencumbered with tools, wood storage and accumulated debris. The bench is essentially a clamping station, designed to hold work in as many different configurations as possible. It's a rigid skeleton, usually freestanding and often designed to knock down for portability.

The second group views the base as the most convenient place in the workshop for tool storage. In the tradition of the Shakers, they view the workbench as a self-contained workstation. It may incorporate a large bank of drawers, shelves or cabinets. It is sometimes freestanding, sometimes built into tight quarters and almost always very heavy.

Of course, a wide range of compromise exists between these two points of view. On occasion, partisans of the open base will allow a single shelf between the stretchers or a small drawer hung on the underside of the top. And those who prefer a base cabinet may design a space between the cabinet and the benchtop to allow for clamping (as on the benches at Sheridan

College). It is important to realize that whether the base is a stick-built skeleton or a massive tool chest, it must provide a rigid foundation for the top. While the open-frame base relies upon careful joinery, the enclosed base usually accomplishes this by virtue of its mass and a truckload of mechanical fasteners. The enclosed base also provides an opportunity for considerable elaboration.

The base on Jim Mattson's bench (see p. 97) is a perfect complement to the twin tops. It consists of three mitered plywood frames covered with tempered Masonite skins. (Mattson sheathes all the frames with 1/8-in. Masonite, except for the two outside faces of the end panels, on which he uses 1/4-in. skins. The thicker material gives the exposed ends of the bench a little more protection.) Sandwiched between the panels are two birch-plywood cabinets screwed into the panels from the inside. Adjustable maple bearers, 1 3/4 in. by 4 in., are mounted on top of each panel to support the tops.

The skins are glued to the frames and thickness-sanded like the tops. Also like the tops, the base frames are edged with maple strips, which are connected at the corners in unglued finger joints. This allows Mattson to remove the maple slats on the front and back of the bench to gain access to the leveling nuts.

To assemble large casegoods in the shop that later would be installed on location, Mattson wanted the tops level. The height of the bench is fully adjustable in two places—on the bottom of each panel and between the panels and bearers. Mattson admits to overkill in the leveling department. "I probably should have put handles on here," he says, as he struggles to pry off one of the maple slats to show me how the nuts work. "But I've never had to get at them except when I'm moving the bench." The bearers separate the tops from the cabinets, allowing ample room to clamp work to the tops.

Double doors, with 1/4-in. tempered Masonite panels set in mitered maple frames, provide access to the interior space from both sides of the bench. The interior is open: "I was going to put shelves inside," Mattson says, "but once you get a bunch of junk in there, inertia sets in." And when the whole base is glued and screwed together, it's about as inert as a workbench can be.

Mattson base

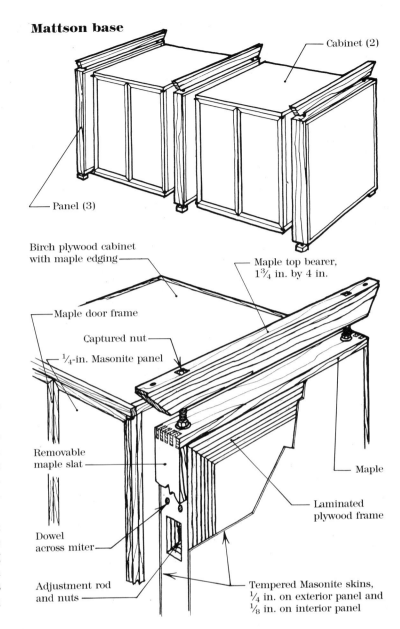

Cabinet (2)

Panel (3)

Birch plywood cabinet with maple edging

Maple top bearer, 1 3/4 in. by 4 in.

Maple door frame

Captured nut

1/4-in. Masonite panel

Maple

Removable maple slat

Laminated plywood frame

Dowel across miter

Adjustment rod and nuts

Tempered Masonite skins, 1/4 in. on exterior panel and 1/8 in. on interior panel

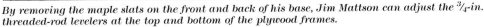

By removing the maple slats on the front and back of his base, Jim Mattson can adjust the 3/4-in. threaded-rod levelers at the top and bottom of the plywood frames.

To make room in his cabinet shop, Lewis Judy designed his heavy workbench to roll. Most of the time it perches securely on 14-in.-long feet mounted at the bottom of each leg. But the feet pivot on a single bolt at the top to expose heavy-duty 4-in. wheels, and lock in place with a steel pin near the middle.

The drawers on Judy's bench pull right through the carcase, allowing access from either side—a handy feature when work is clamped across the front or back of the bench.

Lewis Judy's workbench, shown above, also is designed with a substantial base cabinet. Judy, of Jefferson, Oregon, added a built-in bank of drawers that pull out from either side of the bench. Instead of a traditional, wooden tail-vise system, Judy mounted a large, quick-action Record vise on each corner of the top. He uses this 'multilateral' bench mostly for routing, chiseling and sanding cabinetwork, and the four vises present numerous clamping options. But with work clamped along one side of the bench, the drawers in the base would be obstructed if they didn't open from both sides.

The dovetailed carcase, which Judy made separately from the base, contains five graduated drawers hung on $\frac{1}{2}$-in.-square oak splines. The carcase and drawer fronts are made of walnut; the drawer sides and dividers are all $\frac{1}{2}$-in.-thick oak. Drawer bottoms are oak-veneer plywood. The rest of the base and the top are robusta eucalyptus, milled from logs cleared off a Hawaiian highway. The wood is extremely hard, dense and unstable (although it has finally settled down after four years of benchlife). It also eats carbide router bits. The density of the robusta, in combination with the mass of wood and its 36-in. overall width, makes for an exceptionally sturdy workbench.

Before leaving the discussion of bases and tops, it's worth noting the methods used to connect them. I've seen several, but they all have to account for one basic problem: a solid-wood top expands and contracts with changes in humidity, while a rigid base structure does not. Michael Fortune installs a single lag screw through the center of each trestle into the underside of the top. Frank Klausz simply lets the considerable weight of the top rest on a bullet-shaped dowel, glued into the top bearer of each trestle. Ian Kirby prefers a more positive attachment and through-bolts his benchtop to the rails of the base. The bolts ride in oversize holes drilled in each rail.

An effective but more time-consuming method is the one employed by Peter McMahon, a cabinetmaker in Washington Grove, Maryland. McMahon joins the top to the upper rails of the base with a 2-in.-wide sliding dovetail. He cut the female slots in the underside of the top by hogging them out first with a straight bit, and then finished them in two passes with a 1-in. dovetail bit. The top is lag-bolted to the center of the rails so that it expands outward toward the edges of the bench.

Accessories

Regardless of the kind of bench you have (or decide to build), a few simple accessories will greatly extend its usefulness. Benchdogs, for example, when used in conjunction with a face vise or tail vise, can be as good as another pair of hands or an able apprentice. Shooting boards, bench hooks, wedges and holdfasts are among the other useful items you will want to consider folding into your workbench ensemble.

If you're just starting out, keep in mind that a few accessories may be all you need to transform a slab of wood on four legs into a practical, functional workbench. In fact, the simpler the bench, the more important these accessories will become. (The Roubo bench in Chapter 2, and Ian Kirby's bench in Chapter 6, could hardly function without them.)

I've described a number of accessories elsewhere in the book, many of which seemed to be intrinsic to a particular workbench. Here I'll introduce accessories that can be used with almost any bench. While a few can be purchased, they are usually so easy to make that most woodworkers I know prefer to do so. It saves money and it gives you one more chance to tailor your bench to your own needs.

Ian Kirby makes extensive use of bench hooks and shooting boards, and these are described on pp. 84-85. Other bench hooks worth noting are shown at right. The narrow bench hooks were common in 19th-century America. Each is made of a solid piece of wood, stepped top and bottom, and the pair can be spread as far apart as necessary to support a board of any length along the edge of the bench. The bottom drawing shows a combination bench hook and miterbox. By making three slices—two at 45° and one at 90°—in the top cleat of an ordinary bench hook, you can obtain accurate guides for your saw.

Perhaps the most common bench accessory is the benchdog, or stop. Every bench with a tail vise has at least two—one in the vise and one in the benchtop. Benchdogs make it easy to trap wood of all different lengths on the top for planing, mortising and a variety of other operations. Some commercial metal face vises also incorporate a small dog in the movable jaw.

There are two basic types of dogs—metal and wood. Those woodworkers who prefer metal dogs appreciate them for their weight, strength, stiffness and the secure grip of the crosshatched face. This type of dog comes with most commercially made benches, but can be bought separately from woodworking supply companies. These dogs usually have a thin metal spring riveted to one side, which makes them adjustable in height.

Most woodworkers I met, however, like wooden dogs (a preference I share). Perhaps the most eloquent, and vehement, champion of wooden benchdogs is James Krenov, a furnituremaker, teacher and author. He has no patience with the metal dog, which is notorious for leaving its signature in the once-keen edge of an errant plane blade. In his book *The Fine Art of Cabinetmaking,* Krenov responds to the complaint that a wood-

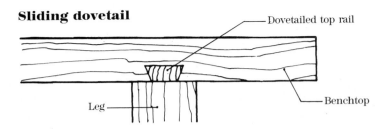

Sliding dovetail

Dovetailed top rail

Benchtop

Leg

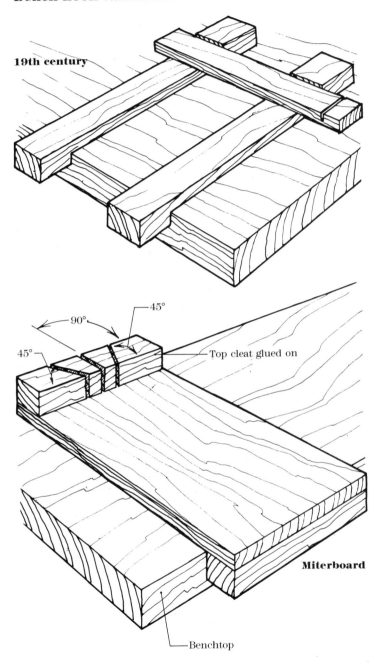

Bench-hook variations

19th century

90°

45°

45°

Top cleat glued on

Miterboard

Benchtop

James Krenov uses two springs on his benchdogs—a thin strip of wood screwed to the bottom of the dog, and a metal spring from a ballpoint pen behind it.

Edge dogs, or edge vise clamps, are used to clamp a board against the front of the bench.

These simple edge dogs were made by students at The College of the Redwoods by gluing a dowel in a wooden jaw and lining the contact surfaces with cork.

en dog may break: "Well, yes. If you are big and strong and brace yourself against the wall and give the screw all you've got—they are apt to break. But then, so is the piece of wood between them!" Krenov's dogs are made of hornbeam, and have a slim wooden spring backed up by a small metal spring taken from a ballpoint pen. He recommends setting the dogs low, because only the bottom edge of most pieces of wood needs to be grabbed to hold the piece securely.

Like Krenov's dogs, most rectangular, wooden dogs incorporate some type of spring, as shown in the drawing below. The drawing also shows a selection of dogs with both round and rectangular shafts, a further distinction between wooden benchdogs. Those who like rectangular dogs complain that the round ones rotate in their holes and loosen. But those who like round dogs usually consider that an asset, as it makes it possible to swivel the dog's head to fit whatever is being clamped. And, since it's easier to drill a round hole in a benchtop than to chop a rectangular mortise, the round dogs are much quicker to install.

Several of the benches in Krenov's classroom are fitted with edge dogs, shown at left. With the dogs' round shafts inserted into holes, one in the benchtop, one in the tail vise, a wide board can be held edge-up against the front of the bench. Michael Fortune uses the commercially available edge dogs shown at top left on the facing page. To overcome the loose fit of the round shafts, he has made rectangular doghole liners.

The toothed aluminum stops, shown at top right on the facing page, are commercially made. They are inlaid flush in the benchtop, and the teeth can be adjusted up or down with the small thumbscrew in the top of the stop. The teeth grip the end of a board without much pressure, preventing the work from slipping, which is all right if you don't mind toothmarks in end grain. Because of their grip, one toothed stop is sometimes deployed alone, without a tail vise, for planing operations. They clog easily with sawdust, however, and are best kept clean with a blast of compressed air.

The wrought-iron dogs shown at the bottom of the facing page are a wonderful example of an early style of handmade benchdog. The points on the two lower legs are hammered into the bench and the work is set against the pointed head. The dogs may be arranged at both ends of the work and braced against an edge for additional support. While most of us would shudder at the thought of pounding them into a well-kept benchtop, having a few in the job-site toolbox could prove invaluable by enabling a 2x10 to be turned into a makeshift workbench.

Benchdog variations

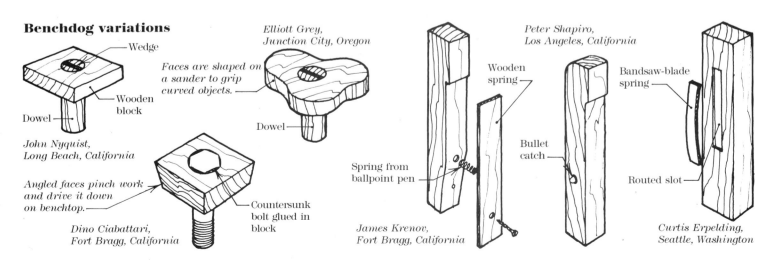

Wedge

Wooden block

Dowel

John Nyquist,
Long Beach, California

Elliott Grey,
Junction City, Oregon

Faces are shaped on a sander to grip curved objects.

Dowel

Angled faces pinch work and drive it down on benchtop.

Countersunk bolt glued in block

Dino Ciabattari,
Fort Bragg, California

Peter Shapiro,
Los Angeles, California

Wooden spring

Bullet catch

Spring from ballpoint pen

James Krenov,
Fort Bragg, California

Bandsaw-blade spring

Routed slot

Curtis Erpelding,
Seattle, Washington

These 1½-in. by 2½-in. adjustable aluminum bench stops may be used with or without a vise.

Michael Fortune adapted these store-bought edge dogs to match the holes in his benchtop. The shaft of the edge dog fits in a hole bored in a rectangular wooden dog, which keeps it centered.

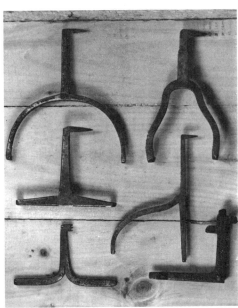

Wrought-iron benchdogs can transform almost any board into a worksurface. The collection above is a measure of their diversity. The commercially made "Portable Bench Dog" at lower right in the photo above was patented by Stanley in 1909 and sold for 30¢ in 1929.

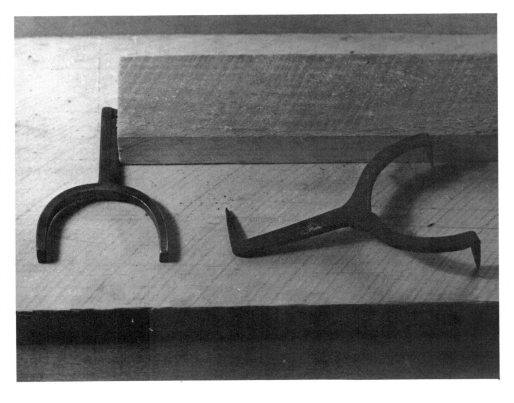

Don McKinley's hooked wedge and bench-fixed rails are mainly used for planing. Stock is trapped between the wedge and rail as it is pushed forward. Photos by Donald Lloyd McKinley.

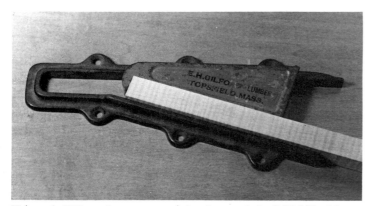

This early 20th-century commercial wedge, made by E.H. Gilford Lumber of Topsfield, Massachusetts, is a lot less versatile than McKinley's homemade alternative.

Wedging systems are probably as old as, if not older than, the workbench itself. I've seen a couple of variations, but they all follow the same principle. Stock jammed into a receptacle with tapered sides is held fast to the benchtop, without aid of a vise, clamp, or any other mechanical device. These systems are among the simplest, most effective holding devices I can think of that still free both hands for the work.

I found the most versatile of these systems to be Don McKinley's hooked wedge with bench-fixed rails, shown above. The force of planing pushes the end of the work against the hook of the wedge, jamming the work and wedge against the converging rails. The rails are located with dowels in a double row of holes across the width of the benchtop. Moving the rails apart accommodates larger work. The system holds the work securely, will not mar softwood stock, and is instantly released with a tap on the hook end of the wedge. McKinley has no tail vise on his bench, and only one toothed metal stop and a face vise, so this wedging system is almost indispensable to him.

McKinley devised his hooked wedge-and-rail system out of frustration with a commercial, metal version, similar to the one shown at left. This requires a more permanent fixing to the benchtop, is much more limited in the stock it can handle, and requires that the workpiece have sides that taper to match the sides of the wedge. (If it doesn't, McKinley found that the work will become "abused to this condition whether you wanted it or not.") What's more, as with a metal benchdog, if you nick it with your plane blade, you will have undone any convenience it may have offered in the first place.

McKinley's wedges and rails are made out of varying thicknesses of hardboard or particleboard, from $\frac{1}{8}$ in. to 1 in., to accommodate different-size work. Two dowels are glued and wedged in each rail. On the last batch of rails he made, McKinley located the dowels to one side of the centerline. This provides three different spacings for each set of rails if he simply turns one or both rails end-for-end in the holes. The wedge is tapered about 1 in. in 4 in., or 15°, and the edges of the wedge and rails may be planed to a bevel to alter the pressure on the workpiece.

Another benchtop clamping device of ancient origin is the holdfast, or hold-down. In the 18th century, the typical cabinet bench had at least one of these hand-wrought, iron gooseneck holdfasts (see p. 27). The holdfast passes through a hole drilled in the benchtop and is set by striking its head with a hammer. The spring in the iron shaft and neck holds the work. A scaled-down version is commercially available today from a few tool dealers, although you would probably do better by having the holdfasts custom-made by a local blacksmith, as did Rob Tarule when he built his reproduction of the Roubo workbench (Chapter 2).

Record makes a screw-operated holdfast that I found on many benches during my travels. This is usually set in a cast-iron collar, which is inlaid and screwed in the benchtop. The collar protects the benchtop from being chewed up by the ribbed metal shaft. This holdfast has a maximum reach of about 7 in. and can accept work up to 7½ in. thick. Some woodworkers buy a bunch of extra collars and install them all over the benchtop, while others prefer one or two well-placed locations. Furnituremaker Curtis Erpelding, of Seattle, Washington, installed only one collar near his face vise (photo at right), a setup that allows him to hold work firmly in two directions. A carver, Bob Bessmer of West Sand Lake, New York, omitted the collar, choosing instead to inlay a ¼-in. by 2-in. steel bar the length of his benchtop. A series of 1-in.-dia. holes on 5-in. centers in the bar accommodates one or more holdfasts.

Jorgenson calls their smaller version of a holdfast (photo at far right) a hold-down. It is much less expensive than the Record holdfast, but has a maximum opening of only 3 in., with a 1½-in. throat. The neck slips over the head of a small bolt, which drops into a countersunk hole in the benchtop when not in use. This hold-down is among the favorite accessories of David Powell, who runs the Leeds Design Workshops of Easthampton, Massachusetts. Powell has eight bolts spaced around his benchtop, and he told me that the only thing he'd change on a future bench would be to add more bolt holes. "They never seem to be where I need them," Powell says.

Apart from these standard commercial holdfasts, I found several unusual wooden ones that you can easily make yourself. John Leeke adapted half of a wood handscrew clamp to make the holdfast shown in the top drawing at right. Leeke bored a series of holes in his benchtop, through which he passes a wood screw threaded into a single wooden clamp jaw. A shoulder turned on the head of the screw bears against the underside of the benchtop, which functions like the missing jaw of the clamp. A second screw is cut off short and adjusted up or down against the benchtop to keep the jaw parallel.

Leeke devised the clever holdfast in the bottom drawing to solve a specific problem. He was making a quantity of Shaker-style fly swatters—about a hundred of them in one crack—and needed a way to hold the round shaft firmly to the bench without marring it while pinning the swatter head to the handle. Leeke's solution couldn't be simpler. He passed a loop of rope through a hole drilled in the benchtop and slipped the swatter handle through it. Pressure is applied by stepping on a wooden pedal tied to the bottom of the rope.

Curtis Erpelding installed the collar for his Record holdfast (above left) near his face vise, allowing him to clamp a joint in two directions. The bolts that attach the 3-in.-high Jorgenson hold-down to David Powell's benchtop (above right) are housed in countersunk holes when they are not in use.

Handscrew holdfast

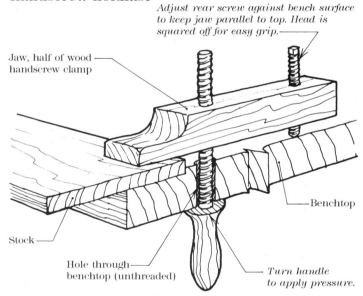

Adjust rear screw against bench surface to keep jaw parallel to top. Head is squared off for easy grip.

Jaw, half of wood handscrew clamp

Benchtop

Stock

Hole through benchtop (unthreaded)

Turn handle to apply pressure.

Rope holdfast

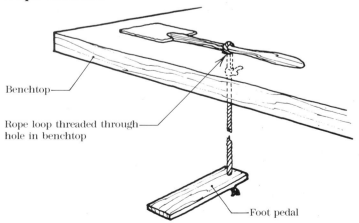

Benchtop

Rope loop threaded through hole in benchtop

Foot pedal

Curtis Erpelding uses unusually large oak go-bars to clamp work firmly to his bench. It takes only a few seconds to spring them between the ceiling and the pads that hold the work.

The French goberge is an early ancestor of the modern go-bar. Shown here in a 17th-century joiner's shop, the goberge is at least as convenient as the traditional iron holdfast. It can be used anywhere on the benchtop, and requires no holes bored in the surface. From Des Principes de L'Architecture, *André Félibien, Paris, 1676.*

Without a doubt, the most interesting holdfasts I found were the 'go-bars,' shown above. Curtis Erpelding built these 9-ft. 3-in.-long flexible springpoles for all kinds of quick-clamping applications. Sprung into place between the ceiling and the work on the benchtop, the go-bars hold securely and leave all edges free. With go-bars it's easy, for example, to cantilever work off the edge of the bench, or to clamp a straightedge to a piece of plywood—a difficult task with traditional clamps.

The modern go-bar, such as Erpelding's, undoubtedly evolved from the 18th-century, French *goberge*, pictured at left. It is also well known among musical-instrument makers. I've seen much smaller go-bars used to glue braces to the soundboard of a guitar, or to attach the soundboard to the body. This is usually done within an open box that is large enough for the instrument and tall enough for short go-bars. In all applications, the go-bars cost little to make, are quick to use and allow for easy access on all sides of the point of pressure.

Erpelding has three poles, all made from air-dried oak, about 1¼ in. in diameter, shaped to an octagon using a chamfer bit on a router. (White ash or any straight-grain springy wood would do as well, or you could laminate three thinner pieces together.) The octagonal grip is easier to make than a round pole, and the crisp facets provide a better gripping surface.

The length of the go-bars depends on the distance between the benchtop and ceiling, the thickness of the work being held, and the amount of flex in the bars. Curtis' poles are about an inch shorter than the distance between the benchtop and the particleboard box he has mounted on the uneven ceiling above his bench. At this length, they develop just the right amount of pressure on 1-in.-thick work. Anything over 2 in. thick places the poles at too sharp an angle for safety. Lower ceilings require thinner bars. To accommodate a variety of thicknesses, make go-bars of different lengths or use shims. Curtis always places a protective pad of cork-faced plywood between the bar and the work, cork face down.

Setting a go-bar takes longer to describe than to do. Plant the bottom first against the work and slide the top into position. To remove the pole, reverse the process, placing pressure against the top while you slide the bottom out. Curtis has used his go-bars for years without incident, but a 9-ft. oak pole crashing down on the bench is something to be avoided. Be particularly careful if the work you're doing causes a lot of vibration. And take note that when you place two go-bars close together they must be the same length, or the shorter one will loosen when you tighten the other one.

Small is portable

When Scott Grandstaff of Happy Camp, California, started making miniatures, he tried just about every way to hold the small parts. Nothing worked. Vises and clamps crushed them and were too big to work around—they obscured the part being carved.

Grandstaff's solution was the miniature workbench shown at right. The bench is 13½ in. high, with a 1-in. by 3½-in. by 8-in. top, sporting a row of nine ⅛-in. by ⁵⁄₁₆-in. dogholes. The whole rig can be clamped in a vise or to the surface of a larger bench, or, as shown in the photos, you can sit on the splayed wings that form the base.

Two bolts in the top serve as screws for the wooden end vise. The metal vise is made of angle iron and small scraps of rod and brass.

Partly because he likes things small, and partly because a portable bench made a lot of sense for his restoration and repair business, Grandstaff came up with the two sawhorse/workbenches shown below. "What I really needed," he says, "was a plane stop out in the field." The larger of the two hybrids has a small vise mounted on one end and 16 dogholes along one edge. The rolling bench, which is the same height, can be pulled into service to support the other end of a board. Both benches fit easily in the back of

Grandstaff's station wagon. They hold a surprising number of tools and are easily set up and moved on a job site. Plus, Grandstaff points out, the small benches look pretty nice in someone's living room if they've got to be there awhile.

For added convenience and yet another level of portability, a removable toolbox fits above the drawers on the larger bench, with two doors that flap open on either side of the handle. The outside of the doors was wasted away from solid stock to leave two thicker sections resembling raised panels. These were then routed out from the inside and covered with small, hinged lids to provide handy pockets for drill bits and other items that get lost in the clutter of most toolboxes.

The larger bench is a smorgasbord of found woods—cherry, birch, oak, ash, pine, walnut and mahogany—while the rolling bench has a tan oak top and legs. After living with these for a couple of years, Grandstaff decided to make a heavier version of the stationary bench for his shop. It's got a wider, thicker top, a heavier vise and angled back legs.

What's the ideal height for these portable benches? Grandstaff recommends that you "stand on one leg, bend the other one at the knee and measure the distance from the bent knee to the ground."

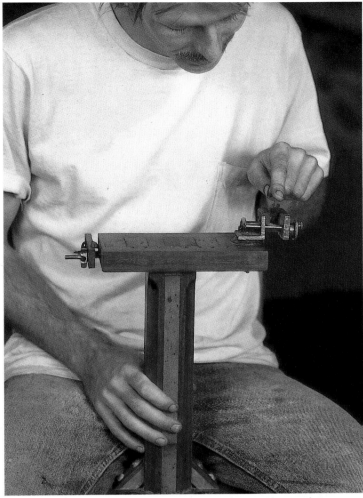

Grandstaff sits on top of his bench to carve fine details. Here, he carves a 1¼-in.-high cabriole-leg assembly for a Philadelphia Chippendale arm chair.

Grandstaff's small bench combines the features of a workbench, a sawhorse and a toolbox. The toolbox at right is stored above the tray in the base.

Portable combination

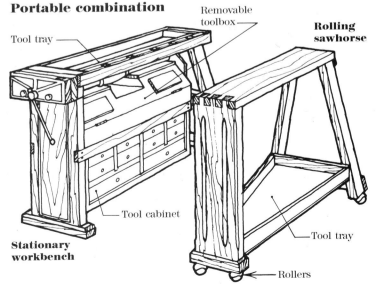

Tool tray

Removable toolbox

Rolling sawhorse

Tool cabinet

Stationary workbench

Tool tray

Rollers

Acorn Design has been one of the more successful North American benchmaking ventures. Here, Tony Moreno routs the dogholes in a stack of Acorn workbenches.

Benches to Market

fter considering all the features you can incorporate into your new workbench you might feel overwhelmed. Making a cabinetmaker's bench can require a lot of machinery, materials, space, time and hard work. While most woodworkers I know have built a bench, many others—professionals and amateurs alike—figure it's just not worth the effort. Like the homeowner who would rather buy oil than split wood, they reason that the time saved in buying a bench is better spent using one. Indeed, if you paid yourself by the hour, your homemade bench would probably cost as much as the most expensive store-bought model.

Commercial benchmaking is not new. It was well established in the United States by the middle of the 19th century. Old catalogs of Hammacher, Schlemmer & Company (see p. 16) and Ohio Tool Company show an impressive variety of workbenches, vises and accessories, pitched (as commercial benches are today) both to professionals and the much larger market of serious amateurs. Perhaps due to changes in Americans' leisure-time activities, or because modern cabinetmakers began to work more with sheet laminates than solid woods, U.S. bench manufacturing declined precipitously in the first half of this century. A cursory glance at any modern tool catalog will reveal that most of the cabinetmaker's workbenches sold in North America are now made in Europe. (Two of the major competitors are shown on p. 119.)

I find this hard to understand. The best European benches are expensive, costing over $1,000. These prices are at least partly inflated by trans-Atlantic shipping costs and the double mark-up, or 'two-step,' added by the U.S. importer and the retailer. And despite their pricetags, many European benches exhibit typical construction short cuts: exposed bolt heads, end caps lag-screwed into end grain, stapled-on tool trays, knotty wood, lightweight stock in the base, end-joined strips in the lamination and so on. Even the most expensive benches look better in the glossy catalogs than they do in the flesh.

Workbenches are mainly wood and a few bits of steel: why not make them here? A small shop, with its lower overhead, might be able to build in the quality of materials, construction and finishing which the larger producers—on any continent—can't begin to compete with. It can offer more options, perhaps even do custom work.

I've met several North American furnituremakers who have been seduced by similar thoughts into the workbench business. The most prolific is undoubtedly Peter Shapiro of Los Angeles, California, who built more than 700 benches under the Acorn Design trademark. When I first spoke with Shapiro in the winter of 1986, Acorn Design was five years old, and he was considering a major financial commitment that would expand production. Six months later, Shapiro was working for the J. Paul Getty Museum in nearby Santa Monica and making benches only on special order in his spare time. Shapiro's benchmaking ups and downs provide some valuable insights into the bench business.

Peter Shapiro's Acorn workbench has a full-width end vise and a double row of dogholes. A face vise of the same design is mounted flush with the opposite end of the bench. Photo by Lee Peterson.

The benchtop in the original Acorn kit came in 10 major pieces. The customer had to glue on the doghole strips, bolt on the end caps, glue up the end-vise jaw and attach both vises.

One of the strongest factors working against the success of a commercial workbench is the woodworker's deeply held conviction that he should build it himself. As a woodworker's totem, the bench rallies some visceral instinct: you not only *can* build your own, you *should*. Aware of this reality, Shapiro decided to allow the customer to participate in the bench construction by selling it as a kit. The customer could point to the finished project with pride and say, "I built that bench." In consultation with Jesse Barragan, who was working for The Cutting Edge tool store in Los Angeles, Shapiro reasoned that the kit also would enable him to bring the cost of a quality workbench below that of the European imports.

Having decided on a kit, Shapiro and Barragan set out to compose the right package of features. The overall dimensions—24 in. wide by 79 in. long by 34½ in. high—were based on standards in the industry. "Benches tend to look a particular way because that's what works," Shapiro says. That's also what the public has come to expect and to make any drastic alterations is asking for trouble. The other features came from benches Shapiro had built for his own shop, based roughly on one of the German Ulmia/Ott workbenches.

The single most important design decision they made was to incorporate a full-width end vise. "I wanted the double row of benchdogs," Shapiro says, "although I wasn't really sure why. Maybe it was that if one row is good, two rows have to be better." He had worked on a bench with a narrow worksurface and a tool tray and found it wanting. The bottom line, as Barragan put it, was that "if you're going to spend six hundred bucks on a bench, you want worksurface—not tool tray." It didn't hurt that both the tool tray and the traditional *L*-shaped tail vise are more difficult to build. Other specifics were determined by eye, and by what could be obtained from available lumber. The base was built out of 12/4 stock and made as large as possible, given the limitations of the vise hardware.

The top of the standard kit was shipped out in pieces—a pre-laminated body, doghole strips, end caps and vises as shown at the bottom of the facing page. The customer also had to glue and assemble the mortise-and-tenon base trestles and trim the tenons on the ends of the stretchers to fit the mortises in the legs. The trestles were attached to the stretchers with knock-down, double-wedged through tenons as shown below, secured with pre-cut tapered wedges made from 1-in. dowels. The customer's last responsibility was to sand and oil the top and the base, two steps that Shapiro was chagrined to discover many customers omitted. Four benchdogs accompanied each kit and these Shapiro made in pairs. He cut the recessed shaft of the dogs on a router table, then sliced them apart and installed a bullet catch in the side of each one.

It was obvious that kit manufacture would require greater consistency than Shapiro's previous one-off woodworking. The parts had to be interchangeable—each bench had to go together without a hitch, regardless of where it was assembled. I visited Shapiro when he was still in production at his Acorn shop in San Gabriel to find out how he dealt with this challenge.

As I drove to the shop with Shapiro in a February downpour, the first thing on my mind was moisture. It had been raining for days, storm sewers looked like slalom kayak courses, houses and mud along the coast highway were becoming shore frontage and the door in the apartment where I was staying would no longer shut. Everyone acknowledges the importance of using well-seasoned wood and allowing for its movement in the design of a workbench. But what happens when you build a bench kit in a Los Angeles monsoon and ship it to Arizona to be assembled? I pictured a nightmare.

In reply, Shapiro gave me a thumbnail introduction to local meteorology. Despite the deluge we were driving in, the ambient moisture content in Southern California normally fluctuates between 25% and 40%—somewhat drier than the East Coast, somewhat wetter than the Southwest. Every now and then, however, a high-pressure center develops over the desert and pushes its way toward Hollywood Boulevard, reversing the normal flow of humid air off the ocean. The hot dry wind, called a Santa Ana after the mountain range from which it descends, can practically turn grapes into raisins. It's not uncommon, Shapiro explained, for the humidity to drop from 40% to 15%, or even 5%, overnight. And it can stay there for weeks.

When Shapiro built his first batch of workbenches in July of 1981, a Santa Ana blew in—raising the temperature in his second-story shop to about 105° in a few hours. Shapiro stood in the middle of 15 workbench kits listening to the wood dry. "It sounded like popcorn," he recalls. Cracks 1/16-in.-wide appeared in end-grain surfaces all over the shop. "I panicked," Shapiro says. He closed the windows and poured buckets of water all over the wood floor. Within 15 minutes, the place was a steam bath, with a slurry of soggy sawdust on the floor. Shapiro figures he had raised the humidity to at least 80%. He closed the shop for the day and went home. Now, when a Santa Ana hits, he slaps a quick coat of cheap varnish on the ends of all the rough bench stock and hopes for the best.

What happens when the finished parts are sent off to the Northwest rain forest or the desert? Surprisingly, not much. Occasionally Shapiro hears about a cracked top, which most often is damaged in shipping, and he replaces or fixes it. But there have been few problems. In part, Shapiro credits the wood he uses. In the early days he built the benches from rough lumber—16-ft. lengths of maple "with hair all over it." It took three men a week just to mill it into thin strips for laminating. The lumber alone cost about $200 per bench (roughly 100 bd. ft.) and much of that was waste. So Shapiro began to look for ways to cut costs. One of his first steps was to have the wood milled into strips before delivery. The cost jumped about $20 on each bench, but it saved shoptime. Later he devised the handy clamp below for keeping the tops flat during glue-up.

The best solution to the problem of cost was to go to the source. He located a maple producer in Wisconsin that was able to deliver laminated tops for less money than Shapiro had been paying for rough lumber in Los Angeles. He found that he could specify no cracks, knots, checks, warp, wane—nothing but clear rock maple, kiln dried to 6½%, sanded to 120-grit and wrapped in plastic. He was saving money at every turn: he had cut out the middlemen, he was no longer shipping water or waste and he saved two days a week in gluing and sanding. The retail price of the Acorn kit dropped by $100 to $599. (By the time of this writing, the price had gone up to $750.)

Acorn wedged tenon

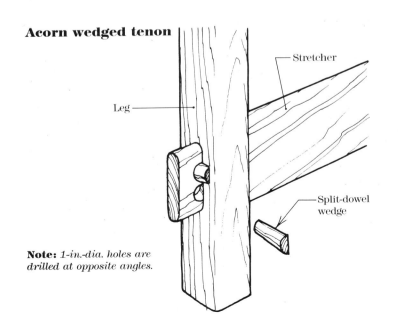

Stretcher

Leg

Split-dowel wedge

Note: *1-in.-dia. holes are drilled at opposite angles.*

Benchtop gluing clamp

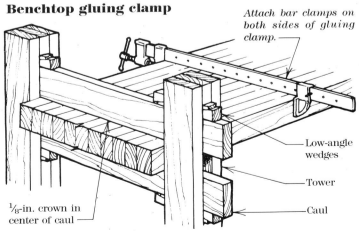

Attach bar clamps on both sides of gluing clamp.

Low-angle wedges

Tower

Caul

1/8-in. crown in center of caul

Note: *Tower is three-piece laminate, with middle strip 1/16 in. thicker than caul to leave room for glue build-up. Length and width of cauls are varied for different stock. To use clamp, tighten wedges by hand, then tap home with mallet. To release wedges, strike top of tower with rubber mallet.*

The vises, of all the components, are at once the bench's strongest feature and its greatest limitation. "If the vises don't work, people get really upset," Shapiro says. "It's like having a chest of drawers where you can't open the drawers." The wide jaws on the Acorn vises offer a large clamping area, and the convenient four-point benchdog connection of the end vise seems to be increasingly popular among woodworkers. The problem with such wide vises is that clamping rarely occurs directly over the screw and the vise racks quickly with off-center pressure.

Both face and end vises are essentially the same simple design (I'll describe the end vise here). Aside from the screw and jaws, the guts consist of two parallel, solid-steel guide rods that pass through the jaw and end cap, and a third steel alignment rod that is stationary beneath the benchtop. A wooden drag bar runs along the underside of the bench to prevent sagging. Steel bushings epoxied into the end cap and the drag bar are honed to fit the guide rods to a close tolerance, thereby reducing racking. (On several benches I saw, the steel bushings had become dislodged, a problem Shapiro later resolved by using a slow-setting epoxy.) The vise will never be as tight as a well-made tail vise, but it's a lot easier to build.

To bore the vise holes with the accuracy required for interchangeable vise parts—tolerances within three thousandths of an inch—Shapiro uses a custom-made mortiser with two 3-hp, 3-phase motor heads mounted on gibbed, dovetailed ways. It used to take one man on a Taiwan drill press four days to bore all the holes in 15 benches. Now it's done in a day. Even the doghole-cutting operation has been streamlined with the router jig shown on p. 114. The 3½-in.-thick doghole strips are twice as thick as the rest of the top, which lends an *I*-beam-like rigidity to the structure and economizes on wood.

Early Acorn practice was to glue the benchdog strips to the top with aliphatic resin glue, but one day Shapiro ran out of clamps and realized "we could buy 24 pairs of corner clamps or one wood-welder." The welder he bought rotates the glue molecules, causing them to rub against each other, creating friction and heat, which sets the glue. Each top spends only five minutes in clamps—exactly enough time for his assistant, Tony

Moreno, to cut the slots for Lamello alignment biscuits in the next top and strips. Later, the top is drum-sanded smooth and boxed for shipment. At the height of production Shapiro and Moreno were able to turn out a workbench kit in five hours.

Aside from the quality of the materials and the accuracy of the vise parts, I suspect that the kits work because not much is left to chance. Shapiro has streamlined the process so that 90% of the customers pay about $70 more to receive a fully laminated top, rather than attach the doghole strips themselves.

While he was striving to become more efficient, Shapiro continued to take the time to make his benches a little bit different, or a little bit nicer than the rest of the commercial herd. The standard kits included vise handles and benchdogs made from exotic woods like gonçalo alves or padouk. "They keep coming out cleaner, stronger, with fewer defects," Shapiro told me as he was finishing what was to be one of his last production runs. What's more, after 700 benches, he still looked at both sides of a piece of wood to put the pretty side out.

When I left Peter Shapiro that day, it looked like his five years of benchmaking might begin to pay off. He'd gotten material costs down, streamlined production and was building a nice bench. He had a backlog of orders for kits. So I was surprised to hear only 6 months later that Shapiro had decided to pack it in. He had elected to exchange one kind of certainty—that of many more years of hustling benches—for another, more secure variety. He'd accepted a full-time job as the woodshop supervisor at the Getty Museum, an offer he couldn't refuse.

Shapiro's experience was not unique. The handful of other North American companies that were making benches in the early 1980s are no longer active. Besides Shapiro's Acorn bench, Hunter Kariher's Liberty Hill workbench, shown below, was also recently discontinued. Even the inexpensive worktables made by Garden Way in Charlotte, Vermont, and by Tennessee Hardwoods in Woodbury, Tennessee, are no longer in production. To my knowledge this leaves only one commercial producer of cabinetmaker's benches in the United States—M. Chandler & Company of Houston, Texas—and two in Canada—Alaska Wood Industry of Ashton, Ontario, and Cambridge Tool Company Ltd., of Cambridge, Ontario.

Acorn end vise
(bottom view)

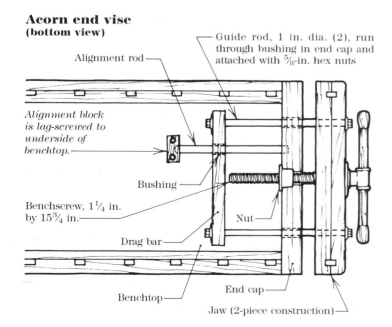

One of 22 Liberty Hill benches Hunter Kariher sold to College of the Redwoods in Fort Bragg, California. Massive 3-in. finger joints hold the shoulder and tail vises together. Vise hardware is custom made.

The one handicap all North American bench businesses seem to share is a poor life expectancy. Robert Larson, a San Francisco tool importer, isn't convinced that there is a workbench manufacturer in North America that will be around in ten years. Larson told me that he figured a North American company could build a good business, but it would take five years to turn a profit and a lifetime to really make it work. Given their recent history, it's no wonder that distributors and consumers don't trust them like they do the Europeans, who have been flogging benches for over a century.

In the last two decades, European benches such as the Ulmia/Ott in the photo at bottom right and the Lachappelle at top right have set the standards against which all commercial benches have had to compete. (Several other imported benches are sold in North America, and their distributors are listed in the Sources of Supply.) Ulmia and Lachappelle benches are the best made and most expensive of the overseas crop; they range in price from $300 to $400 for the smallest hobby and carving models, to around $1100 for the largest cabinetmaker's benches. Despite these prices, the benches continue to appeal to a number of North Americans. While some of their success is due to their affiliations with major U.S. tool dealers such as Garrett Wade Company and Woodcraft Supply Corp., several other factors have helped.

Both companies evolved from old family businesses and now ship benches around the world. Ulmia/Ott was founded in 1877 in Ulm, Germany, by cabinetmaker Georg Ott. Lachappelle Ltd., of Kriens, Switzerland, has been selling benches since 1840. Both gained their reputations in Europe, selling mainly to professionals. Both offer benches featuring a heavy redbeech top with a traditional tail vise and wood/metal face vise (see pp. 128-129), standard features since the 19th century.

One of the benefits of size and experience is that it enables Ulmia and Lachappelle to offer a selection. Both companies sell several models in the United States, with others available on special order. The Ulmia catalog shows 16 different benches, including two- and four-sided school benches and benches with a large tilting tool drawer in the base. Periodically, both companies add new models or make changes to the line in an effort to reach more people. Recent additions are the full-width end vise with two rows of dogholes (now available on both Ulmia and Lachappelle benches) and a $1495 Ulmia sculptor's bench, which features a tool chest and a tilting top. The novel adjustment feature shown in the drawing at right is now standard on all Lachappelle tail vises.

No matter how many options companies like Ulmia and Lachappelle offer, I suspect they will continue to face stiff competition in North America. This won't come from American or other European manufacturers, or even from African imports (several companies are now offering small benches built in Zimbabwe). The strongest challenge to any commercial workbench is the one identified by Peter Shapiro and Jesse Barragan when they designed the Acorn kit. Many woodworkers will never feel at home at a store-bought bench.

Benches made by two of Europe's major manufacturers. Top: Lachappelle's 'Improved Versatile Craftsman's Bench.' Bottom: Ulmia's 'Ultimate' workbench, with tilting tool drawer.

Lachappelle tail vise

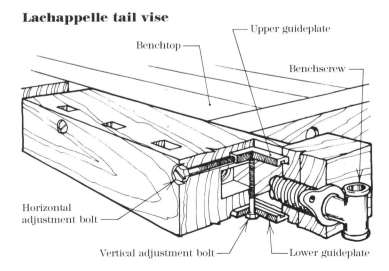

Note: Horizontal adjustment bolts screw into edge of upper guideplate, allowing vise to be kept tight to bench. Vertical bolts secure upper and lower guideplates.

This 19th-century leg vise, built by Aaron Kratz of Plumsteadville, Pennsylvania, extends above the benchtop to facilitate the shaping of curved carriage parts. Kratz's workbench is in the collection of Fonthill Museum in Doylestown, Pennsylvania.

Shop-Built Vises

Chapter 9

Without some way of holding the work, a workbench is hardly more than a table. The most sophisticated tools and hard-won skills are of no use if you can't keep the work from scudding across the benchtop. These days, most of us look to a vise to do the job.

Woodworkers tend to be conservative creatures—we cling to what we know and we know what we have been taught. But it's worth noting that while the vise has one purpose—to secure stock while it is worked—there are a lot of different ways this can be accomplished. The dozens of benches in this book suggest a vast range of possibilities. No two are alike in their selection or deployment of vises. And probably more than any other single feature, the vises you choose will determine the nature of your workbench.

In general, a tail vise traps work in two positions: down on the surface of the bench between dogs or clamped between its jaws. Leg and face vises are designed to clamp work to the front or, occasionally, to the end of the bench. Various specialty vises fix the work between jaws above the surface of the bench for three-dimensional shaping.

Today, with a hardware store down the street and a mail-order catalog at the bedside, it's easy to take the lowly vise for granted. We can no more imagine a workbench without at least one vise than we can picture a hand without its prehensile thumb; without them we'd be lost.

This wasn't always so. Compared to the workbench, which has been around in one form or another for a couple of thousand years, the vise is a relative newcomer. While the form may have originated as early as the fourth century B.C., there is no evidence that a vise appeared on a woodworking bench before the Middle Ages. This is surprising, considering that the operative part of the vise—the screw—was well known to the ancient Greeks. Archimedes (287-212 B.C.) is credited with having developed the so-called 'endless screw' for drawing water, and the wooden screw was in common use in fruit presses and other machinery as early as the first century A.D. Archimedes recognized the screw as the circular analogy to the inclined plane, as can be demonstrated by wrapping a right-angled paper triangle around a cylinder.

Why wasn't the screw immediately embraced by woodworkers and incorporated into a vise? In part, it may be due to that familiar craftsman's trait, stubborn conservatism. But I suspect that the simple efficiency of more primitive workholding methods kept our precursors happy (and their hands full) for a couple of millenia.

The wedge played a fundamental role at the workbench long before the evolution of the benchscrew. Next to the human body itself, the wedge has been the primary workholding de-

Vice   Old and modern French: *vis* (screw); *vit* (speed). Latin: *vitis* (vine, vine stem).

—*Oxford Dictionary of English Etymology*

vice for about as long as men have worked wood, Indeed, for certain tasks, many of the woodworkers I've met prefer simple, pre-industrial devices to the more formidable (and expensive) modern alternatives. These range from simple interlocking wedges to an assortment of ropes, levers and wedges (see Chapters 7 and 13), and shaving horses (Chapter 12). Don't fix it if it's not broken, the saying goes, and the hypothetical woodworker of antiquity may have considered attaching a screw press to a bench, only to reject it because it would take too much time and be no better than a good wedge anyway.

Progress, or at least change, is inexorable. Once the wood vise was introduced—perhaps copied from a blacksmith (who would have had a hard time holding a hot iron with his knee)—it was here to stay. By the beginning of the 16th century, German woodworkers were using highly developed tail vises and face vises on their benches (see center photo, p. 9), although it seems to have taken the rest of Europe and the New World almost two more centuries to catch on.

By the late 19th century, mass-produced metal vises were widely available through stores and mail-order catalogs. An ample selection of their 20th-century descendants is available to the modern woodworker (see Chapter 10). For strength, speed of operation and ease of installation, it's hard to beat a store-bought vise. But just the same, quite a few woodworkers I visited chose to make their own. A shop-built wooden vise may not take as much torque as its iron and steel counterpart, but making your own vise allows you to tailor it to your needs. While it takes time to make a vise, it usually costs much less than buying one, and like building your own bench, the dividends in satisfaction return every time you turn the screw.

Wedging devices

Wedged face vise

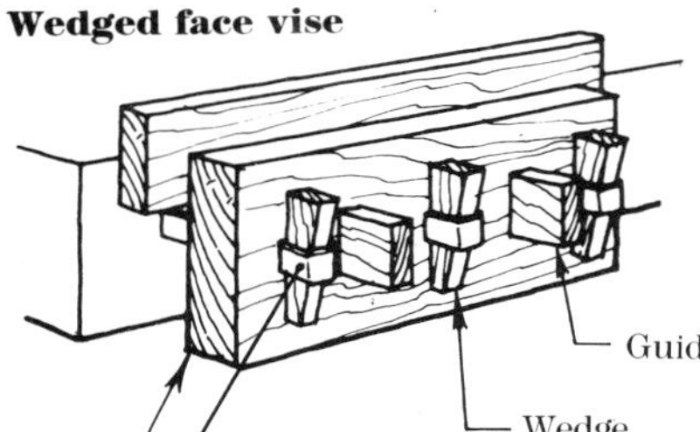

Japanese carpenter's vise

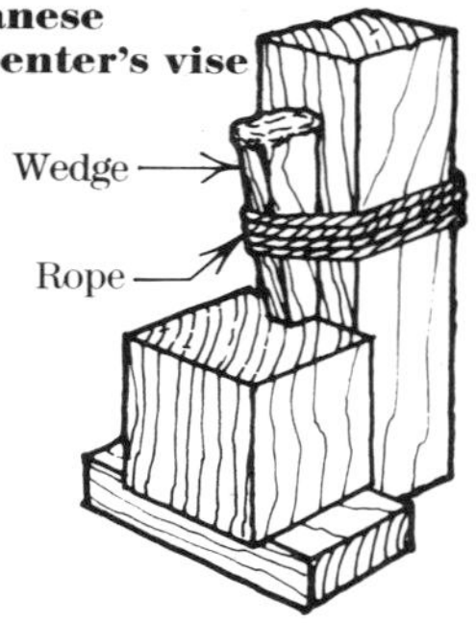

Stop and wedges

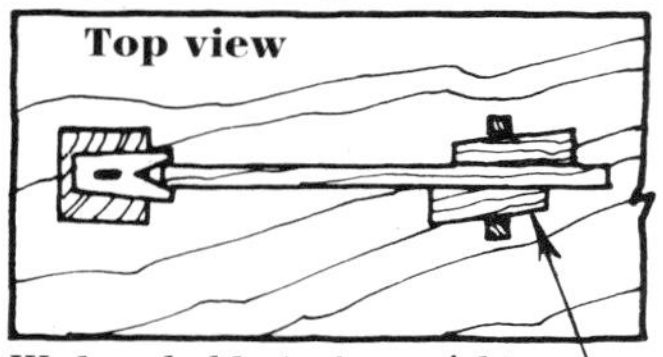

Wedges hold stock upright.

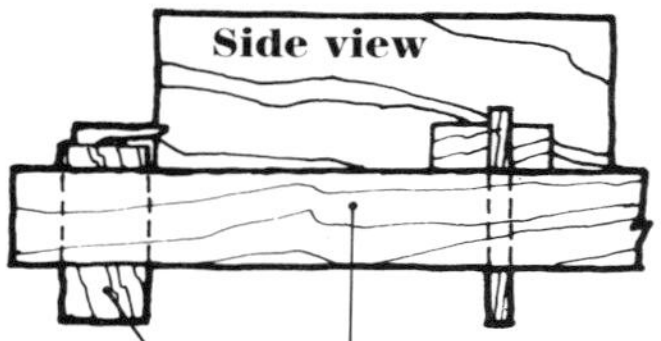

V-blocks

Angled dowels may be used to secure V-block on benchtop.

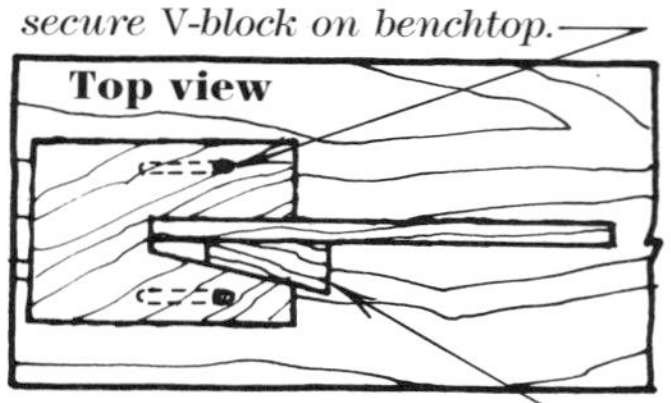

Wedge holds stock tight.

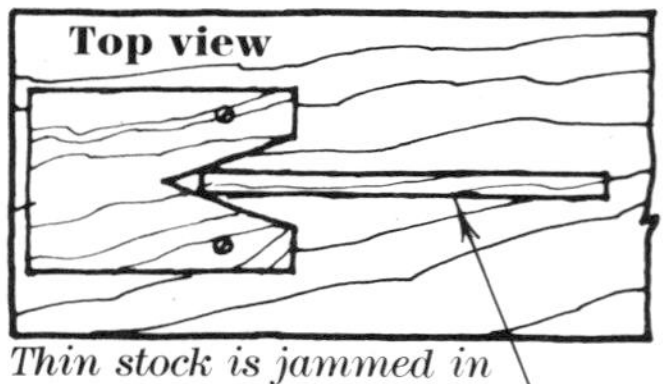

Thin stock is jammed in mouth of V-block.

A history of threads

Without threaded wood or metal screws, the modern woodworking vise wouldn't exist. Richard Starr, a woodworking teacher and writer in Vermont, has long been fascinated by wooden threads. Here he examines their history and how they're made.

The first person to make a screw probably did it by hand the way the Eskimos did. Historical photographs suggest the Eskimo's technique: holding a piece of antler, bone or wood in one hand, they'd twist it past a knife grasped in the other. With the blade at an angle to the shaft, the knife would scribe a helical mark (a spiral) on the material, resulting usually in a left-hand thread because most people are right-handed (try it!). Then, whittling toward the incision, they produced a buttress-shaped thread that could hold a spear tip to its shaft.

That this isolated aboriginal society had threads is a glitch in the history of technology, since most researchers believe every screw on earth had direct ancestors in ancient Greece. Though helices appear in nature and in decorative arts worldwide, we know of no practical application of the shape until the first century B.C. in the land of Plato and Aristotle. The pyramid-building Egyptians never thought of it; Chinese machinery did without screws until the 17th century. So if the Eskimos did come up with the idea on their own, they share the pride of invention with a rather sophisticated culture.

By the first century A.D., screws of wood and metal were common in Hellenistic technology. A press for flattening cloth has survived at Herculaneum (covered by Mount Vesuvius's eruption in 79 A.D.), its wooden screw in fine condition. At the surgeon's house in neighboring Pompeii were found dilating instruments (specula) operated by metal screws, as are modern ones. A twin-screw press appears in a wall painting in that doomed city.

How were screws manufactured in antiquity? Fortunately, we had a reporter on the scene: Hero of Alexandria, who lived during the first century A.D. He created several tools of fundamental value, including a basic surveying instrument, but he is best remembered for his simple steam turbine, which was only a toy. An early engineer who wrote broadly about the mechanical technology of his time, Hero described the evolutionary improvement of screw presses used to produce olive oil. Machines identical to the ones he knew survived into the 19th century. He also explained how screws were made in both wood and metal.

Until quite recently, historically speaking, large wooden screws, up to 12 in. or more in diameter, were cut the way Hero described. After laying out a helix on the surface of the cylinder (he used a metal template) you would saw a notch along the mark to the depth of the threads. Then you'd chisel the *V* shape into the sawkerf. I've tried this; it's easy.

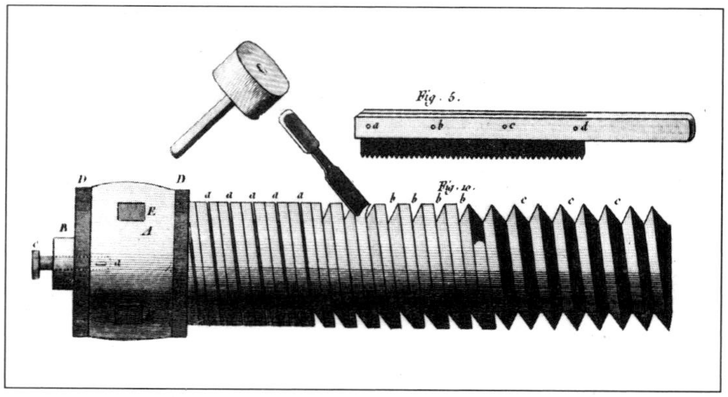

Almost 2,000 years ago, Hero described how large wooden screws were cut in Greece. The same saw-and-chisel technique was demonstrated by Louis E. Bergeron (pseudonym for L.G. Salivet) in this detail from his Manual du Tourneur *(Paris, 1792).*

Making the nut was a problem. The earliest method was to use a bare hole with one or more dowels intruding into it to engage the threads. This worked, but lacked strength. Another method was to carve the nut in two halves, then fasten the halves together. This was stronger than the dowel method, but its strength was limited by the integrity of the fastenings, which might have been glue, rivets or bindings of some sort. Besides, fitting the female thread to the male was incredibly tedious. Despite these shortcomings, the practice survives today, as shown in the photo at right.

Finally, Hero described (and possibly invented) a mechanical tap that etched a thread in a hole, working a little like a modern machine lathe. This gadget, shown in the drawing below, remained in use for almost 2,000 years until hydraulic presses made the wooden machinery obsolete.

In Hero's time, if you needed a small-diameter metal screw, you'd probably cut it with a file and use the dowel-in-the hole method for the nut. It was also possible to cast a nut around an accurately filed screw. The worm drive, where a male screw engages a gear rather than a nut, is said to have been developed by Archimedes in the third century B.C.

Blacksmiths had a technique where inside and outside threads were made at the same time. First the smith would forge a ribbon of iron, square in section, and fold it back on itself, then he would wrap the doubled strip around a metal rod. Sliding the rod out, he'd separate the pair of helices, then solder one to the rod, the other inside a hole. Large screws for presses or vises were made this way and jewelers could use the method on tiny work.

Threading taps for metal and wood, similar to the design common today, were described by da Vinci in the 16th century and probably were in use much earlier. Usually these amounted to notches filed on the corners of a square rod, very simple to make but capable of cutting a decent thread.

Dies, the female-threaded devices designed to cut male screws, are probably as old as the metal-cutting tap needed to make one. Screw boxes, the wood-cutting equivalents of the die, use a *V-gouge* cutter positioned against a nut. I imagine this tool to be very old, although I doubt they existed in antiquity or Hero would have described them. Da Vinci sketched a tool that may or may not be a screw box; if it is, it's the earliest representation I've been able to find. The 18th-century screw box and tap are almost identical to those available today. Several devices are now available that use a router to cut screws in wood very neatly.

After Hero's wood-threading tap it was probably twelve or fourteen hundred years before people resumed the search for new methods of cutting screws quickly and accurately. Most methods were adaptions of the lathe, a tool that had been in worldwide use for thousands of years. The challenge of threading and, later, of turning screw-like ornamental shapes, stretched mankind's ingenuity and eventually evolved into the machine-tool industry upon which our modern technology is based. As woodworkers we owe a nod to the early inventors who made possible our labor-saving machinery. And when we cut a screw in wood for a child's toy or a workbench vise we are a lot closer to our roots than we may think.

To carve the female threads, the ancient Greeks commonly cut the nut in half. Robert Yorgey, a Pennsylvania farmer, makes screws for his vises the same way. Yorgey fits the threads by using lampblack on the crests, the way a dentist locates a raised filling with carbon paper. Photo by Richard Starr.

An ancient screw box

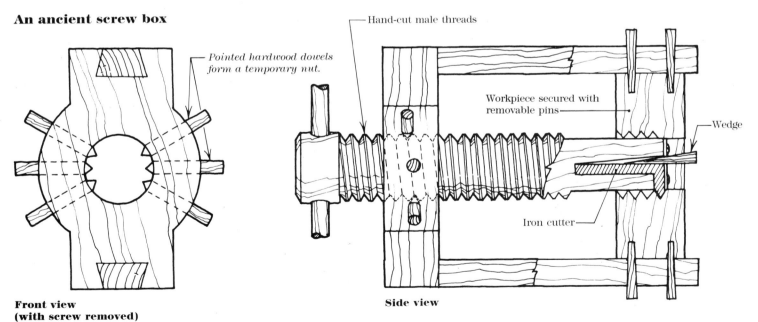

Hand-cut male threads

Pointed hardwood dowels form a temporary nut.

Workpiece secured with removable pins

Wedge

Iron cutter

Front view (with screw removed)

Side view

Note: *Hero described this thread-cutting screw box in the first century A.D. Male screw threads are hand-cut on one end of a shaft. These run in a temporary nut, above left, formed by pointed dowels inserted in drilled holes. At the other end of the shaft, above right, an iron cutter is wedged in a slot. The cutter is propelled forward when the screw in the temporary nut is turned.*

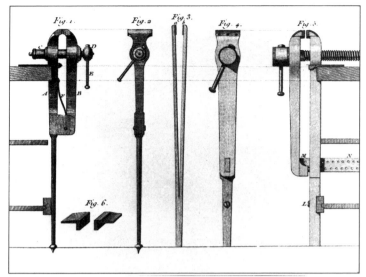

Leg vises, from Jacques-André Roubo's L'art du Menuisier.

Leg vise

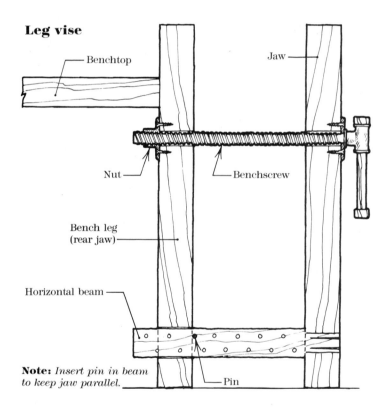

Benchtop

Jaw

Nut

Benchscrew

Bench leg (rear jaw)

Horizontal beam

Note: *Insert pin in beam to keep jaw parallel.*

Pin

Foot-adjustment mechanism
G.L. Gilmore

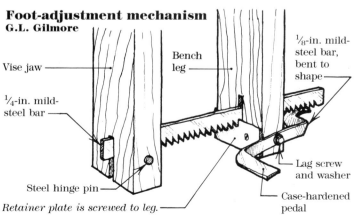

Vise jaw

Bench leg

1/8-in. mild-steel bar, bent to shape

1/4-in. mild-steel bar

Steel hinge pin

Lag screw and washer

Retainer plate is screwed to leg.

Case-hardened pedal

Leg vises

If you are considering building your own vise, the leg vise shown in the drawing at center is a good place to start. With either a store-bought metal screw or an old wooden one, it is probably the simplest woodworking vise—and one of the strongest—you can make. These two features made the leg vise very popular in 19th-century America, and versions of it can be found on many of the benches that survive.

In principle, leg vises are all the same. Their vertical jaws are operated by a single screw, placed somewhere below the clamping surface and above the center of the jaw. If you think of the jaw as a lever, the vise derives much of its strength from the fact that the fulcrum (the horizontal beam) is near the floor—far away from the point of pressure. Accordingly, the closer the screw is to the top of the vise (and the farther from the fulcrum), the stronger the grip, although the clamping area will be reduced. In Roubo's 18th-century engraving at left, the screw is placed as close to the jaw as possible. Roubo also depicts the close relationship between the woodworker's leg vise and the blacksmith's iron post vise, from which it probably evolved.

Leg vises typically use the left front leg of the bench as both the rear jaw of the vise and the nut. (Either the leg itself is threaded, or a separate nut may be installed behind it.) Occasionally, a leg vise may be entirely independent of the base of the bench. The top of the jaws may be made flush with the top of the bench, or raised above it to permit easier clamping and carving of irregularly shaped objects.

The major problem presented by the leg vise is how to keep its jaws parallel. When the screw is tightened, both the top and bottom of the jaw are pressed against the leg. Without some method of supporting the bottom of the jaw, the clamping surface will be angled. The mild angle produced when clamping thin stock probably won't matter, but the radical angle that develops when you clamp thick stock will make it impossible to get a good grip. (This can work to your advantage, of course, if you're clamping tapered objects.)

The most common solution to this problem is the one shown in both the Roubo engraving and the leg-vise drawing at left—a horizontal beam attached near the bottom of the front jaw. The beam passes through a mortise in the rear jaw of the vise, or in the leg of the bench to which the front jaw is screwed. By inserting a pin in one of a series of holes drilled in the beam, the bottom of the front jaw may be held away from the rear jaw. The pin is easily moved to another hole to keep the jaws more or less parallel when clamping thicker or thinner stock. (In an earlier volume Roubo described a simpler method—inserting a block of wood at the foot of the vise to hold the jaws apart, thus replacing the horizontal beam as the fulcrum.)

A host of creative minds have been kept busy over the years trying to design a more effective alternative, or one that does not require bending over to adjust. Patent offices around the world are chock full of the ingenious, over-engineered results. I've seen a few successful variations that are worth noting. A screw-adjusted mechanism that appears on some Shaker leg vises (see p. 46) allows for precise parallel adjustment. The foot-operated mechanism shown at left, built by G.L. Gilmore of Wilmington, Delaware, is more involved but does the job quite nicely. (Gilmore adapted the design from *The Boy Mechanic*, published by Popular Mechanics in 1915.)

The leg vise at right, from a 19th-century workbench, is reinforced with twin horizontal guides at the bottom of the leg and an iron adjustment screw to keep the jaws parallel. A wooden roller has been fitted to the bottom of the vise leg to make it easy to adjust.

Most leg vises I have found on old benches are vertical, like the one shown in Roubo. In workshops and among tool dealers I visited in Pennsylvania and Maryland, however, it is at least as common to find an angled leg vise, like the one in the photo at right. The angle of this vise is only slightly off the vertical, while others I have seen approach an acute 60° angle to the floor. In some cases, only the vise itself is angled, while on other benches such as this, the entire left leg assembly of the bench is skewed. I have no idea why this quirk of construction gravitated to the mid-Atlantic states, but I have never seen one north or east of the Delaware River.

From conversations with woodworkers who have used this type of vise and from what I can glean from old texts, I suspect that this striking feature has several advantages. First, the angle of the vise opposes the natural force of planing. If the entire left leg structure is also angled, the bench tends to resist racking or sliding. Perhaps more to the point, when a piece of wood is clamped in the vise, it is aligned with the center of the jaw, as shown below. This feature, coupled with the enlarged clamping area of the jaw, allows pressure from the screw to be applied directly against the work, rather than to one side of it, as is common with a vertical leg vise. In addition, a simple horizontal beam can be installed on the centerline of a skewed vise without requiring a mortise-and-tenon joint.

The vise shown on p. 120 extends 9 in. above the benchtop, which is characteristic of a 19th-century wheelwright's bench. This feature allows the craftsman to work the curved felloes of a wheel rim and shave the spokes without busting his knuckles on the top. The bench, which is in the collection of Fonthill Museum (the home of Henry Chapman Mercer in Doylestown, Pennsylvania), was acquired from a local carriage builder, Aaron Kratz, who used it from 1856 until his retirement in 1916.

Although it is perhaps the oldest form of woodworking vise, the leg vise isn't altogether a relic. I've met boatbuilders, luthiers, period furnituremakers and carvers who use them. Several of these leg vises are shown elsewhere in this book, with details of their construction; check the index to find them.

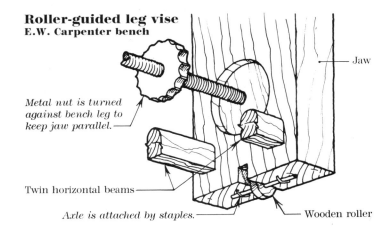

Roller-guided leg vise
E.W. Carpenter bench

Jaw

Metal nut is turned against bench leg to keep jaw parallel.

Twin horizontal beams

Axle is attached by staples.

Wooden roller

The angled leg vise and base frame on this 19th-century bench help resist movement during planing. The first several inches of threads on the leg vise and the matching female threads in the leg have been stripped—a frequent sign of old age among wooden vises. A new nut has been added behind the leg to keep it working.

Leg vises

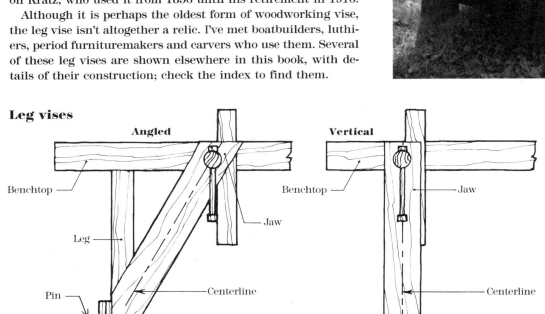

Angled

Benchtop

Leg

Jaw

Centerline

Pin

Horizontal beam

Attached beam guide

Note: *With leg vise angled, work can be clamped more securely along centerline of jaw.*

Vertical

Benchtop

Jaw

Centerline

Mortised beam

Face vises

In time, the leg vise was largely supplanted by the wooden face vise. The need to adjust the beam, and the narrow width of the jaws, greatly restrict the leg vise's applications. The early face vise in the drawing below, which is modeled after one on a 17th- or 18th-century French workbench in the J. Paul Getty Museum in Malibu, California, appears to form a transition between the leg vise and two divergent types of face vise. It is essentially a leg vise turned on its side—a guide bar and adjusting nut similar to those of a leg vise keep the jaws parallel. It's but a short step from this vise to the twin-screw face vise found on the 18th-century Dominy bench in Chapter 1, or the shoulder vise in Chapter 4. By eliminating the adjustment nut and adding a second guide on the opposite side of the benchscrew, it's only a slightly larger step to the full-fledged face vise.

The traditional wooden face vise (the one in the photo below was by cabinetmaker Ray Creager) solves the problem of angled jaws by flanking the screw with two guide bars. The wide jaw also opened up many clamping possibilities to the cabinetmaker who had previously relied on a leg-vise jaw perhaps one-quarter its width. The guide bars have two main functions: they aid in tracking the jaw in and out and, by fitting

snugly against the underside of the top, they prevent the jaw from becoming skewed when work is clamped along its top edge. As with any vise, the closer work can be clamped to the screw, the less stress is placed on all its parts. As is common with wooden leg vises, the front edge of the bench serves as the rear jaw of the face vise. Today's commercially made face vise is nothing more than a mass-produced, metal version of the traditional handmade wooden vise.

Creager used a wooden screw when he built the face vise below. A number of craftsmen I met chose to do the same thing, which isn't difficult if you can make or find the screw and nut. (Thread-cutting methods have a long history, as Richard Starr points out on p. 122.) Creager's 2-in.-dia. screw and the tapped board, which he lag-screwed to the underside of the bench, cost him $2 at an auction. Wooden screw boxes and taps are readily available up to about 1 in. in diameter. These are expensive, however, and larger sizes, such as the one shown in the top left photo on the facing page, are hard to come by. Screw-threading router jigs are less expensive and at least one source, the Beall Tool Company of Newark, Ohio, will cut larger-diameter threads to special order. These are lathe-cut, square-threaded affairs (see photo, p. 129). The square threads are not nearly as smooth in operation and are much weaker than the traditional V-shaped screws, cut with a screw box.

In its construction, Creager's vise is straightforward. Its success depends mainly upon the parallel orientation of the guide bars and screw. This is accomplished by the ¾-in. plywood brace screwed to the back of the guide bars. A nub turned on the end of the benchscrew fits in a hole drilled in the brace, halfway between the guides. Creager cut a 3⅜-in.-wide jog in the plywood brace to clear the base's top bearer. To make up the difference between the 3½-in.-thick front section of the benchtop and the 3-in.-thick main section, he added ½-in. shims to the underside of the thinner portion. This enables the guide bars to slide smoothly against the top, effectively restricting vise racking. The guides slide in tight-fitting notches cut in the bottom portion of the vise's rear jaw, and their front ends are joined to the front jaw by double-wedged through tenons. The screw is secured in the jaw by a garter, a friction-fit wooden key that is another relative latecomer to the workbench vise.

French face vise

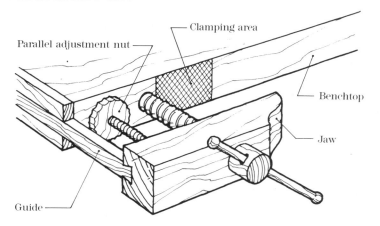

Clamping area

Parallel adjustment nut

Benchtop

Jaw

Guide

The jaw on Ray Creager's wooden face vise is a massive 3¼ in. thick by 9 in. high by 22⅝ in. long. (Creager, a lefty, mounted the vise on the right end of the bench.)

A 2-in.-dia. tap and screw box. New wood-threading kits are commercially available up to 1 in. in diameter.

The two horizontal guides on Creager's vise are screwed to a plywood brace, which slides along the underside of the benchtop, preventing the vise from sagging.

Wooden face vise
Ray Creager

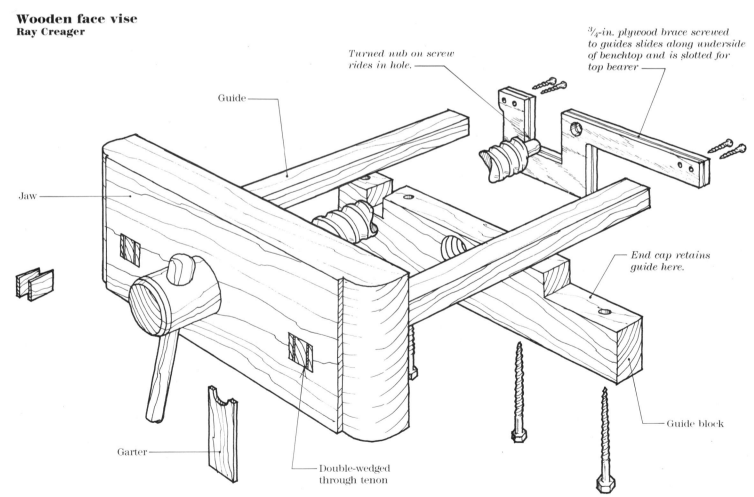

Guide

Turned nub on screw rides in hole.

¾-in. plywood brace screwed to guides slides along underside of benchtop and is slotted for top bearer

Jaw

End cap retains guide here.

Guide block

Garter

Double-wedged through tenon

Note: Benchtop (not shown) and guide block form rear jaw.

Tom Nelson outfitted his wooden face vise with German-made hardware. The steel guide rods slide in a cast-iron harness screwed to the underside of the bench.

For those less enamored of wood and more interested in convenience, there are several brands of metal face-vise hardware available. Tom Nelson, whose tail vise is described in Chapter 5, built the hybrid face vise at left using wooden jaws and commercial hardware. In principle, the design is identical to that of both the traditional wooden face vise and the modern metal vise. It includes an Acme-threaded benchscrew, flanked by two steel guide rods, running in a cast-iron harness, which holds the unit together and is mounted to the bench.

The installation of such a vise is largely a matter of boring accurate holes in the jaws and screwing the harness squarely to the underside of the benchtop. The process is similar to that described for the Record vise in the next chapter, with a few additional considerations.

Nelson's 4⅝-in.-thick benchtop presented a problem. If he simply attached the harness to the underside of the bench, the vise screw would be positioned quite low in the jaw and subject to a lot of torque. Instead, he routed into the underside to inset the harness.

The metal guide rods slide in cast-iron sleeves that must be installed in the front of the bench (not all brands of hardware come with these sleeves). Nelson fitted a block to the underside of the bench along its front edge to provide a mounting surface for the sleeves. (Also, the block was relieved for the doghole slots that it covers.) At the back of the harness, he attached a spacer to provide a flat mounting surface on the underside of the bench.

Nelson bored the holes in the jaw and in the front sleeve block at the same time. He clamped the two together and bored them on the drill press, making the holes slightly oversize so that the jaw could be lifted about 1/16 in. proud of the top surface of the bench when the front casting was attached. After boring the holes, he routed recesses for the sleeves in the block that screws to the bench.

Routing the recesses can be bothersome, so for his next bench Nelson plans to substitute bronze sleeves for the sleeves that come with the hardware. The jaws should make contact

Wood/metal face vise
Tom Nelson

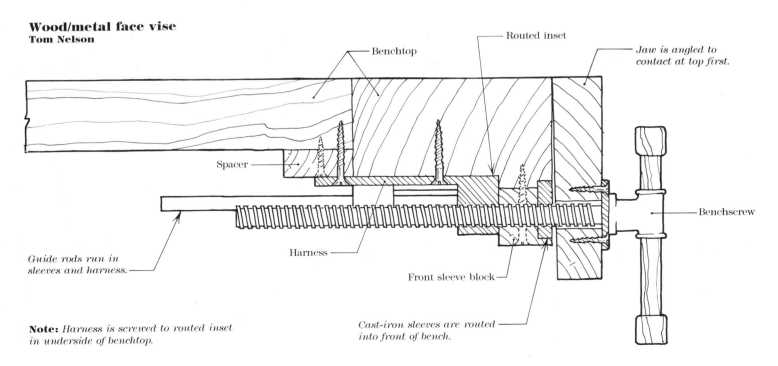

Benchtop

Routed inset

Jaw is angled to
contact at top first.

Spacer

Benchscrew

Guide rods run in
sleeves and harness.

Harness

Front sleeve block

Note: *Harness is screwed to routed inset
in underside of benchtop.*

Cast-iron sleeves are routed
into front of bench.

first along the top edge, closing gently to parallel as pressure is applied. This can be accomplished by simply planing a taper on the jaws. After the vise has been attached and is working smoothly, the jaw's top can be planed flush with the benchtop.

Over the years, Nelson has helped his students install more than 30 of these face vises. He admits that, to one degree or another, the vises all have the same problem: the jaw racks if you don't center your work over the screw. Even with the most careful mounting job, the racking is so predictable that Nelson has come to accept it as "the nature of the beast."

Most of the vise hardware (for both face and tail vises) currently available in the United States and Canada comes from one of three sources: Gebruder Busch of West Germany, makers of the Hirsch trademark; Cambridge Tool Company Ltd. of Cambridge, Ontario; and Record Marples Ltd. of Sheffield, England. Distributors for these and other companies are listed under Sources of Supply.

It costs roughly half as much to buy a plain vise screw and nut than to buy one already affixed to guide bars and castings, and this allows more flexibility in designing your own vise. Drew Lansgner's vise, shown at right and below, is another hybrid, a homemade replica of a store-bought face vise.

Langsner, of Marshall, North Carolina, built the vise with wooden jaws, a 16-in. Record benchscrew and two 1-in.-dia., cold-rolled steel guide rods he got from a steel fabricator. The rods slide within standard pipe sections threaded into pipe flanges at both ends. They are friction-fit in the front jaw and each is secured from below with a single screw. The nut for the Record screw is attached to the back of the rear jaw, and both the rear jaw and brace are bolted to the underside of the bench.

Langsner used a drill press to ensure the accuracy of all the guide-rod holes in the wooden jaws. He also advises having the supplier cut the steel rods to length with a power hacksaw to save time and a lot of frustration. Langsner's vise takes more time to build than Tom Nelson's, but even at its maximum opening of about 8 in. it hardly wobbles.

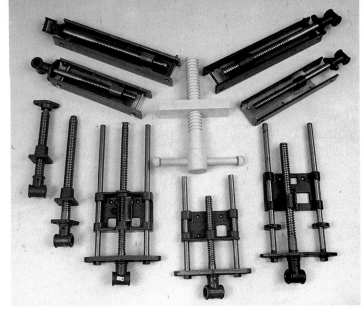

A sampling of bench-vise hardware: tail vises (top), face vises (bottom), shoulder-vise screw (far left), tail-vise screw (second from left) and wooden benchscrew (center). The black hardware is made by Gebruder Busch of West Germany. The blue hardware is made by Cambridge Tool Co. of Canada. The wooden screw is no longer in production.

Drew Langsner built this basic face vise using a Record benchscrew and locally available steel rods and pipe fittings. The guide rods slide within the short pipe sections, keeping the vise rigid.

Pipe-fixture face vise
Drew Langsner

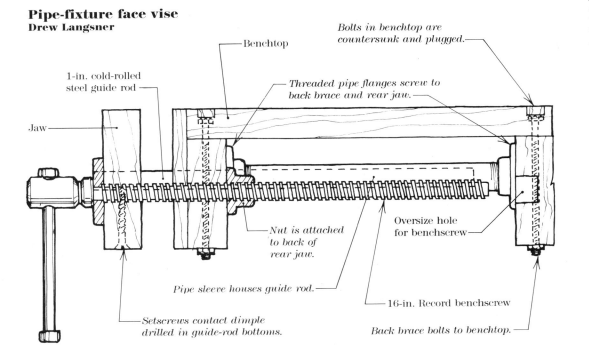

Benchtop

Bolts in benchtop are countersunk and plugged.

1-in. cold-rolled steel guide rod

Threaded pipe flanges screw to back brace and rear jaw.

Jaw

Nut is attached to back of rear jaw.

Oversize hole for benchscrew

Pipe sleeve houses guide rod.

16-in. Record benchscrew

Setscrews contact dimple drilled in guide-rod bottoms.

Back brace bolts to benchtop.

Tail vises

The tail vise may be the most distinctive feature of the cabinet-maker's workbench; I've described several versions of the traditional tail vise in Chapters 3, 4 and 5. The following two tail vises, which were devised by their makers to meet particular needs, depart from the norm. I include them here to indicate the variety that exists in even as traditional an object as the tail vise.

I introduce them, however, with the following caution. When I adapted Tom Nelson's vise to fit Michael Fortune's bench (Chapter 5), I learned an important lesson about the integrity of the tail vise. Once I began tampering with tail-vise dimensions for the adaptation, I discovered I had to alter many other measurements in the benchtop to match. Be forewarned. If you adapt a vise from one bench design to another, be sure to check its relationship to all parts of the top and base before you cut. To a lesser extent, the same holds true, of course, for all the vises and other design features throughout this book.

Traditional tail vises hold the work in several ways. The most obvious, and predominant, is between the benchdogs. In addition, work can be held between the front jaws (those opening along the front edge of the bench); you can clamp a curved part in the jaw to work on one end, while the rest of it trails off under the bench. And, finally, work is sometimes clamped between the end of the bench and the rear jaw of an L-shaped tail vise. Most woodworkers object to this last method, contending that the rear jaw exists principally to strengthen the vise and to provide a place to attach the guides that keep the vise square and running parallel. Work clamped that far away from the centerline of the screw will eventually distort the vise. (On some commercial benches the bolt that attaches the end cap is not countersunk; the protruding bolt head is meant to discourage those who would use the rear jaw as another vise.)

For its principal use—holding work between dogs—the traditional tail vise performs admirably. But it has one limitation: the stock is invariably clamped over the space created by the open jaw. This may not matter if the stock is thick and the gap isn't too large, but with thin stock and the vise open wide the vibration and general lack of support can be a problem.

Faced with this problem, David Powell, who operates Leeds Design Workshop in Easthampton, Massachusetts, designed the tail vise shown below and on the facing page. Powell's vise has a single benchdog housed in a wooden block. The block travels within the enclosed right corner of the bench and functions as a 'trapped' jaw. On this vise, the gap of the standard tail-vise jaw is greatly reduced, providing firm support for most thin stock. On the other hand, it's impossible to clamp a piece vertically, and the Record face vise mounted at the bench's other end has a limited application in that regard. "But I haven't felt the need for it so far," Powell says.

Powell built the cavity for the trapped jaw into the bench when he glued on the doghole strip and dovetailed the front laminate to the end cap. He bored a hole in the end cap for the 16-in. Record benchscrew and flipped the benchtop over to fit the nut to the sliding wooden block. The wooden block is mounted to a ¼-in.-thick steel plate, which slides in grooves milled in two ¾-in.-square steel guide bars. The bars are screwed to the underside of the bench on both sides of the cavity. To keep the guide bars parallel, Powell screwed and pinned a metal crossbar near both ends. At one end of the block a piece of ¼-in. steel angle bolted to the steel plate provides a mounting surface for the benchscrew nut. Powell mortised the block, plate and angle for the benchdog.

Feeling his way through the fitting process, Powell hacked away with a drill and chisel until there was room for the sliding block assembly and the guide bars. When the hardware was installed, he inserted the benchscrew and secured its remaining collar.

The vise construction looked fairly complicated and unforgiving to me, so I wondered if he would do it again. "I might," Powell said, adding, "I'm not going to make another bench." Interestingly, some months later I discovered the remarkably similar Löffelholz vise, described on p. 9. Drawn in 1505, it is the earliest reference to a workbench tail vise I have found.

Powell's bench is 8 ft. 8½ in. long, and he can clamp work as short as 3 in. or as long as 8 ft. between dogs. The vise operates contrary to conventional practice—that is, a clockwise turn on the screw opens, rather than closes, the dog. I found this awkward, but I suspect I'd get used to it.

Enclosed tail vise
David Powell

Note: *Section is taken inside front laminate.*

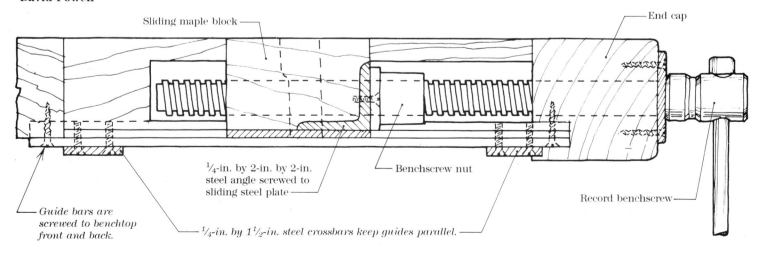

Sliding maple block

End cap

¼-in. by 2-in. by 2-in. steel angle screwed to sliding steel plate

Benchscrew nut

Record benchscrew

Guide bars are screwed to benchtop front and back.

¼-in. by 1½-in. steel crossbars keep guides parallel.

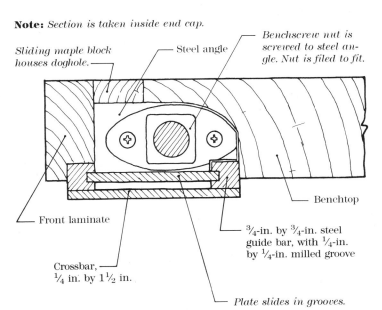

Note: *Section is taken inside end cap.*

Sliding maple block houses doghole.

Steel angle

Benchscrew nut is screwed to steel angle. Nut is filed to fit.

Benchtop

Front laminate

3/4-in. by 3/4-in. steel guide bar, with 1/4-in. by 1/4-in. milled groove

Crossbar, 1/4 in. by 1 1/2 in.

Plate slides in grooves.

David Powell's enclosed tail vise requires less space to operate and offers good support for thin stock. The benchdog rides in a wooden block that slides on a steel plate and parallel steel guide bars.

Apart from being equally unusual, John Nyquist's tail vise is about as different from David Powell's as it could possibly be. What Powell's vise is missing is precisely what Nyquist was looking for: the capacity to clamp work vertically in the front jaws of the vise. Because Nyquist builds more chairs than anything else, he is more often shaping three-dimensional stock than planing thin boards. Although the front jaw of his vise is only $2\frac{1}{2}$ in. wide at the benchtop, it is 6 in. deep and extends for a full 12 in. under the top. This extra clamping surface enables Nyquist to hold curved chair parts securely.

Nyquist also frequently clamps work between the rear jaw and the end cap—with no apparent weakening of the vise. The width of the front jaw and the careful fit of the waxed, teak runners and guide certainly contribute to the sturdiness of the vise, as does the $\frac{5}{8}$-in.-square, cold-rolled steel bar that guides the front jaw. Fastened into a mortise in the jaw, the bar slides in a corresponding slot in the edge of the benchtop. To beef up the benchscrew connections, Nyquist attached the flange to the outside of the vise jaw and the nut to the inside of the end cap with flat-head bolts instead of wood screws. The bolts are secured with *T*-nuts, countersunk and plugged, as shown in the drawing below. The plugs present a smooth clamping surface on the mating faces of the jaw and end cap.

"The whole secret to building a good jaw," Nyquist says, "is to build it on an armature." His armature is a plywood form, made to the exact dimensions of the tail-vise notch on the benchtop and assembled dead square. To this, Nyquist clamped the vise parts, ensuring an exact fit. The result speaks for itself—at full extension ($12\frac{5}{8}$ in.), Nyquist's vise is one of the tightest I've seen. Using well-seasoned hardwoods certainly contributes to the vise's smooth operation and stability. Nyquist used walnut for the vise and iroko for the 3-in.-thick benchtop, air-dried in the rafters of his Long Beach, California, shop for 20 years. The simple fixture shown in the drawing at right below clamps in the front of the tail vise to hold chair parts above the benchtop for shaping.

The front jaw on John Nyquist's tail vise is 6 in. deep and extends 12 in. beneath the bench, making it easy for Nyquist to clamp curved chair parts. The walnut vise has massive finger joints at the corners. It runs on two waxed teak guides beneath the benchtop and a $\frac{5}{8}$-in.-square steel bar mortised into the inside of the jaw and the front edge of the bench.

Tail vise
John Nyquist

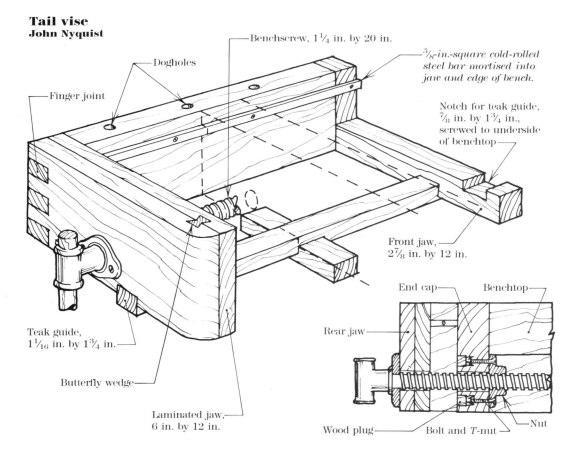

- Dogholes
- Finger joint
- Benchscrew, $1\frac{1}{4}$ in. by 20 in.
- $\frac{5}{8}$-in.-square cold-rolled steel bar mortised into jaw and edge of bench.
- Notch for teak guide, $\frac{7}{8}$ in. by $1\frac{3}{4}$ in., screwed to underside of benchtop
- Front jaw, $2\frac{7}{8}$ in. by 12 in.
- Teak guide, $1\frac{1}{16}$ in. by $1\frac{3}{4}$ in.
- Butterfly wedge
- Laminated jaw, 6 in. by 12 in.
- End cap
- Benchtop
- Rear jaw
- Wood plug
- Bolt and *T*-nut
- Nut

Chair-clamping fixture

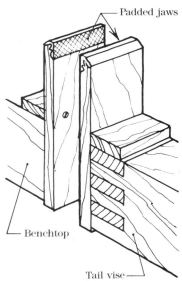

- Padded jaws
- Benchtop
- Tail vise

Before leaving the subject of unorthodox tail vises, it's worth noting that a standard face vise can be installed on the end of a workbench. Face vises are far less complicated than most tail vises, and with very little modification, a face vise such as Ray Creager's or Drew Langsner's may be enlarged to cover the entire end of the bench.

That was precisely what Peter Shapiro did when he designed the vises for his Acorn workbench, shown at right. Shapiro's end vise has the ability to clamp a wide carcase or frame the full width of the bench, with a four-point grip. Its primary disadvantage is that, unless you work exclusively on wide stock, most planing, chiseling, etc. takes place along the front edge of the bench. Because the screw is in the center of the jaw, the vise skews and may eventually be destroyed by repeated clamping on one side. Dino Ciabattari tackled this problem by building a full-width end vise with two screws, one at each end of the jaw, as described below.

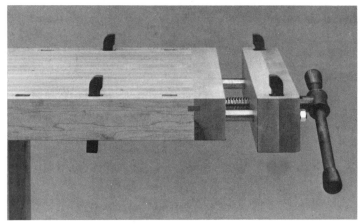

To reduce racking on the Acorn end vise, the guide rods travel in steel bushings in the end cap. A third rod and a drag bar beneath the benchtop also help keep the jaws parallel. Photo by Lee Peterson.

A chain-drive end vise

It's hard to resist the appeal of a sleek workbench such as Nyquist's or Powell's, or many of the others described in this book: dovetailed end caps, exotic hardwoods, frame-and-panel cabinets, self-locking joinery. But I've always been as attracted to the unusual, even oddball, as I am to the elegant. So I was pleased when, on a trip to Ft. Bragg, California, I came across a most unusual tail vise.

Students descend on Ft. Bragg from all over the country to spend a year learning to woodwork at The College of the Redwoods and at the feet of James Krenov. Few realize that ten blocks away, in a nondescript one-story garage, another woodworker has been plying his trade for at least as long as Krenov. Dino Ciabattari has been woodworking for the last 50 years. Born in California but raised in Italy, Ciabattari learned the trade as an 11-year-old apprentice in a furniture factory in Florence. He set up his Ft. Bragg shop in 1940, and for the last 20 years has specialized in cabinets, large carcase work and built-ins.

In the middle of a cluttered workshop crammed with homemade machinery sits Ciabattari's workbench. Lit by a panel of skylights, the 3-ft. by 11-ft. behemoth is the focus of the shop. It is built of softwood, with the middle of the structure supported by twisted wire and turnbuckles—a miniature Golden Gate bridge. When I arrive, the benchtop is a hopeless jungle of tools, hardware and coils of yellow airhose. There are two metal vises on opposite corners of one end of the bench (one is for woodwork, the other for metalwork). But Dino lavished his attention on the other end.

At first glance, the tail vise is an ordinary affair—a softwood jaw spans the end of the bench and is closed with two steel benchscrews. Two dogholes near both ends of the vise jaw correspond to parallel rows of holes in the top to provide four-point clamping pressure for large panels or frames. But at this point Dino's vise departs from the ordinary. To solve the perennial predicament of a twin-screw vise—how to open or close the jaw in a parallel orientation—Ciabattari joined the heads of both screws with a bicycle chain that travels on a small sprocket on each screw. Turning only one handle opens or closes both screws simultaneously.

As an added feature, only one of the sprockets is welded to the head of the benchscrew. The other slips over an extension of the benchscrew shaft and is engaged when an oversized, round nut is turned against a leather washer, which exerts pressure against the sprocket. The screw can be placed in 'neutral' simply by backing off the nut and leather washer 'clutch' that hold the sprocket under pressure. In this manner, the jaw can be re-oriented to conform to an out-of-square object. When the clutch is re-engaged, the vise opens and closes with whatever skewed orientation is required.

In its own way, Ciabattari's workbench reflects the pragmatic ingenuity he applies to his business. It was one of Krenov's students who introduced me to Dino. Dino had told the student: "You let them [College of the Redwoods] teach you how to do it—I'll teach you how to make a living at it."

Dino Ciabattari custom-built this twin-screw end vise for clamping wide panels or frames (above left). The action of the vise is controlled by a bicycle chain that runs on a sprocket mounted to the head of each screw. The screws can be operated in tandem, or they can be adjusted independently by backing off the large nut and leather washer on the right-hand screw (above right).

Dovetailing vise
Joel Seaman

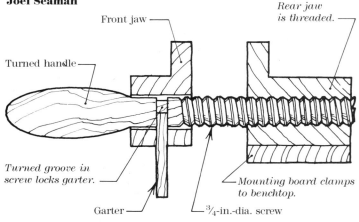

Front jaw

Rear jaw
is threaded.

Turned handle

Turned groove in
screw locks garter.

Garter

¾-in.-dia. screw

Mounting board clamps
to benchtop.

*Joel Seaman built this modified bookbinder's vise to clamp wide
boards vertically for cutting dovetails.*

Specialty vises

No matter how good your leg, face or tail vise is, chances are
you will occasionally need to clamp something that just won't
fit in any of them. There are commercially made holdfasts and
all kinds of specialty vises you can buy that might do the trick.
Or you can make your own.

To cut dovetails, Joel Seaman (cabinetmaker at Hancock
Shaker Village in Massachusetts, described in Chapter 3) made
the small vise shown at left, which resembles a traditional
bookbinder's press. Clamped anywhere along the front of the
bench, Seaman's vise allows him to secure a wide board vertically, something he can't do in either his tail vise or leg vise.
The vise screws are keyed to the front jaw with a garter, fitted
in a slot below the jaw.

David Welter, of Fort Bragg, California, devised the small
vise shown at the top of the facing page for chopping dovetails.
The work is clamped between a wooden cleat and the benchtop
by two long dogs, which pass through matching dogholes in
the cleat and the bench. The dogs hook under the bench, and
are cut away at an angle to compensate for the slope in the
doghole slots in the benchtop. The wedges that tighten the
cleat are adjusted easily with thumb pressure, or by gently tapping them with a mallet.

To use the vise, Welter aligns the edge of the cleat with the
bottom of the dovetails. The cleat itself serves as a guide for
the chisel when trimming the bottoms of the joint. Welter
made the dogs and cleat of oak, but later faced the cleat with
maple to keep from chipping out the open grain of the oak.
Two pieces of fine, wet-or-dry sandpaper are glued to the underside of the cleat to keep the work from slipping.

All of the vises described in this chapter, with the exception
of some of the leg vises, have their jaws adjacent to the benchtop. That's fine for most cabinetry, but you might find them
wanting if you have to carve a three-dimensional object such
as a canoe paddle or a curved chair part. There are vises you
can buy that are intended for this purpose (see p. 149), but you
can also make one yourself, following a design like the one at
the bottom of the facing page. This plan, which was circulated
several years ago by Record Marples Ltd., requires only that
you buy a metal screw. All other parts are made of wood and
their dimensions may be varied to suit yourself.

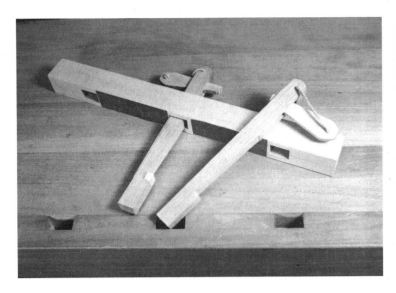

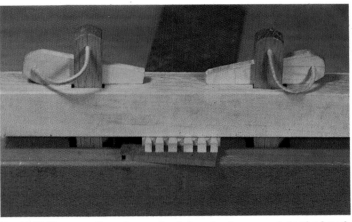

David Welter uses these two wedged dogs to clamp work to the bench while chopping dovetails. A hooked shoulder at the bottom of the dogs catches the underside of the bench, and three dogholes in the cleat accommodate different-size work.

Carving vise

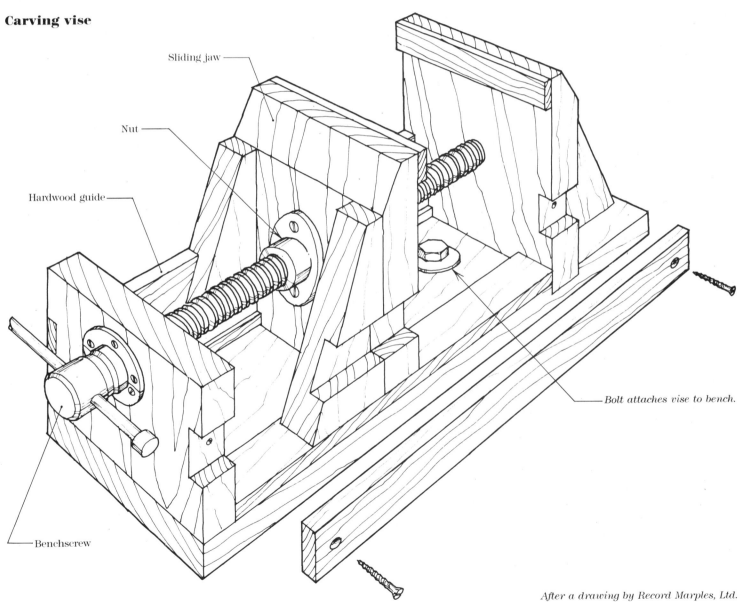

Sliding jaw

Nut

Hardwood guide

Benchscrew

Bolt attaches vise to bench.

After a drawing by Record Marples, Ltd.

The Emmert Roto-Vise, a classic patternmaker's tool, has unparalleled versatility. Seen from above, this large model (K-1) has been lifted so that its jaws parallel the benchtop.

Off-the-Shelf Vises

For centuries, if you wanted a vise (or a workbench), you had to make it yourself. Wedges and levers, wooden screws and vise jaws are all relatively easy to fabricate on the spot. As craftsmen became more specialized and sophisticated, however, their workholding requirements followed suit. In the Western world, that specialization peaked at the end of the 19th century and the beginning of the 20th, when industry flexed its manufacturing muscles in Europe and North America. With the combined technological advancements and economic demands of the Industrial Revolution, vises began to be mass-produced to meet the particular needs of virtually every branch of the wood and metal trades.

It is no surprise that both professional and amateur woodworkers of the period happily turned to these ready-made products when they became available. Industry was capable of producing hardware that the individual woodworker could not hope to make in his own shop, and 19th-century tool catalogs evidence the enormous selection of vises being manufactured at the time. With the attachment of one of these store-bought vises, any slab of wood on four legs might become a workbench of sorts. And, in the end, the time saved by not making a vise could be better spent in getting on with the job at hand.

The most popular commercially made vises used by today's 'post-industrial' woodworker were born around the turn of the century. Dozens of different types of woodworking vises are still around (cast iron must have a half-life approaching that of plutonium), but only a handful are still being manufactured. Two of the most interesting are the Emmert, the classic American patternmaker's vise; and England's Record vise, perhaps the most popular vise among professional woodworkers in North America.

Emmert

In early catalogs the Emmert Universal Pattern Maker's Vise was promoted as the 'Iron Hand' and 'The Peer of All Woodworker's Vises.' For once ad copy was not hyperbole—the Emmert is not your everyday woodworker's vise. From its elaborate, cast-iron fins to its unparalleled flexibility, the Emmert is exceptional. I am always delighted by eccentric 19th-century machinery—two of my favorites being the Velocipede saw and the Singer sewing machine—where function is wedded (but not sacrificed) to a Victorian design aesthetic. To my eye, the Emmert is an upstanding member of the same family.

Among woodworkers, patternmakers are a breed apart. As one former patternmaker described the trade, "The point of wood patternmaking is a very practical one—to produce the forms that create the molds for metal castings—but the work is scarcely less than elegant in its best expressions." Patternmaking is exacting work. Tolerances are measured in thousandths of an inch. The patterns are often mammoth, rarely rectilinear and almost always unique. To meet these requirements, a vise has to be unusually strong and versatile. The Emmert must have been indispensable to the turn-of-the-century pattern shop.

In its 'natural position,' the Emmert hangs from the front of the workbench like any other metal vise, its jaws at a right angle to the top surface. But the likeness stops there. By turning one or two screws, or shifting a cam-activated knob on the front of the vise, the angle of the front jaw can be changed to accommodate a non-square piece of wood. A lever below the vise allows the jaws to be rotated a full 360°, and locked in eight different positions along the way. At 90° the rectangular jaws are vertical and protrude well above the top of the bench—handy for shaping an elaborate pattern or a curved chair leg. At 180° two small jaws stick up like tongs to grip metalwork securely. Another lever below the bench can be adjusted to flip the whole vise up so that the back jaw is parallel to the bench-top. In that mode, the vise can be locked in an infinite number of postions. Four dogs are included, and wooden pads or an additional pivot plate ('tilt jaw' in the parts list) can be added to enhance the clamping options.

Joseph F. Emmert was an inveterate tinkerer. Trained as a farmer and patternmaker, Emmert generated a host of patents and inventions—from an adjustable buggy seat to improvements in grain separators. He began developing vises in the 1870s and continued to fine-tune their design over the next 20 years, a process that culminated in the Universal, which was patented on August 11, 1891.

It was nearly another decade before Emmert was able to turn his invention into profit, with the formation in 1900 of the Emmert Manufacturing Company in Waynesboro, Pennsylvania. He attracted investors, erected a factory and in short order was employing 80 men in the manufacture of more than 50 different sizes and varieties of woodworking and metalworking vises. In the lifetime of the company, Emmert Manufacturing also made wooden handscrews and the Lion Miter Trimmer, and branched out into drafting machines. (Emmert Manufacturing claims the oldest U.S. patent on a mechanical drafting machine.) At its peak, the Emmert Company employed 300 workers.

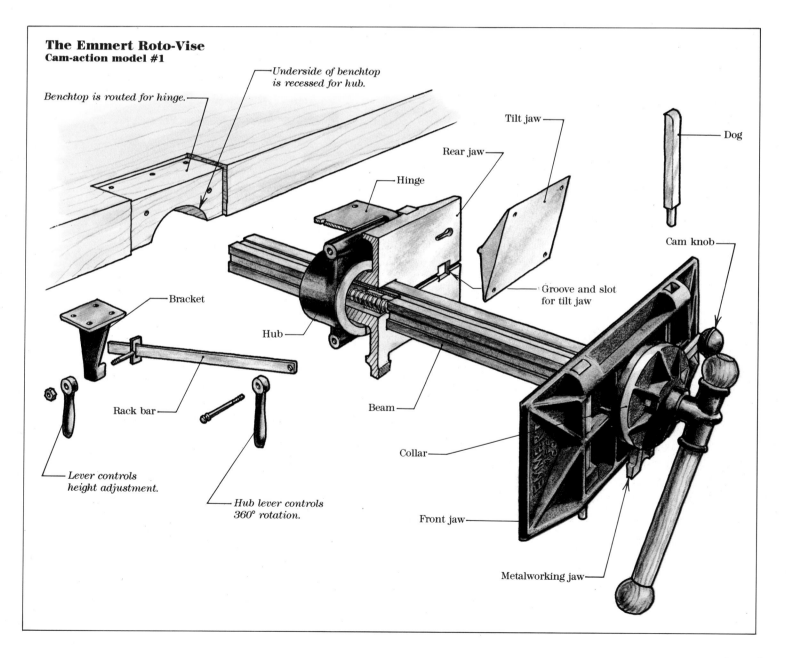

The Emmert Roto-Vise
Cam-action model #1

Benchtop is routed for hinge.

Underside of benchtop is recessed for hub.

Tilt jaw

Dog

Rear jaw

Hinge

Cam knob

Bracket

Groove and slot for tilt jaw

Hub

Rack bar

Beam

Collar

Lever controls height adjustment.

Hub lever controls 360° rotation.

Front jaw

Metalworking jaw

In its 'natural' position (above), the jaws of the Emmert 'turtleback' (Model #82) hang at a right angle to the benchtop like any other woodworking vise. The four dogs housed in the jaws are useful for clamping curved work. The vise can also be adjusted to an infinite number of positions (right).

On this early model, the rack bar locks in one of seven notches; on later models, the bar can be locked in an infinite number of positions.

When the vise is turned 180° (upside down), a pair of small metalworking jaws are on top.

The jaws may be rotated 360° and locked in eight positions to handle unusual clamping needs (above left). The front jaw may be tilted up to 5° to hold tapered work (center). With the release of a lever beneath the bench, the vise can be lifted until its jaws are parallel to the benchtop (above right).

On this turtleback Emmert, the jaw is tilted by means of a knurled knob to the right of the handle. On later models, this mechanism gave way to a retaining collar and cam knob (see drawing, p. 138).

The patternmaker's vise comes in two sizes: the #1, 7 in. by 18 in., and the #2, 5 in. by 14 in. (earlier models were numbered 82 and 96). The large vise opens 15 in., the small one 12 in. The earliest Emmerts were fondly nicknamed 'turtle-back,' so called for the casting that covered its front. The vise featured a 'double-lead' Acme screw with two threads per inch, which added considerable strength and speed to its operation. In 1905, the turtleback was discarded and the thumbscrew was replaced by a lever-activated cam that could be rotated with a flick of the wrist to vary the angle of the jaws. Though other minor alterations were made over the years, the Emmert actually changed very little.

Due in part to the vagaries of management (the last Emmert family member left the business in the early 1920s), the company went bankrupt and changed hands several times in the last 60 years. As the patents ran out, Emmert also ran into competition from G.M. Yost Manufacturing Company, and later the Oliver Machinery Company and the Kindt-Collins Company, all of which have manufactured nearly identical vises. (G.M. Yost managed Emmert for six years before starting his own vise company, promoting an 'Improved Universal Patternmakers' Vise' at the head of the woodworking line.)

If the Emmert vise stood still through most of the 20th century, it was at the eye of a storm. The patternmaker's vise was designed to suit the highly specialized iron-casting industry that flourished at the turn of the century. By the 1930s the Emmert was being used by squadrons of naval and automotive patternmakers to fuel the industrial buildup of America. By World War II there were more than a quarter million in use across the country. But by the end of the war, much iron casting (and hence patternmaking) had been replaced by welded fabrication, which was usually cheaper and quicker to do and required less skill. Sales of the Emmert plummeted in the late 1940s and '50s to a mere thousand vises per year (with only four employees dedicated to making them). In half a century, the face of North American industry had changed dramatically. By the 1960s and '70s much of the heavy manufacturing in the United States had either moved abroad or folded completely.

Still, patternmaking will survive as long as some manufacturing remains. What will likely seal the fate of the Emmert vise as the backbone of industry is that patternmaking has moved away from wooden castings. Ceramic materials now are used to produce precisely molded castings. These tend to be smaller and lighter, and because they are molded instead of shaped, they do not require a behemoth like the Emmert to hold them. What demand remains for a vise such as the Emmert may be pretty well met by the reserve of old vises that can be found in bankrupt pattern shops around the country.

The Emmert has never been cheap, carrying a price tag commensurate with its mass. Sold for $15 around 1914, the #1 Emmert cost $40 in the 1940s, $100 in the 1960s, $300 in the 1970s and has recently been fetching more than $800. Kindt-Collins, of Cleveland, Ohio, the last manufacturer of an Emmert clone, sells about a dozen a year for over $1,000 apiece—mainly to pattern shops in the Navy or to General Motors. If you can locate a second-hand Emmert (or one of the clones) from an antique-tool dealer or a pattern-shop liquidator, you can expect to do a lot better.

In the spring of 1984, when patternmakers were about as scarce as solvent railroads, the Emmert Company was sold once more. The duplex mill (used to machine the hollow channel in the sides of the beam), the 40-ton Lucas press (used to broach the square hole in the rear jaw for the beam) and the last of the castings and patterns were bought for $4,500 by Robert Kinslow of Hagerstown, Maryland. Along with the machinery and the parts for about 150 vises, Kinslow acquired the rights to the Emmert name, which he hopes to revive in the woodworker's market. As of this writing, Kinslow has had new patterns made for the vise jaws and is just beginning to manufacture parts for Emmert vises.

Over the years, Emmert vises have fallen into the hands of a few fortunate cabinetmakers. Indeed, at 86 lb. (56 lb. for the #2), they're probably a lot more vise than most woodworkers really need. Despite their size and cost, however, Emmerts are highly prized. Many of the woodworkers I spoke to who own one would have nothing else. I know one man who owns seven of them of assorted vintage and condition. In another workshop, the lead screws of two Emmerts have been replaced with foot-operated pneumatic cylinders, as shown at top left on the facing page. Some time back, when Don McKinley found himself stranded in Tasmania without his trusty Emmert, he modified a Record vise to mimic the Emmert, as shown at top right on the facing page.

Perhaps the largest collection of Emmerts, about 200 vises, is at the M.P. Möller organ company in Hagerstown, Maryland. I saw fewer than half that many in a brief tour I took around the sprawling 125,000-sq.-ft. factory, where 150 employees build about nine organs a month. The others, left over from the days when Möller had 600 workers on the payroll, are held for spare parts or as a hedge against inflation. There is at least one Emmert on every workbench in the building, every one mounted in its 'natural' position. A pity, I thought. It's a little like keeping a Jaguar in second gear. What's more, every vise had a piece of pipe instead of the original wooden handle, and many of them were missing chunks of casting, the result of over-torque and general abuse. My guide at Möller, a Mr. Peck, told me: "As far as we're concerned, there is no other vise." And I can attest to that; at least in the cavernous workshops of M.P. Möller, I never found one.

A worker in Thomas Moser's furniture factory in Auburn, Maine, shapes a headboard using a modified Emmert. The beam is bolted to a floor-mounted stanchion; a pneumatic cylinder controls the jaws. Photo by Paul Bertorelli.

If you can't find (or afford) an Emmert, you can come pretty close by souping up a standard woodworking vise. Don McKinley's Record #52½ vise does just about everything an Emmert does. To tilt up, the vise is mounted on an L-shaped block, made from scraps of eucalyptus. The block attaches to the bench with two standard door hinges, their barrels let into the surface. A long, 1½-in. by 2-in. wooden arm, bolted to the inside of the block, pins to the crossbrace in the bench base at the desired angle. A removable pivot block fitted to the front vise cheek holds tapered work. To make the pivot, McKinley screwed one leaf of a butt hinge to the back of the block and bent the other at 90°. The bent leaf fits in a slot cut in the middle of the cheek. The vise has a maximum opening of 11½ in. without the pivot block, and 9½ in. with it. Of course, it still retains the quick-action mechanism of the original Record vise, a feature not even the Emmert can claim.

Installing an Emmert

New Emmerts are expensive and hard to come by, but second-hand vises (an original Emmert or one of the Oliver or Yost clones) can still be found in old pattern shops or antique-tool stores. If you're lucky enough to find one in good shape, you can count on many generations of active service. The following installation guidelines were gleaned from Donald McKinley and Stefan Smeja of the wood studio at Sheridan College in Mississauga, Ontario. In the course of reconditioning the shop benches, Smeja remounted about 12 Emmert #1 vises. (The process is the same for the #2.)

• Turn the screw to open the vise and remove the front jaw and beam. This reduces the weight of the hinge plate and rear-jaw assembly, which are mounted on the bench.

• Trace the position of the hinge plate on the top and front edge of the bench and rout or chisel away the two adjacent surfaces until the plate can be mounted flush.

• If the benchtop is more than 1¾ in. thick, you will have to relieve the underside of the top to clear the hub. (This won't be necessary if the bench is thinner, but the weight of the vise generally requires a thick benchtop.) This can be done by hand or with a router, but if your benchtop can be removed, it's easy to cut a neat trough using the radial-arm saw, as shown below. Turn the saw to its ripping position. Place the benchtop upside down on the saw table and raise the blade to clear the wood surface. Mark the centerline of the trough. Remove the wood by lowering the blade in ¹⁄₁₆-in. increments as you push it back and forth against a stop clamped to the arm of the saw. (Using an 8-in. blade, Smeja made the trough 5 in. wide at the bottom of the benchtop.)

• Mount the hinge plate on the bench with 1½-in. #14 flat-head wood screws. If the top is less than 1½ in. thick, use machine screws and nuts (or stove bolts) to bolt the hinge plate to the top of the bench and 1½-in. #14 wood screws at the front edge.

• With the rear jaw and hinge-plate assembly mounted on bench, reassemble the front jaw, beam and lead screw.

• Install the rack bar (or equivalent device on your model), which controls the 90° tilt action. Bolt the rack bar to the bottom of the hub, assemble the bracket, nut and lever, and screw the bracket to the underside of the benchtop. Be sure that the rack bar is parallel to the beam and that the bracket is positioned to allow full 90° movement of the vise. On the Emmert #1, the front of the bracket should be 5 in. back from the bench's front edge and about 2¾ in. to the left of the beam's center.

Making the trough

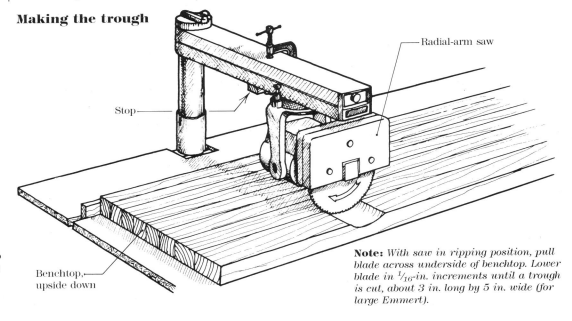

Stop

Radial-arm saw

Benchtop, upside down

Note: With saw in ripping position, pull blade across underside of benchtop. Lower blade in ¹⁄₁₆-in. increments until a trough is cut, about 3 in. long by 5 in. wide (for large Emmert).

Record

At about the same time that Joseph Emmert was tinkering with the design of his patternmaker's vise in Pennsylvania, the forerunners of the Record woodworking vise were taking shape in England. While the Emmert arouses strong support among a handful of dedicated devotees, Record vises have become the virtual standard among the majority of professional cabinetmakers I visited—with good reason. The Record vise is strong and accurate, like the Emmert, but comes without all the bells and whistles, at less than half the weight and at a fraction of the cost. Priced close to $100, the top-of-the-line Records (#52, #52½ and #53) are not cheap. But they're cheap enough for several woodworkers I know to have installed four of them on a single bench—one mounted on each corner.

Alongside the Emmert, the Record is what I'd call a conventional metal face vise. In its basic design, it resembles many woodworking vises manufactured in the United States since 1900. Columbian, Abernathy and Milwaukee are a few of the better-known North American products. All are built with heavy cast-iron jaws, solid-steel guide bars (one on either side of the center screw) and a quick-action mechanism. They are available with an adjustable dog on top of the front jaw, which allows the vise to function as a tail vise when mounted on one end of the bench. All of these companies, Record included, offer smaller, cheaper vises for the hobby market.

Record's quality, in combination with a strong U.S. dollar, has enabled the British company to dominate the high end of the woodworking market in the United States. Woodcraft Supply Corp. of Woburn, Massachusetts, one of the largest North American outlets for Record vises, has been selling them for years. David Draves, Woodcraft's merchandise manager, told me that the company originally picked up the Record vise because of its quick-action design. But Woodcraft has continued to carry it long after vise companies in the United States came out with their own version.

Milwaukee Tool & Equipment Company, of Wisconsin, has been making vises for at least as long as Record, and the company takes pride in its longevity. You can still get parts for the original Morgan vise, introduced by Milwaukee in 1873. They make a quality product, which in weight, features and retail price closely resembles the Record vise. Milwaukee's #200A rapid-action vise has 10-in.-wide jaws, opens 12 in., weighs 30 lb. and retails for about $100. Record's #52½D is 9 in. wide, opens 13 in., weighs 36 lb. and costs about the same. But when Milwaukee approached Woodcraft Supply about a year ago, hoping to crack Record's near monopoly, they lost out in a price war with a Canadian company, Cambridge Tool Company Ltd. of Ontario. Cambridge was able to offer their vises to Woodcraft for about 30% less than Milwaukee's, the difference being about the same as the difference between the U.S. and Canadian dollar. That advantage earned Cambridge a slot next to Record in the 1987 Woodcraft catalog and a large entry into the U.S. market.

If you are used to shopping the tool-store catalogs, Milwaukee (or other American-made) vises are hard to find. This reflects a marketing dilemma that is not unique to woodworking vises. The cheapest vises, sold in small hardware stores and the national chains, have traditionally been American-made, but they are steadily being edged out by imports from Taiwan. Companies like Milwaukee have continued to sell mainly to

The Record #52½D quick-action vise is standard equipment on many woodworking benches. The 9-in.-wide jaws open 13 in. (without the wood cheeks).

schools and industry, where traditional buying patterns and a 'buy American' philosophy are prevalent. Meanwhile, their sales have languished among consumers—particularly professionals and serious amateurs.

According to Dick Berger, Milwaukee's sales manager, the failure of companies like Milwaukee to anticipate the phenomenal growth that has taken place at the high end of the consumer market was an opportunity missed. When they awoke to the potential, there were strong foreign competitors like Record and an unfavorable exchange rate to contend with. They also faced several other obstacles, which have so far proved insurmountable. Record's complete line of woodworking tools makes it easy for retailers to carry their vises—one-stop shopping from a single distributor saves the consumer time. And, not to be discounted, there's a certain caché to the European import—Record's 'old-world quality' sells well in America. "But more than anything," Woodcraft's Draves suggested, "name recognition is it." American woodworkers are familiar with Record products. Many of their workshops are equipped with Record planes and Marples chisels, so they are naturally receptive to the Record vise.

As I've explained in earlier chapters, most metal woodworking vises—the American-made models as well as the Record—have one major drawback. It is impossible to clamp a piece of wood vertically so that it is centered directly behind the screw, as you can in a tail vise or a traditional shoulder vise like Frank Klausz's in Chapter 4. Otherwise, when you clamp a wide board vertically on one side of the guide—when cutting dovetails, for example—you may want to block the opposite side of the vise or support the other end of the board with a bar clamp across the bench, as shown in the photo on the facing page.

In its design, the Record vise has several distinguishing features. The jaws toe in slightly at the top so that they close there first and snug up elsewhere as the vise is tightened, guaranteeing a good grip on small work clamped in the top edge. The quick-action mechanism is engaged by pressing a

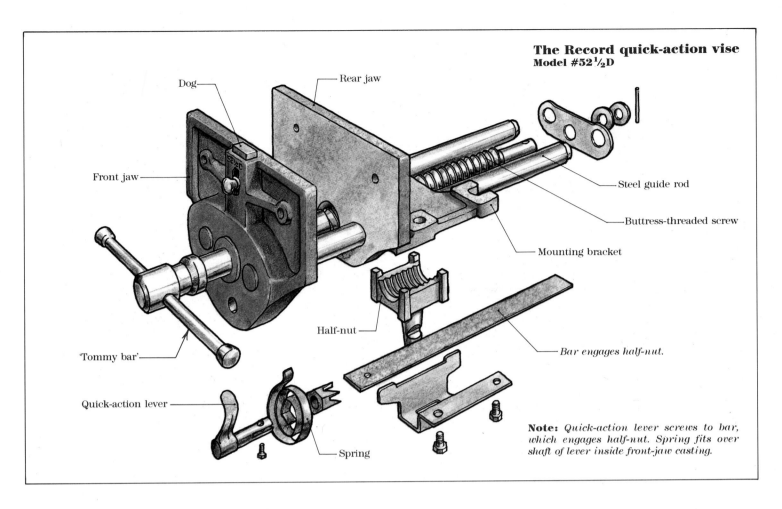

Dog

Rear jaw

Front jaw

Steel guide rod

Buttress-threaded screw

Mounting bracket

'Tommy bar'

Half-nut

Bar engages half-nut.

Quick-action lever

Spring

Note: *Quick-action lever screws to bar, which engages half-nut. Spring fits over shaft of lever inside front-jaw casting.*

spring-loaded lever, which trips a half-nut to release the screw. The lever wraps around the right side of the lead-screw head and may be operated easily with a thumb or forefinger when the vise pressure is released. It's simple and quick and, if kept reasonably clean, not likely to fail. Quick-action vises made in the United States are pressure-activated, requiring the screw to be backed off a half-turn to release a hinged portion of the nut and allow the jaw to be repositioned. I find this slightly more time-consuming and occasionally irritating as the screw threads become worn and prone to slipping. One old 'quick-action' Sears vise I own has a groove machined in the underside of the screw, which must be jiggled until it is aligned with a matching tooth in the nut—a tedious process at best that becomes even more so as the vise and I get older.

The Record screw is turned with a wedge-like buttress thread, undercut by one or two degrees, while its American competitors typically use an Acme thread. The Acme is plenty strong and is standard on a lot of heavy machinery, as well as on all the benchscrews that are designed to be used for leg or tail vises. But when you tighten a buttress-threaded screw it locks down in the deep grooves, and is less likely to jump out of the nut. It continues to work well when it is worn, while an Acme screw may begin to slip. The Acme thread is easier (and therefore cheaper) to produce, which may account for its predominance on American-manufactured vises.

The guide rods on all conventional woodworking vises prohibit clamping stock vertically behind the screw. You can compensate for this by clamping the other edge of a wide board across the bench.

I've used at least three different brands of British-made vises—Woden, Record and Paramo. They look almost identical and, in fact, they are closely related. In the process of reorganizing themselves over the last century, British manufacturers multiplied and divided many times, not unlike the way Emmert begat Yost and Oliver. I visited England to try to sort out the differences myself and to see how a modern vise is made.

According to Tony Hampton, a retired fifth-generation vise maker who lives in Hampshire, in the south of England, Woden was one of the first British manufacturers of a parallel-jaw vise. Hampton's great-great grandfather, Joseph Hampton, started the Woden Company around 1870 in Birmingham. When Hampton's grandfather, Charles, and great-uncle, Joseph, moved to Sheffield in 1898 to start C. & J. Hampton Ltd. (later to become Record), they brought the Woden tool line with them and began making vises patterned after Woden's. Woden tried, unsuccessfully, to sue, but the vises were unprotected by patent. The rest is history. Record absorbed Woden in the 1950s.

In the 1930s, when the company was well established in England, Record ventured into North America, exporting first to Canada, and later to the United States. The Canadian market has always been more receptive, according to Hampton, partly because of its historic Commonwealth connection. Record also discovered that Americans preferred a wooden handle to the steel 'tommy bar.' (All of the American-made woodworker's vises I've seen are available with wooden handles.) I had to admit to the same prejudice—the steel is noisy and can give a nasty pinch if you're not careful. "Well, a wooden one will do just the same," Hampton replied, "if you drop it hard enough. These are old wives' tales that go 'round the world."

In 1940, Sheffield was bombed and the Record factory hit. According to Hampton, Record was the only vise manufacturer in the country, and the wartime Ministry of Supply reasoned that it would be hard to build and repair RAF fighters and Lee Enfield rifles without vises. Record was instructed to train another company in their manufacture. A stove-grate foundry, F. Paramore & Sons, had been making Record's castings and was the obvious choice. "We had to go along and show them how to do it," Hampton recalls. "We were annoyed." Forty years later, F. Paramore & Sons has reorganized as The Paramo Tools Group Ltd., and is one of several companies giving the Record vise its first real challenge for the North American professional market. My visits to both Record and Paramo turned up some interesting differences.

If you've leafed through a tool catalog in the last 20 years, you'll find many Sheffield names familiar. Rabone, Chesterman, Joseph Tyzack, Sorby, Footprint, Paramo, Record, Marples and Ridgway are to Sheffield what Ford, GM and Chrysler are to Detroit. For centuries, Sheffield has been the center of the British hand-tool industry. Situated in the industrial heartland of Yorkshire, Sheffield grew up close to coal and at the confluence of two major rivers, the Don and the Sheaf. Only a few miles from town five other tributaries feed into these rivers. If moving water was important for transportation in the city's early days, it was crucial for energy. In 1770 there were 161 cutler's waterwheels along the banks of Sheffield's waterways—an average of five of them in every mile, powering the trip hammers and grinding wheels essential to the knife-making industry.

The Record vise takes its shape from these metal patterns (above). The automated molding and casting machine (right) packs sand around the patterns and then removes them, retaining their impressions in a long, continuous mold. Molten iron is ladled into the mold, producing more than 100 tons of cast-iron vise and tool parts every week.

In the 20th century, Sheffield experienced a pattern of economic booms and busts similar to that of its manufacturing counterparts in North America. Like Detroit, Sheffield was especially hard hit by the recessions of the 1970s and early '80s. Many companies went under and those that survived reorganized. Record changed hands several times and is now climbing out of the trenches—with six company names carved in its masthead and 1,200 fewer employees than it had ten years ago. It now operates its own foundry—a capital burden that I've been told almost put the company under. In a partial role reversal, Paramo now farms out its castings—but not to Record.

A vise, before it becomes a vise, is pig iron and scrap metal. Great volumes of the stuff are hauled in boxcars and trucks to Record's Parkway foundry, where vise-making begins. The foundry is a remarkable place. Titanic cauldrons of molten metal tower over me as I stand with my face seared by intense heat and my brain addled by an orchestrated cacophony. I can't help but feel immediate empathy for millworkers around the world. I imagine my finger on the sooty pulse of industry. At Record, the fingers of the foundry operators rest on a set of remote-control levers, protected behind a Plexiglas heat shield. Glowing like lava, cooked to 1400°C, molten iron is ladled into holes in a writhing snake of hard-packed sand. I am watching an automatic, assembly-line casting machine called a DisaMatic. State-of-the-art, I'm told, and it's hard not to be impressed. At the end of the line, an enormous vibrator shakes freshly molded cast iron out of the sand, 120 to 140 tons per week. If vise castings are not held back for inspection or to fill a large order, the entire process, from molten metal to painted castings, will take only a couple of hours.

Once in the machine shop, a few hundred yards away, it takes even less time to turn the castings into Record vises. There are only a few major parts in each vise—the cast jaws and the threaded screw and nut being the most important. By manufacturing these parts themselves, Record maintains control over the quality of the critical components. With the aid of Japanese-built CNC (computer numerically controlled) machinery, Record produces more vises with only half as many workers as pre-CNC. And when the whistle blows on the shop floor signaling lunch, the machines continue to work—double shifts, 16 hours a day, producing "nothing but the best," according to my guide.

The CNC machine is a marvel of modern engineering. An operator (yes, they still need them) loads a carousel with four or eight vise castings. He runs a quick 'tool-change program' indicating which of the 16 or 32 different cutter heads in the magazine will be required and in what order. Next he punches the X, Y and Z coordinates into the keyboard, along with cutting speeds, depth of cut, approach and retraction speeds. Then he sets the beast in motion and, while the machine drills perfect holes in the castings and mills the outside edges to spec, the operator loads another carousel. When it's finished, any front jaw can be married to any rear jaw with an unparalleled degree of accuracy.

Before CNC, each operation would have been performed by a separate 'dedicated' machine. Before, the quality of the product depended upon the skill of the operator. Now, it's governed by the skill of the programmer and AMT (advanced manufacturing technology). Record, as I am often reminded on my tour, is "at the front end" of things.

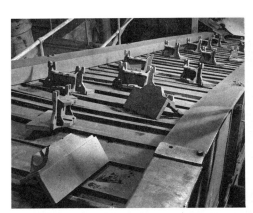

The making of a Record takes place in the foundry and the machine shop. Top left: After being poured, the fresh castings pass through a cooling tunnel and the sand mold is removed on a vibrating conveyor. Bottom left: One end of the steel benchscrew rod is heated and upset in a die to form the head. Top right: The cast-iron half-nut is cut with the characteristic buttress thread. Bottom right: The handle is secured by cold-forging both ends. Right: Cast jaws are loaded in a CNC machine, where they are automatically bored for guide rods, lead screw and quick-action lever. Vises are then assembled, dipped in a rust inhibitor and packed for shipping.

The only element in noticeably short supply is people: there are very few around to slow down production. At the end of the woodworking-vise line, however, a worker checks each vise and files off any recalcitrant burrs. Lead screws, which have been turned the old-fashioned way on an automatic metal lathe, are mated with castings and guide rods. A quick test ensures that the vise is operable. It is moved around the room to a conveyor belt where it is merged with the line of metalworking vises. Then it is washed, dipped in rust inhibitor (from now on, you won't find your Record vise slathered in grease), blown off, dried and perhaps sent to America.

When Record was restructured a few years ago, they dropped many items from their catalog in an attempt to 'rationalize' the product range. They planned to hold less 'work in progress' (stock) and fewer components. Some of the eliminated products—like the large Record benchscrews, which were popular with people making their own bench vises and other fixtures— are now being returned to the line.

Looking ahead, there are dozens of ideas on Record drawing boards. At the world's largest annual tool show in Cologne, Germany, the company recently previewed a reversible-jaw vise (for woodwork and metalwork). A range of quick-action C-clamps and a new quick-action woodworking vise are in prototype. Some of these products have been in development and patented for years. Others were designed and in production within 24 hours.

Record likes to look ahead. When I expressed interest in printing an old photo of a fellow in a smock with his sleeves rolled up to test vises, Record squirmed. The company doesn't want to appear stodgy or linger too long in its own old-world tradition. Earlier, I had seen a photo of a float the company had entered in this year's Lord Mayor's parade. (The Lord Mayor of Sheffield happens to work in the Record plant.) The float was based on the theme 'Shaping the Future,' and won first prize.

Across town, the people at The Paramo Tools Group Ltd. take a certain pride in looking back. The way Record used to make vises, I'm told—and the way that company built the reputation it now enjoys—is pretty much the way Paramo makes them today.

Setting a Record

There's more to hanging a Record (or similar vise) than simply bolting it to the bench. To work properly it must be straight, level with the top and secure. At the very least, once the vise position is decided, you must accurately bore four holes, attach the mounting bracket (which is a single casting with the rear jaw) and add wood cheeks. But there are several fine points and a variety of mounting options to consider, as shown on the facing page.

The rear jaw may be mounted onto the edge of the benchtop (Fig. 1), inset flush with the edge (Fig. 2), set behind an apron (Fig. 3), or mortised into the underside of the bench (Fig. 4). If the working surface of the rear jaw is the front edge of the benchtop (Figs. 3 and 4), it will be easy to add additional clamps to secure a long board to the bench. On the other hand, if the cheek protrudes (Figs. 1 and 2), irregularities in the stock won't strike the benchtop edge and make it difficult to close the vise jaws. Which vise-mounting method you choose depends on the thickness of your benchtop, the shape of the edge and your own preference. Here are some other considerations to make vise installation easier and vise operation more effective:

• When positioning the vise, make sure that when the vise is closed the screw and guide bars will not interfere with any dogholes or with the legs of the bench.

• Fitting the rear jaw/bracket to the bench will be easier if you turn the benchtop upside down or on its edge. If this is not possible, you can remove the front jaw of the vise along with the lead screw and guide bars to reduce the weight.

• Unless your benchtop is unusually thick, you will have to insert a spacer between the mounting bracket and the underside of the bench. This can be made of hardwood, plywood or fiberboard, or built up of $1/4$-in. or $1/8$-in. tempered Masonite.

• Size the spacers to position the top of the rear jaw about $1/2$ in. to $3/4$ in. below the top surface of the bench. This allows for periodic resurfacing of the benchtop. (The wooden cheeks should be flush with the top.)

• If you let the rear jaw of the vise into the front edge or underside of the bench, allow a $1/16$-in. gap above the casting. The spacer is bound to compress when you attach the vise, and this gap will close. Without the gap, the wood may buckle above the jaw and have to be planed off. (A snug fit on the sides of the rear jaw helps position the vise.)

• To hang the vise, use either $3/8$-in. bolts or lag screws. Bolts provide a more positive fixing (Fig. 1), but their heads must be countersunk beneath the top surface and the holes should be plugged. (The square shank beneath the head of a carriage bolt will strip the wood after several installations, so I prefer to use machine bolts and lock washers.) Lag screws work well (Fig. 2), but make sure that you size and bore the pilot holes carefully, and don't remove the vise more often than is necessary. Lag screws and machine bolts may be combined using an enlarged spacer (Fig. 3), which strengthens the fixing.

• Metal vise jaws should always be covered to protect your work and the edges of your tools; $3/4$-in.- to 1-in.-thick hardwood is fine. You can make these cheeks wider than the metal jaws to extend the clamping capacity, but bear in mind that the farther you clamp away from the center screw the more the vise will rack out of square. For a neater job (and more protection), the wooden cheeks can also be routed to fit around the top and sides of the front jaw (Fig. 1). Allow about $1/2$ in. of space between the tops of the guide rods and the bottom of the cheeks so that veneer edges or moldings can fit between them.

• If you let the rear jaw into the front edge, wood must be routed away to the exact thickness of the casting. If too much wood is removed, the wooden cheek will dish. If not enough is removed, there will be a gap at the top between the cheek and the front edge of the bench. Sawdust will work its way in and wedge the cheek away from the bench.

• The Record and Paramo vises are designed to make contact first along the top edge of the jaws. This 'toe-in' should be retained for a better grip. If your vise jaws are parallel, you can create your own 'toe-in' by tapering the wooden cheeks.

• To make it easy to align work vertically in the vise, inlay thin pieces of veneer in the top of the front cheek. These should lie at a right angle to the outside edges of the guide rods. Work can be quickly installed in the vise by pushing it against a guide rod and aligning it with the veneer on top.

The vises of the two companies illustrate their common lineage. At a quick glance, Paramo's vises could be mistaken for Record's. The shape is the same, even the familiar 'Record-blue' paint job looks the same. On closer inspection, the Paramo shows some of the signs of a handmade product—slightly irregular castings and a few off-the-shelf washers and springs. There are minor disparate details: the jaws on the Paramo vise are tapped to accept two screws for attaching the wooden cheeks, while the Record vise jaws are just drilled, and the Paramo has a heavier casting. Altogether, there's not enough of a difference between them, I think, to make one vise much more desirable than the other.

Of greater import is the fact that Paramo's vises retail in the United States for slightly more than Record's, a difficult position for a challenger. And the Paramo vise is a relative newcomer in the high end of the retail tool market, where Record is firmly ensconced. Like Record, Paramo offers a selection of woodworking and metalworking vises, as well as assorted other tools—from chisels to screwdrivers—which helps to diversify and support their line.

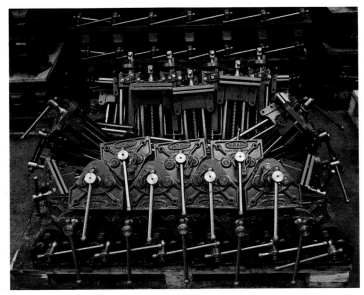

A pallet-load of Paramo woodworking vises ready for packing.

Setting a Record

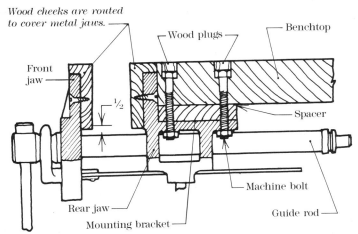

Wood cheeks are routed to cover metal jaws.

Front jaw

Wood plugs

Benchtop

Spacer

Machine bolt

Rear jaw

Mounting bracket

Guide rod

½

Fig. 1: Edge mount

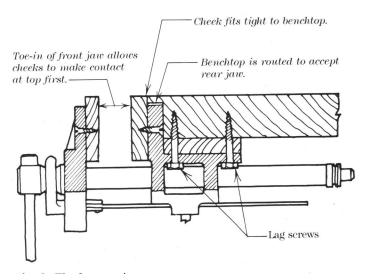

Cheek fits tight to benchtop.

Toe-in of front jaw allows cheeks to make contact at top first.

Benchtop is routed to accept rear jaw.

Lag screws

Fig. 2: Flush mount

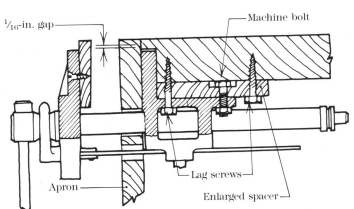

1/16-in. gap

Machine bolt

Apron

Lag screws

Enlarged spacer

Fig. 3: Flush mount behind apron

Note: *Remove apron to rout recess in edge of benchtop. Replace apron, with holes drilled for guide rods and screw. Vise will have to be dismantled to be installed.*

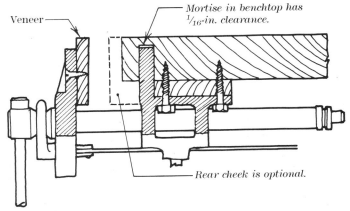

Veneer

Mortise in benchtop has 1/16-in. clearance.

Rear cheek is optional.

Fig. 4: Mortised mount

Note: *Mortise may be routed, drilled or chopped after benchtop is assembled, or sawed before top is laminated. Veneer inlaid in top of cheek aligns vertically with outside edges of guidebars.*

At Paramo, the 2° undercut buttress-threaded vise screw is cut on an automatic machine lathe (top). Heads are welded on threaded shafts (bottom), achieving a breaking strength of about 17 tons (for the large screws).

Most of the distinction between the two vises takes place in their manufacture. Paramo's Rotherham works, a short drive northeast of Sheffield, is a lot more cluttered than Record's plant—with machinery, the grime of a typical machine shop, and people. My guide through the 13,000-sq.-ft. plant is its manager, Arthur Dorling, who started working for Paramo when he was 14.

When vise castings arrive from the supplier, Dorling explains, "I inspect them on the lorrie before we unload. I'm there with me straightedge and if they're not straight I send them back." Inside, they are run through about 15 different operations, from one end of the shop to the other. Holes are drilled, one at a time, on dedicated machines. Edges are milled clean.

Screw heads are welded to the buttress-threaded shafts by hydraulic pressure, achieving a breaking strength of about 17 tons. (They used to turn the whole screw down from the larger diameter of the head.) Heads of the smaller screws are attached using little riveters, which only recently replaced a man with a hammer. "I had a chap there with muscles just like Garth," Dorling recalls. At Record, the heads of the lead screws are heated and then upset into a die by a hydraulic ram. At the end of the line two experienced workers assemble each vise. Because of the variables of Paramo's lower-tech production, parts must be matched carefully. "I don't put any Tom, Dick or Harry on it," Dorling says. "It's got to be just right."

According to Dorling, Paramo could turn a vise off the line every five minutes if he had an operator at every machine. In fact, Paramo produces only about 200 to 300 large woodworking vises a week, and the operators wear lots of different hats. "We muck in," as one employee puts it, meaning that when they see a job, they do it. One cheery woman demonstrated three different machines to me and I was told that she can do any job in the place. There are 29 employees at Rotherham and another 30 more in the office and small tool factory back in Sheffield—a fraction of the 750 Record employees. In addition to phone calls and visitors, the receptionist handles shipping and helps with exports.

Because of the similar product lines and proximity of Record and Paramo, I wondered whether labor flowed between them. It does occasionally, according to Paul Allen, Paramo's export manager. "But in each case, they left Record to come here," he adds with a laugh. While Paramo didn't seem to be on everyone's mind at Record, clearly Paramo has its eye on the leader. "Record is the brand-name manufacturer," Allen admits. But if Paramo can build as good a product, the theory goes, they might just snatch a corner of the market.

Although Paramo has made a few inroads in the last year or two among major suppliers to the North American market, Record remains solidly out in front. While Paramo turns out about 10,000 to 15,000 large woodworking vises a year, Record was decidedly tight-lipped about their vise sales figures, which must be quite a bit higher. Both companies, however, felt that they had a lot more room to grow in the United States, particularly in the hobby end of the market, which has traditionally been dominated by the old American companies.

In the long run, the stiffest competition for both companies won't come from Sheffield, Canada or the United States. Allen hinted as much when he told me that "every year at Cologne an overseas manufacturer [from the Far East] comes up and says 'nice vise, that—we can make it for you cheaply.'" So far, every year they turn him down.

Specialty vises

For those occasions when a regular woodworking vise won't do the job, there are all kinds of elaborate, articulated 'second hands' that might tickle your fancy. Although I am predisposed to conventional hardware, I've come across several unusual vises that are worth considering. Two are manufactured, and two you can make yourself (I've discussed these in the previous chapter). All of them are designed to facilitate three-dimensional shaping and carving by holding the work away from the surface of the bench.

The Zyliss vise, shown in the bottom photo, was commissioned 28 years ago as a field vise for the Swiss army. While it hasn't quite achieved the currency of their little red pocket knife, it almost approaches the Emmert in its versatility. The standard Zyliss fixes on the edge of a worktable in about eight major clamping positions, which cover most of the work of a face vise and tail vise. With a few extra attachments it can be contorted to perform about as many other tasks. It is made of a cast-magnesium alloy, with all stress points reinforced with Swedish steel. A child could twirl it around in one hand, so don't plan on pounding bar stock flat on the jaws or clamping large timbers. While I've only used one at a demonstration, it seems handy for a wide variety of lightweight projects.

The Universal vise (also called a patternmaker's vise in most catalogs), shown in the top photo, was popular among several woodworkers I visited. It mounts on the benchtop (up to 4½ in. thick) by bolting through a hole drilled in the top. The body of the vise swivels 360° around the base, and the wood-lined jaws, which open 6 in., also pivot to allow further flexibility. It's much heavier than the Zyliss vise and the center of its jaws are 8 in. above the bench surface, providing a convenient height for intricate drawknife and spokeshave work. I discovered only one drawback. Because the center of pressure in the jaws is 4 in. above the single Acme lead screw, the jaws are inclined to distort as pressure is exerted on the screw, thus loosening their grip. (The short screws that attached the wooden cheeks of the vise I tested also stripped out when I clamped work between the top edges.) I judge it satisfactory, however, for most light-to-medium purposes. The Universal vise in the photograph is made by Cambridge Tool Company Ltd. in Ontario, Canada. Similar vises are manufactured in both West Germany and Taiwan and are available in North America.

There are, of course, many other manufactured vises, too numerous to describe here. There are miter vises, vises that turn into lathes, flexible ball-and-socket vises. And new ones appear all the time—like Black & Decker's 'Quick Vise,' which is based on the same twin-screw principle as their Workmate. Vises also come and go. One excellent product, the Versa Vise, was discontinued during the writing of this book, but six months later it was back in production by another manufacturer (see Sources of Supply). I wouldn't be surprised if several others sprouted up at the same time. A new vise may not make a great difference in the quality of your work, but it can sometimes make life a bit easier.

The Universal vise is ideal for projects such as this canoe paddle. The jaws pivot and provide maximum clearance for shaping the oval shaft with a spokeshave.

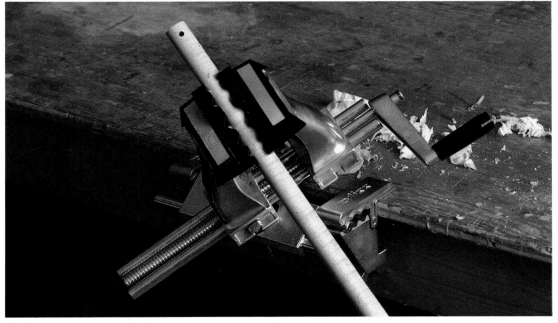

The Zyliss vise is lightweight, portable and about as flexible as the Emmert.

Toshio Odate surrounded by tools and benches in his new workshop. Left to right: The planing beam, small horses and planing board. "The workshop has to be clean," says Odate. "It's a very sacred place."

Japanese Beams and Trestles

Of all the tools of the Japanese woodworker, the workbench may be the most foreign to the North American. Japanese planes, saws and chisels are functionally different from European planes, saws and chisels, yet no one would question their identity. But a Japanese workbench? Looking around most Japanese workshops, you might easily mistake the bench for a piece of stock, perhaps an incomplete project, or a pair of sawhorses.

Little more than simple frames with heavy boards on top, Japanese benches also confound Westerners by their lack of vises. Instead, stops and the woodworker's body substitute. By sitting on the work, you can immobilize it for chiseling. By stepping on it or clamping it against the bench with one hand, you can hold the work firmly in position for sawing. Since the Japanese pull both planes and saws, the body is a natural stop.

Why devote an entire chapter in a book on workbenches to objects that are so patently simple? In the first place, I believe it is true that the simpler the tool, the more skill it takes to use. Anyone can surface a timber on a thickness planer if they can lift it to the in-feed table. Smoothing it with a hand plane (or a broadax, for that matter) is another story.

Simplicity aside, however, the Japanese workbench qualifies as a workbench according to any meaningful definition I can construct. Like its most elaborate European counterpart, it has a worksurface (a beam or a heavy plank), a means of support (a frame or one or more trestles) and a workholding system (a stop and/or the craftsman's body). In fact, there are several distinct species of Japanese workbench—some at floor level, others approaching Western bench height. Some are angled, others perfectly level. Each one is well suited to a particular kind of work as well as to specific tools and their method of deployment. Each one reflects the training of its maker—there are workbenches for screen makers, carpenters and furnituremakers, and in 20th-century Japan there are even benches for chairmakers.

It is a curious twist that the workbench may provide an even better perspective on Japanese woodworking than do any of the more popular tools. It is true that a Japanese plane, saw or chisel may be used effectively on a Western workbench, and I know many woodworkers who do. But the Japanese workbench is quite meaningless when divorced from the tools, the techniques and the sensibilities of the tradition. As Toshio Odate observes in his book *Japanese Woodworking Tools*, "...the *shokunin*'s [craftsman's] art is difficult, if not impossible, to separate from his workspace, his tools and his equipment. The craft is not apart from his life so much as a heightened detail of life." Understanding this important part—the bench—can lead to a better understanding of the whole.

"...perfection is finally attained not when there is no longer anything to add, but when there is no longer anything to take away..."   —Antoine de Saint-Exupéry, *Wind, Sand and Stars*

Talking with Toshio Odate in his Woodbury, Connecticut, studio, it occurs to me that Toshio talks about his tools the way some people talk about their children, with a mixture of pride and responsibility—and at some length. Toshio can haul out a saw or a plane blade to share with a visitor as easily as he does a newspaper clipping of his son Shobu's last swim meet. Toshio became a *shokunin* in post-war Japan, after training as a *tategu-shi* (sliding-door, or *shoji*, maker) in a rigorous, traditional apprenticeship with his own stepfather.

Toshio Odate seems to relish contradiction. He moved to the United States from Japan in 1958 and has now spent exactly half his life on each side of the Pacific. Trained as a craftsman, he is now a sculptor, teacher and author living in rural Connecticut—in many respects, an American. But, as Toshio told me, "You can't change a Doberman pinscher into a French poodle." He is what might be called a 'non-practicing' *shokunin*. Toshio's sculpture reflects this tension, mixing masonry with silken wood timbers, translucent rice paper with cordwood. His sculpture studio is a cavernous aluminum hangar with a concrete floor, upon which Toshio spreads the *tatami* or *mushiro* mats as he did 40 years ago as an apprentice to delineate quiet workspace from the chaos around it.

In the teaching workshops that Toshio now conducts throughout North America on the subject of Japanese woodworking tools, he likes to begin by matter-of-factly stating that he can always tell the amateurs from the professionals in the crowd. The amateurs, he explains to the riveted assembly, are the ones who fondle his tools. The worst among them run their thumb across the edges of the chisels to test their sharpness—implying that they might be something other than perfect. This preemptive announcement works. No one would be caught dead shaving the hair off their forearm with one of Toshio's chisels. I, too, constrain myself to stealing sidelong glances at the tools in the shop. In his book, Toshio expands on this theme: "The bond between the *shokunin* and his tools is not only practical but emotional and spiritual; this bond, once experienced, is not easily ignored or changed. I feel this way not only about my hand tools, but about all of my machinery, and though I can intellectually chide myself for this attitude, I am helpless to feel otherwise."

I was curious to know whether Toshio felt the same about his workbenches as he did about his other tools. He hadn't considered the question before and paused to think. In his shop, Toshio uses three different benches—a planing beam, a pair of low horses and a planing board—each for different operations. They are important, he says, very important to the way a *shokunin* works. But they are somehow different from his planes, chisels or saws.

As an apprentice in Japan, Toshio and his master traveled the countryside, carrying their tools in a box and setting up shop for as short as a week or as long as a few months in spaces provided by their customers. Anything from an empty cow shed to a corner of a house or an open field could be transformed into a workshop. On every job their first activity would be to prepare the planing beams and horses, usually out of materials provided on the site. If it was a new building, spare construction timbers, often cedar or fir, would be turned into planing beams. These materials are relatively stable and easily squared and kept flat. At the end of the job, the beams would be stored with the customer, something Toshio and his master never would have considered doing with their other hand tools.

No *shokunin* would dream of using another's plane, although Toshio would not feel violated if a qualified craftsman used his beam or horse. But if he cut into or otherwise abused it, that would be different. "Is it like chipping a blade?" I ask. "Not chipping," Toshio pounces, "touching. I get goosebumps if somebody touches my tools...he has invaded my soul."

Like the workshop, the benches should command respect. "The workshop has to be clean," Toshio explains. "It's a very sacred place...the bench surface has to be very, very flat." Like Japanese toolboxes, benches must be strong and neat, but simple and functional. "Tools," he continues to ruminate, "are an extension of one's own body—mentally and physically. Tools are an extension of the hand. The planing beam and sawing horses belong to the floor."

I wanted to watch Toshio work, but there was something holding him back. It finally became clear after three visits that he was not anxious to demonstrate the use of his planing beam and horses until he had completed the new 'Japanese-style' workshop he was preparing. When I returned several months later, Toshio said proudly, "For my fifty-sixth birthday I gave myself a good workshop...one of my dreams."

The new shop is in a converted one-car garage adjacent to the old studio. It was as Toshio predicted it would be—pure and peaceful, as though "everything is very calmly waiting for you." He likes machinery and uses it frequently, but it has its place—next door.

"How big is it?" I ask as Toshio sweeps out the shop and clears a place for me to sit down. "I don't know," he chuckles, "but we can measure." On top of the concrete garage floor Toshio built a 7-in.-high pine platform that measures about 11 ft. wide by 14 ft. long. At the front of the shop, between the platform and the garage door (which will eventually be replaced by a sliding door), is a 4½-ft.-deep cement foyer where Toshio plans to install his sharpening area and a stove. At the bottom of the back wall of the shop is a shallow ledge upon which are stored miscellaneous small tools that don't fit on the wall, as well as the short block that supports the butt end of his planing beam. The space seems small compared to most Western shops, but not to Toshio. "Are you kidding?" he asks. "This is a luxury you'd never have in Japan." Like the beginning of a new friendship, there is something special about a fresh workshop. "From here," Toshio says, "it will start aging."

"The beam is the center of the workshop," Toshio explains. "All the tools are available around it." Toshio has a favorite planing beam, made of cherry and kept flat and square, with the edges lightly broken. It measures about 5 in. square—a thinner beam might warp or flex under pressure. Almost any straight-grained wood will suffice, but Toshio prefers a medium-dense hardwood like cherry or maple to lend mass to the beam without being too slippery. A hefty portion of wane edge remains on one lower corner, recalling an observation made by Edward S. Morse a hundred years ago in *Japanese Homes and Their Surroundings*: "Whenever the Japanese workman can leave a bit of Nature in this way he is delighted to do so. He is sure to avail himself of all curious features in wood."

The beam is elevated at one end by a thin trestle built of three pieces of 1-in. by 1½-in. oak, fastened with a single nail at each intersection. The trestle is angled slightly against a half-lap notch cut in the underside of the beam. As planing pressure is applied to the top surface of the beam, it is resisted by the trestle and transferred along the length of the beam to the short post at the other end. This tripod works as well on uneven ground as on the floor boards of the workshop. In Toshio's shop, the short post simply leans against the wall, but outdoors it would be lashed to a tree or to another post driven into the ground. Because all planing motions are taken down the beam, from the trestle to the post, all forces will be opposed if the post is well secured.

The arrangement has several obvious advantages. It is very flexible—simply change the length of the beam or the height of the post or trestle. It is instantly portable—when the top nail of the trestle is pulled, the three pieces collapse and nest alongside each other. Yet there appears to be even less movement than in much heavier, freestanding Western benches, which can slide across the floor.

Like most workbench details, Toshio explains, the angle of the beam is a matter of personal preference, tailored not only to the craftsman's body, but also to the kind of work performed. A shallow angle, which Toshio prefers, is appropriate for fine work. With the aid of gravity and leverage, a steeper angle makes it easier to take heavier cuts. At the beginning of a steep planing stroke, the craftsman's legs are spread wide; at the end of the stroke, they are closer together and his center of gravity is much lower. Toshio's beam is 31 in. high at the top and 24½ in. high at the bottom, falling roughly between his waist and thigh, in the *shokunin* 'strike zone.'

The beam's length also depends on the work you do. Toshio's is about 9½ ft. long, which is just right for the willowy frames of the *shoji*. Although the beam is narrow, he can plane a board up to 12 in. wide simply by moving it so that the portion being planed is centered on the beam. Wider stock could be worked on the floor or he could recruit an assistant to support one edge.

The angled planing beam is the most important bench in the workshop. The beam is supported in the front by a thin, three-slat trestle, and in the back by a short post propped against the wall.

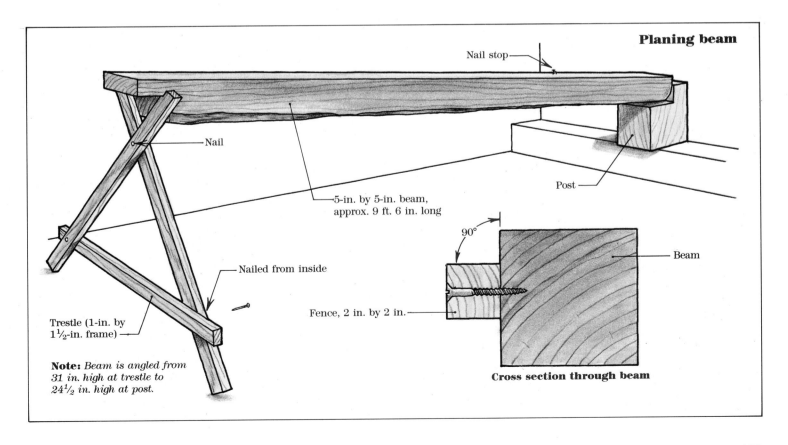

Planing beam

Nail stop

Nail

5-in. by 5-in. beam, approx. 9 ft. 6 in. long

Nailed from inside

Trestle (1-in. by 1½-in. frame)

Note: *Beam is angled from 31 in. high at trestle to 24½ in. high at post.*

Post

90°

Fence, 2 in. by 2 in.

Beam

Cross section through beam

A strip of wood screwed to one side of the beam serves as a fence for the plane when shooting an edge.

Odate stops

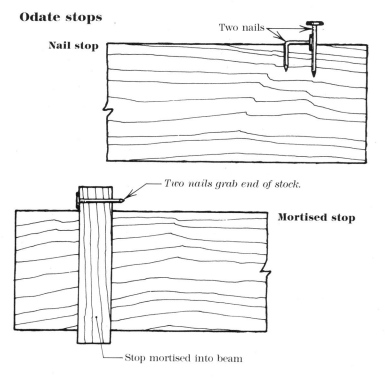

Nail stop

Two nails

Two nails grab end of stock.

Mortised stop

Stop mortised into beam

The beam must be absolutely flat or it will be impossible to produce flat stock. On one side of the beam, a strip of wood (about 2x2) is attached to support the side of a plane when shooting the edge of a board. Toshio's strip is oak and screwed instead of nailed to make it easy to replace. The top of the strip must be exactly parallel to the top of the beam to produce a square-jointed edge on a board. (This works, of course, only if the side of the plane is square to its sole.) The strip should be located so that the center of the plane blade falls at, or slightly above, the top of the beam.

To hold boards on the beam during planing, two nails are driven into the top surface of the beam near its lower end to form a stop. The rear one is driven first, partway into the beam, and bent forward. Then the front one is driven vertically, its shaft supported by the head of the first, as shown in the drawing at left. Alternatively, you could mortise the beam for an adjustable stop, also shown in the drawing. The Japanese mortised stop resembles a typical Western benchdog, except that two parallel nails are driven through the stop near the top. Their points protrude to grip the end of small, thin stock and keep it from slipping while it is being planed. In either case, the stop should be located far enough away from the end of the beam that you will not bump into the wall at the finish of a long planing stroke. A beam with a steeper angle requires more room behind the stop to allow for a longer follow-through. Also note that the side strip must extend well beyond the stop to provide a full bearing surface for the plane all the way to the end of a cut.

Using the planing beam properly is like making a graceful dive. It begins before you reach the end of the diving board and continues after you have hit the water. With his legs spread wide and his weight forward, as shown in the sequence of photos at the top of the facing page, Toshio has already put his plane in motion by the time it first slices into the board. Stretching radically at the waist, Toshio pulls the plane quickly through the wood, shifting his weight to his back leg and drawing his front leg beneath him. In one stroke, he planes half of a 7-ft. board. Without lifting the plane from the surface of the wood, Toshio executes a rapid skip-like two-step. He replaces his feet so that, as before, his right leg is stretched back and his weight is forward over his left leg. The second stroke flows off the end of the board, completing the cut and producing a single shaving.

Performed correctly, it is one long fluid power stroke. But Toshio does it so quickly that I have to ask him to stop at the end of the first stroke so I can photograph his feet. He obliges, but is surprised to find himself stranded in mid-cut. Frozen in position for the camera, he is unsure where to place his next footfall. "Forty years I'm doing that and just now I forget it!" he exclaims. "You can't chop it up."

He also explains that speed is a vital ingredient in the process. Most European planes have enough mass to add momentum and pressure to a stroke. The wooden Japanese plane is much lighter, so it falls upon the sharpness of the edge, the speed of the stroke and the skill of the craftsman to produce the surface for which the Japanese are renowned. Speed is required of the craftsman in all aspects of his work. It is expected that the *tategu-shi*, for example, will be able to make one high-quality *shoji* screen in a day. "The worst insult to a craftsman in Japan is to be called a slow worker," Toshio says. "I can just see my master standing in the studio—laughing."

Coordinated arm and leg motions are the key to a fluid planing stroke. Odate stretches forward to begin the stroke before his blade enters the wood. As the stroke progresses (top left and right), his arms and left leg are drawn into his body. With a quick shuffle, Odate resumes his initial posture (bottom left)—his right leg is stretched back and his arms are forward. It takes only one or two seconds to complete the whole stroke (bottom right), and the plane never leaves the surface of the wood.

For shorter, more delicate work, Toshio uses an oak planing board, as shown at right. A *shokunin* may have several, with various dimensions, depending on the size of the work. It is important that the board be elevated by as much as 6 in. with a crosspiece at each end. These can be fitted to sliding dovetails on the underside of the board—which also help keep it flat—or simply toenailed in place from below. Occasionally, one crosspiece is made higher than the other to provide better leverage for planing, but Toshio also uses his planing board to chop mortises, so it must be level. At one end of the board, a narrow strip of wood is attached to the top surface with a sliding dovetail to provide a replaceable stop. On Toshio's planing board the stop does not go all the way across the width. This allows him to lay a long piece of work flat on the surface.

To use the planing board, Toshio sits on a cushion at the stop end and cradles the board between his extended left leg and his right foot and knee, which support the bottom end. (Outside on the ground, he would sit on either a *tatami* mat or a rice-straw mat.) The board is immobilized in this position by the numerous points of contact with his body. Reaching forward like an athlete stretching a hamstring, Toshio can comfortably take a planing stroke the length of the board.

To plane small work, Odate cradles the board between his legs, stopping its end with his right leg and supporting one edge with his left.

Japanese Beams and Trestles 155

This is the intimate working position for which the Japanese woodworker is perhaps best known. Early paintings of Japanese craftsmen at work invariably depict them seated. While I fear I'd have to do leg stretches and splits before even thinking about working wood from a sitting position, the practical advantages are obvious. Such close physical contact makes heightened control practically inevitable. The speed with which Toshio can change body position and the position of his work is remarkable. In an instant, a piece of wood can be flipped on edge to work a chamfer or a bead, or turned perpendicular to the board for flattening or crosscutting. Moreover, in the sitting position it is easier and less tiresome to focus on a fine detail than it is from a standing position.

Odate sits directly on the work to chop a mortise, a position that affords him maximum control.

To chop a mortise or pare a tenon, Toshio sits directly on the work with his leg tucked under him to maintain perfect control and an excellent vantage point of all layout lines. The actual bench surface may be small, but the floor around it is the set-up table—part of an extended workbench. There's nowhere to drop tools and they're all within easy reach of both hands. When the work is done, the board can be tucked under one arm and carried off, either to another job or to be stacked neatly against the wall until next time.

For marking out and most crosscutting, Toshio uses the small horses shown at the top of the facing page. In the field these may be made of almost anything, but Toshio's workshop horses are built of 2x6 oak, about 7 in. high by 20 in. long, and nicely rounded and chamfered at the corners. It is important that they be a comfortable height for kneeling or squatting and that they elevate the material enough to allow convenient fastening of small clamps or clearance for the end of a sawblade. On occasion Toshio prefers to work standing, in which case the horses can be placed on top of a low workbench in the sculpture studio.

Given the uncommon workmanship of Japanese craftsmen like Toshio Odate, it would be tempting to conclude that Eastern woodworkers are inherently more skillful than Westerners—or that their tools are better. While I'm not sure that such an assumption is altogether unwarranted, Toshio has another explanation. "A lot of people say Japanese planes are better than Western planes," he argues, "but that's the wrong way to look at it. It's a social relationship." The tools and the working habits of the Japanese craftsman are intrinsic to the products of his craft and the predilections of the society he serves.

Within Japanese society, as within our own, there are considerable differences in the way a craft is conducted. I'm told that many modern city carpenters in Japan work exclusively with hand-held power tools and have become as isolated from their hand-tool tradition as their Western counterparts. Toshio is also careful to point out that the workbenches (and many of the tools) he uses are typical for a *tategu-shi*. While he shares a basic tradition with carpenters, housebuilders and furniture-makers, their tools reflect their own special needs. To gain a sense of this variety, I visited several other Japanese, or Japanese-trained, craftsmen.

Planing board

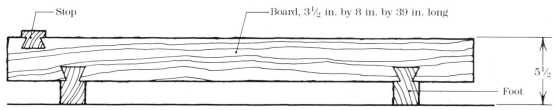

Stop

Board, 3½ in. by 8 in. by 39 in. long

5½

Foot

Note: *Feet and stop are joined to board by sliding dovetails.*

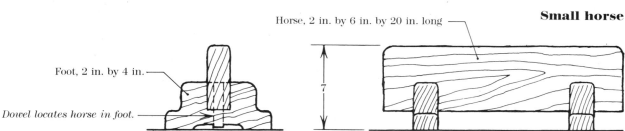

Foot, 2 in. by 4 in.

Dowel locates horse in foot.

Horse, 2 in. by 6 in. by 20 in. long

Small horse

7

The small horses may be used for a wide range of operations anywhere in the shop. About 7 in. high by 20 in. long, they are just right for crosscutting, laying out and ripping.

Sitting vs. standing, pulling vs. pushing

Earliest records of man at work show Egyptian craftsmen seated. Illuminations in medieval European texts frequently portray woodworkers sitting down. Although many 20th-century craftsmen still work in a seated position, for the most part the Western woodworker has risen to his feet. Meanwhile, his Japanese counterpart has remained seated—although hardly immobile.

For a Westerner, sitting down to the job is often the most startling and discomfiting thing about Japanese woodworking. It's as foreign to most of us as karate might be to a boxer. Unpracticed, our limbs just don't bend that way. But for the Japanese, sitting down, whether it's to eat dinner or to make a *shoji* screen, is instinctive. Chairs (like vises) are a relatively modern development in Japan, having been popularized in the 20th century. Even today, most Japanese still opt for the floor.

Japan's many-layered tradition was isolated for centuries from the rest of the world. Within that cultural pressure cooker, many Japanese tools and work habits developed. Early paintings, such as the 14th-century construction scene in the photo at right, show carpenters squatting like water spiders to pull the traditional *yari-kanna*. Until quite recently, Japanese craftsmen pulled this double-edged, spear-like

instrument to smooth the marks left by an adze.

It seems likely that the two most striking features of Japanese technique—sitting and pulling—are closely related. According to Japanese historian Teijiro Muramatsu, the modern plane did not arrive in Japan until the 15th century. It probably came from China by way of Korea, whose influence on Japan was strong, particularly in the southern island of Kyushu. The Koreans, like the Chinese and Westerners, have long used the plane in a pushing motion. Attempting to integrate this unfamiliar tool into their work, the Japanese naturally adapted it to the familiar pulling motion of the *yari-kanna*. Experimenting with both methods, Japanese

craftsmen would have been immediately convinced of the greater efficiency of the pulling motion. When you pull a plane from the sitting position, strength and control are enhanced by the use of body weight and muscles in your back and legs as you uncoil from a forward crouch. A push stroke would be comparatively weak, relying only on the smaller arm muscles.

Evolution is never linear. Japanese hand planes were pushed *and* pulled in some parts of Japan for several centuries. But it is likely that the farther they migrated from Kyushu (and Korean influence), the more thoroughly they were integrated into the Japanese working habit. (I've heard that some craftsmen in Kyushu

continued to push their planes as recently as the 1940s.)

Once the plane was established in Japan, it was probably not long before carpenters moved—albeit reluctantly—from a sitting to a standing position. Only by standing was it possible to execute smooth planing cuts on long timbers. The work of the *tategu-shi* (maker of *shoji*) requires standing to plane the long pieces of the frame and sitting to plane small crossmembers. Whether working on the floor or on a pair of hip-high trestles, Japanese woodworkers adopt the familiar sitting position to chop a mortise. Wrapping one leg over the work, they hold the work steady and get a good vantage point from which to judge the accuracy of their cuts.

This animated scene from a 14th-century scroll depicts the construction of a bamboo forest sanctuary. From Teijiro Muramatsu's A Hundred Pictures of Daiku at Work.

Makoto Imai is a Japanese timber framer in Northern California. Like Odate, Imai had moved recently into a new workshop when I visited him. Dug into the side of the Trinity Mountains, it's an airy, shed-like structure, redolent with the fragrance of Port Orford cedar, which fills the shop. Large windows look out for inspiration across the valley at a blue-gray ridge of mountains beyond.

The 16-ft. by 32-ft. shop is divided into one main work area spanning the length of the building and three smaller work cubicles for Makoto and two apprentices on the sheltered north side. Outside under an extended overhang are the machine tools Makoto uses to prepare stock. Like Odate's, Makoto's workshop is reserved for hand tools, and shoes are left at the door. It's a fitting retreat for a philosophical craftsman. Down the hill, Makoto lives with his wife and children in an aluminum house trailer—"Japanese mobile home," he calls it.

Makoto Imai is 37 years old, wiry and intense, with long black hair pulled back into a ponytail and thinning on top. His skills and attitudes were forged by the same kind of rigorous training as Toshio Odate's. The first two years of his six-year apprenticeship were devoted to sharpening the tools, cleaning the workshop, tending the fire (to warm water for the sharpening stones) and watching his master. In the process, Makoto began to imbibe the spirit of the work—long before he learned how to cut a joint or follow a building plan. Eventually he went on to refine his skills building teahouses, temples and shrines in Kyoto. In this country, Makoto earns his living building exquisite timber-frame structures in traditional fashion.

Chain-smoking Camel cigarettes and squatting on the floor of the shop, Makoto explains that as a timber framer he uses several kinds of workbenches. Low trestles support large beams for chopping mortises or cutting tenons. Most fine planing, sawing, or careful paring is done on a higher bench, shown below. The bench consists of two 25½-in.-high trestles,

Timber framer Makoto Imai does most of his joinery on a pair of Port Orford cedar trestles. To hold the work, Imai relies upon his body and the basic properties of gravity and friction.

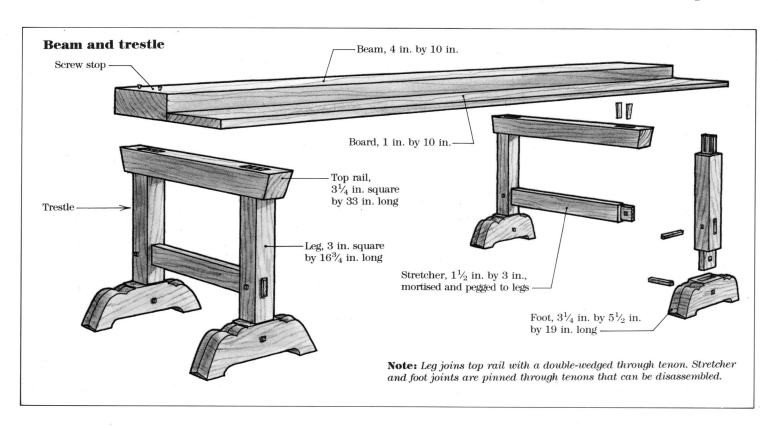

Beam and trestle

Screw stop

Beam, 4 in. by 10 in.

Board, 1 in. by 10 in.

Trestle

Top rail,
3¼ in. square
by 33 in. long

Leg, 3 in. square
by 16¾ in. long

Stretcher, 1½ in. by 3 in.,
mortised and pegged to legs

Foot, 3¼ in. by 5½ in.
by 19 in. long

Note: *Leg joins top rail with a double-wedged through tenon. Stretcher and foot joints are pinned through tenons that can be disassembled.*

upon which he can place one or more beams to provide a work surface. Because the benchtop is level, unlike an angled planing beam, all operations can be performed on the same surface.

The ideal benchtop, he explains, should be level with your hand when your arm is at your side with the palm held face down. At that height, Makoto can produce a single translusent shaving the length of a beam without losing his balance. If the bench were higher, it would raise his center of gravity and make it harder to apply constant pressure; the work might slip and he'd lose control. As with Odate's benches, there is nothing sacred about the dimensions; they are a function of the height of the craftsman and the nature of the work being done.

When Makoto works heavy timbers, they sit directly on the high or low trestles, kept stationary by their own weight. For smaller projects, he rests a 4x10 beam on the high trestles, sometimes with an adjacent 1x10 board to hold tools within reach. Work on the beam is held either with one hand or by one of the stops shown at right. The weight of the beam combined with the wide spread between trestles produces an immobile bench, despite the absence of any fastenings. He joints the edge of a plank in the same manner as Odate, by attaching a small strip to one side of the beam to carry the edge of the plane.

The construction of Makoto's trestles is straightforward timber framing. All joints are either double-wedged or pinned mortise-and-tenons, designed to resist racking and assembled without glue. The bottoms of the feet are relieved for a stable four-point stance. Makoto has several pairs in his shop for himself and his apprentices; all are made of Port Orford cedar. I've seen trestles made of poplar and Makoto tells me that his first 'American' pair, which he sold, was made completely of Douglas fir and had considerably more decoration. Even this relatively simple set of trestles, however, is fancier than those Makoto learned on in Japan, which were comparatively quick-and-dirty. On the job site, Makoto will sometimes nail together a temporary workbench out of available materials, as shown in the bottom drawing at right.

Beams used for the work surface must be stable, clear and dense to absorb the shock of hammer blows. Poplar and white pine are too light and dent too easily. Mahogany is better, but Makoto prefers "Doug" fir for its weight, stability and hardness. It can be resurfaced easily and has a leathery surface texture, which helps to resist sliding during planing. (Quartersawn stock would be ideal for restricting movement in the beam.) The hardwoods used in the traditional European benchtop— typically beech or maple—would be much too hard and slippery to work on. Their hardness would also make them more difficult to keep flat and might mar the work, which is usually of softer woods. The slippery nature of the planed, sanded and well-oiled hardwood tops on Western benches only makes sense in the context of the intricate workholding systems that have evolved along with those bench designs. "Hardwood is slippery," Makoto says, speaking from the floor, where he sits in a lotus-like position sharpening a plane blade. "It's not really practical for no-vise work—that's why Western benches have to have vises."

The beam is hand-planed flat and true in its width and length and square to its edges. It is checked with a square and winding sticks, as any Western benchtop would be. The beam and the trestles are not sanded and are never oiled or finished in any other manner. Even without a finish, the beam doesn't bow or twist much with changes in humidity, but its flatness

is habitually checked before any critical job. Whether it's a workbench, *shoji* screen or temple, the Japanese sensibility prefers a crisp, clean, unadorned surface treatment. As Bernard Rudofsky, an acute observer of the Japanese sensibility, writes in *Kimono Mind*, "The very idea of painting or staining is revolting to [the Japanese]. It is not a matter of taste but a case of vandalism."

Watching Makoto Imai at his bench makes me feel as though all of my previous woodwork has been done with a chainsaw. Whether he's joining building timbers or finishing *shoji* screens, the work is done with consummate skill. The problem I discovered is that, as in Japanese culture itself, the lessons are subtle and not easily grasped by a Westerner on a deadline. If it took Makoto two years to learn how to sharpen, I was not about to decipher his use of the bench in two visits. This is consistent, I think, with his own teaching philosophy. Makoto teaches as he was taught—by example and osmosis. Like his timber-frame structures, he builds his students from the inside out. "Technical is not the most important," Makoto says. "People have to appreciate the process of learning. Some people use their head only to learn, but the body has to learn, too. So it takes more years to become good."

Planing stops

Note: *Two planing stops provide better support for wide stock.*

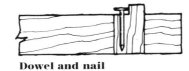

Dowel and nail

Toe nail

Sharpen four sides of screw head.

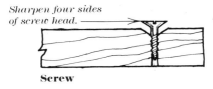

Screw

Job-site carpentry bench

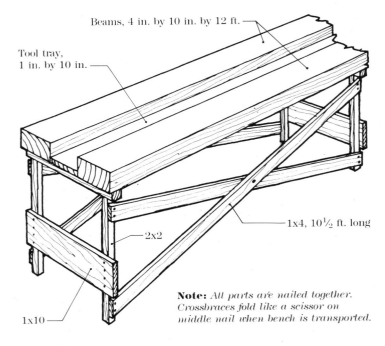

Beams, 4 in. by 10 in. by 12 ft.

Tool tray, 1 in. by 10 in.

1x4, 10½ ft. long

2x2

1x10

Note: *All parts are nailed together. Crossbraces fold like a scissor on middle nail when bench is transported.*

For insight into Makoto's method of working with his benches, I turned to one of his most avid acolytes, Baltimore woodworker Carl Swensson. About ten years ago, while living in Berkeley, California, Swensson enrolled in a workshop on Japanese joinery given by Makoto Imai. Imai walked into the class, set up his bench in the center of a knot of students and, without a word, began working. He cut one half of a dovetail joint and passed it around the group. While they were examining it, he cut the other half—without marking it off the first or resorting to templates. When the first piece came back, he joined them. They fit perfectly. The week before, Swensson had attended another workshop on router joinery in which the instructor discoursed for hours on the subjects of buying a router, using dial indicators to measure its accuracy and making jigs to go with it. Once the jigs were built, Swensson figured a ten-year-old could cut the parts. By comparison, Imai's dovetails were inspired. "Damned if they didn't fit as good as the ones cut by the router guy," Swensson says. He told himself, "This guy [Imai] knows something I want to know."

Swensson took every opportunity over the next couple of years to dog Imai at his public demonstrations, hoping to catch some precious insight unguardedly dropped. Eventually, although he never officially apprenticed to Imai, the master agreed to let him work on his own projects in a corner of the shop. In describing the hollow of a plane sole, Imai would tell his apprentices to scrape it until there was a little line of light. "How much of a line of light?" Swensson wondered. When the others had gone for tea, he stayed and measured the bottom of Imai's plane with a straight edge to find out. He returned to his apartment at night and spent three or four hours practicing what he'd learned that day.

A former tennis pro, Swensson dissected Imai's movements into interactions of muscle and bone—component parts he could understand and methodically reconstruct. Swensson's approach to woodworking is as intellectual as Imai's is intuitive, which makes it that much more comprehensible to many Westerners.

In his basement workshop back in Baltimore, hemmed in by the family laundry room and stacks of aromatic Northwest cedar, Swensson works on a trestle workbench and beam in the style of Imai. "Because the bench is so simple, you have to bring other things to it," Swensson says, as he demonstrates planing on the beam. "The more things I can keep constant...the better." He moves all parts of his body a little—arms, legs, torso—rather than keeping his legs stiff and simply pulling or pushing with his arms.

Swensson's planing stroke resembles the explosive rhythm I'd seen demonstrated by Toshio Odate. Although Swensson's beam is horizontal and Odate's is angled, their technique is much the same. Working from a wide, stable base, Swensson flexes at the knees, shifting weight from front to back. He begins each stroke crouched forward and uncoils through the planing action. For planing, as for most workbench operations, Swensson slides the beam over the front leg of the trestles. (He used a 4x8 beam during my visit, but plans to replace it with a 4x12 for additional worksurface in planing wide boards.) This transfers pressure through the legs directly to the floor, and allows him to stand close to the edge of the beam without running into the top or foot of the trestle.

A large amount of stock can be removed in a series of short, powerful hogging motions, using the body's strongest muscles in the legs and back. With the work centered on the beam against the stop, Swensson works from the dogged end forward until the board is roughly flat. He prefers Imai's simple screw stop (p. 159), which is sharpened below the head for a good bite in end grain. The surface may be smoothed with full-length finishing cuts.

Chiseling falls into two categories: chopping, which relies on the energy of a hammer stroke, and paring, which employs body pressure. Both are done directly over the front leg of the trestle and, as always, Swensson's use of his body is instrumental. To chop, he sits on the work, wrapping one leg over the board, sometimes tucking it beneath his other leg. The lower leg drapes over the bench, establishing contact with the floor. One hand positions the chisel in the cut, the other works the hammer.

To pare, Swensson crouches above the work and locks the top of the handle against his chin or chest with one hand. His other hand stabilizes the blade and guides it through the cut while holding the work on the bench. Swensson's stomach muscles provide the force for the cut, rather than the much weaker arm or shoulder muscles. This posture also affords an extra measure of control as the chisel is operated within the body's sphere of influence. "No need to curl your toes or grit your teeth," Swensson explains. By focusing attention on what you're doing, you can avoid tensing your entire body.

Swensson adopts a stable multi-point base for sawing. Kneeling, with the top of the beam at shoulder height and his left elbow on his upraised knee, he uses his left hand to clamp the work to the bench surface. Weight and leverage lock the work in position, not arm muscles, and he is able to align the sawblade with the layout lines on two adjacent faces of the stock.

For crosscutting, the work is placed on the edge of the beam, so the downward sawcuts will clear the work surface. The benchtop itself functions as a kind of bench hook by supporting and stopping the work; the hand merely holds it lightly in position. For a rip cut, Swensson moves around to the end of the bench and places the stock on the end of the beam, centered as always over the upright leg of the trestle. When the saw approaches the bottom of the layout lines, Swensson pulls it straight down, kneeling over the cut to sight down the blade, as he would a paring chisel.

To rip a longer piece, Swensson leans one end against the trestle, the other on the ground and wedges the bottom end with his foot, if necessary. When he gets to the bottom of the cut, he stands the board on top of the trestle or beam to complete it. Recalling former days on the tennis courts, Swensson advises that you keep your eye and head steady, and your hand, elbow, shoulder and the sawblade all in the same plane. "It's just a little thing," he says, "but you add all those little things together and you've got a joint that fits."

These days, Swensson makes *shoji* screens at home by night and builds organs in a European-style shop by day. The massive benches and Western tools at the organ shop are more of a challenge than an asset, and Swensson has forged his own eclectic brand of woodworking. He shuns the tail vise and works against a fixed dog. He also sits on the bench as he would at home, supporting his dangling leg on a chair. "With Japanese woodworking, you're always working on the edge of your ability," he says. With Western woodworking, the experience is different: "I think my work is as good, but it's like the work's there and I'm here. When I'm sitting on it, it just feels like we're all one."

Swensson centers the beam over the trestle leg and removes stock with a series of short planing motions, working from the dogged end forward. Photos by Paul Bertorelli.

Chisel work also takes place above the trestle leg so that the force is transmitted to the ground. Swensson locks the chisel handle against his chest or under his chin to make a careful paring cut. His stomach muscles drive the tool, providing power and control. Photos by Paul Bertorelli.

Whether crosscutting (above) or ripping (right), Swensson maintains contact between his body, the bench, the work and the ground. The edge of the bench beam resists the downward motion of the saw.

Ted Chase's bench consists of an oak table, with two heavy maple planks on top. The laminated legs are glued and doweled in a lap joint to the apron rails. The oak-plywood top is tacked in a routed rabbet in the aprons; it helps support the beams and provides a tool well between them. Planes are hung upside down within easy reach on a narrow strip of wood screwed across one end of the bench (right).

Before leaving the Japanese workbench, I would like to describe devices used by two other California woodworkers: Ted Chase of Concord and Dennis Young of Petaluma. Chase and Young trained in Japan—under very different conditions—and work primarily in the Japanese style, using Japanese hand tools and lots of body 'English.' Their benches are variations on those used by Odate, Imai and Swensson, but with a few novel twists.

Ted Chase and his wife, Frances, visited Japan in 1977—the first stop in what was to be a two year trip around the world. A serendipitous introduction to a middle-aged woodworker, Kennosuke Hayakawa, led to an invitation to stay and study for a year. Hayakawa had two other apprentices—one had been with him for 8 years, the other 15—and Chase prefers to call his experience a 'study program.' "Look, listen and do, but never ask why," is how Chase describes his tutelage. "When I started, Hayakawa opened a drawer and pulled out a plane. He let me look at it, handle it...then he took it back, knocked the blade out and said 'sharpen it.'" Nine hours a day, six days a week, for three weeks, all Chase did was sharpen the blade. Compared to Makoto Imai's experience, however, Chase was on the fast track.

Hayakawa ran a custom furniture shop, and he used a floor-level planing board for his own detailed work. Ted and the apprentices worked on relatively high (30 in.), stationary workbenches, made of *keyaki*, a highly-prized Japanese wood similar to elm. Before leaving, Ted took the measurements and joinery details off these benches so that he could reproduce one at home.

In practice, Chase's bench functions like Imai's and Swensson's more flexible trestles and beam. For planing, Chase employs an assortment of oak stops, which fit in ¾-in. routed slots near the ends of both beams. The stops extend as much as 1 in. beyond the edges of the beam to support wide work, or can be slipped out entirely for sawing or chiseling. In most operations, one beam is used for a worksurface, the other as a place to spread out chisels. The well holds marking and measuring tools and the assorted detritus of the workshop. The bench frame is hand-planed and finished with three coats of Tung oil. Ted leaves the beams natural, periodically rejointing them if they become roughed up or warped.

Bench and beams

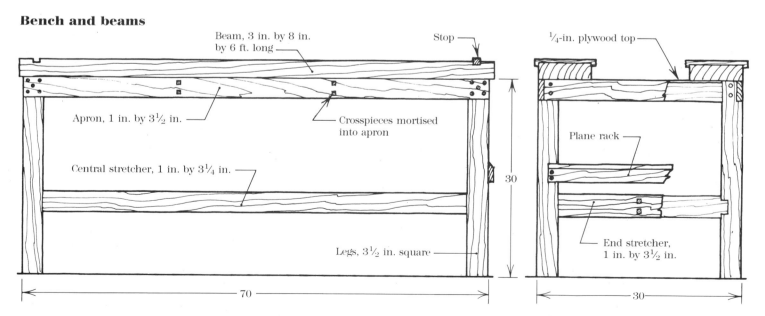

Beam, 3 in. by 8 in. by 6 ft. long

Stop

¼-in. plywood top

Apron, 1 in. by 3½ in.

Crosspieces mortised into apron

Central stretcher, 1 in. by 3¼ in.

Plane rack

Legs, 3½ in. square

End stretcher, 1 in. by 3½ in.

30

70

30

Few Western woodworkers would put up with a Japanese apprenticeship. It's a disciplined, severe environment that galvanizes one's resolve—either to leave or to survive. In no time the long hours, numbing repetition and ego-crunching submission would offset whatever romance may have existed at the start. Dennis Young is an exception. Committed to becoming a craftsman, Young left his native Southern California in 1972 to devote four years to a traditional Japanese apprenticeship. Young served his time at the Matsumoto furniture factory in Nagano Prefecture in central Japan, where a work week was comprised of six, ten-hour days and half a day on Sunday. Sunday afternoon, he was 'free' to work in the company fields.

Three months into his tenure, Young was deposited with an elderly furnituremaker. He was given a small floor-level planing board, a cushion and minimal instruction. After sitting there for two days, ignored by his teacher, Young began to wonder what was going on. Shortly thereafter he was tossed a tool and the learning process ground forward at glacial speed. Like Odate, Imai and Chase, Young also began by sharpening, graduating to tool use and, eventually to his own small projects. After a year and a half, his work—as slow as it was—was judged worthy to be folded into the mainstream production.

"The demands are extreme," Young says, "but once you've reached within yourself to come up with the committment, you're fine...by persevering you realize what it's for." Through the process, Young was not only training his hands and mind, but the basic relationship of his body to the work. "My teacher would get on my case to orient myself to the work," he recalls. With time, and given the planing board Young was riding all day, this became almost inevitable.

At Matsumoto most work was done on low boards. (Occasionally, edge-jointing was done on an angled planing beam, or long, flat surfaces were planed on a table-height bench like Ted Chase's.) Two feet raise the board off the floor, and two thin stops are tacked about 1¾ in. from one end. The board is angled slightly—5¾ in. high at the vise end and 4¾ in. high at the stop end—making it more comfortable for planing. A much steeper angle demands awkward contortions on the part of the Japanese craftsmen, who commonly hold the work with one foot. To hollow chair seats, however, the workers at Matsumoto sometimes propped up one end of their boards by as much as 12 in., bringing the top of the seat into closer proximity to their upper torso.

Young's board is made of *zelkova*, which is dense and stable. The top surface is hand-planed true, when necessary, but Young explains that at Matsumoto the wood was considered "so valuable that you didn't want to take off too much." Planing boards were made in different lengths to suit different purposes—shorter for chairmaking, longer for cabinetmaking.

The most unusual feature of Young's bench is its detachable vise, installed near the upper end of the bench. Vises are uncommon in Japan. Like the chairs that are one of Matsumoto's main products, the screw vise would have been unheard of before the mid-19th century and the 'opening' of Japan to the West. Traditional Japanese cabinetry is rectilinear and I suspect that the ability of the vise to accommodate three-dimensional carving may have been closely linked to the introduction of curved chair parts. In any case, Young didn't receive a vise until his second year at Matsumoto, when he had earned his stripes.

In the factory, Young reports, the vises were used to hold anything that could not be held with the feet—from curved rails and tapered spindles to long boards for edge-jointing. At 15 in. high, Young's vise is quite a bit taller than the others at Matsumoto—"It was hurting my back to lean over all the time," he says. In fact, getting used to the board itself wasn't easy. Young admits "it was difficult on my body."

Now, when the dampness of the cement workshop floor penetrates the board and the folded blanket Young uses for a cushion, he works standing up at a large, cast-iron machinist's vise, or on a 3-ft.-long planing board, which he can place on top of any workshop table.

You might think that after touring so many Japanese workshops, I'd have gotten my fill of Port Orford cedar, green tea, sitting on my haunches and vacuuming my socks. On the contrary, unraveling the nuance of Japanese woodworking is a little like peeling the prickly petals of an artichoke to get to the heart. You can pay good money to get just the hearts—carefully selected and preserved, pickled in glass jars. Or you can begin at the beginning and boil water. There are no shortcuts. To truly appreciate the reward, you must earn it.

Dennis Young still does most of his work at the traditional bench he learned on 15 years ago in Japan.

This vise was common in the chair factory where Young apprenticed. Built like a leg vise, it can be swiveled 360° on its attaching bolt.

Country woodworker, Jason Reed of North Georgia, in the late 1920s. Reed's bench is separate from the seat, and appears to form a transition between a shaving brake and a horse. Photo by Doris Ulmann.

Country Shaves and Brakes

A Swiss milking bucket hangs on the living room wall of Drew and Louise Langsner's North Carolina log home. It isn't very large, and it sits—not exactly hidden, but hardly advertised—beneath the stairs that ascend to the second floor. Its pale white-pine staves undulate gently between the bent wooden hoops that restrain them. When I pause to admire the bucket, turning it over in my hands, Drew points out that it is utterly functional. The bucket is perfectly shaped, he explains, for being pinched between the legs while milking. Indeed, if it had shown signs of hard use, I might even have believed that it was made with straight sides that had been worn hollow by years of knock-kneed milkers. After Langsner has left the room, Louise quietly adds, "More than anything else, that bucket is responsible for Drew's woodworking."

Drew Langsner grew up in Hollywood, California, but since moving to a 100-acre homestead in the Blue Ridge Mountains in 1974, he and Louise have immersed themselves in the culture and crafts of the Southern Highlands. Louise's white-oak baskets and Drew's ladderback chairs and hay rakes—even their hand-hewn log home—are built within a rural Appalachian tradition that they have been instrumental in preserving. The Langsners conduct workshops each summer and have written several books about the traditional country crafts that were once abundant in their valley.

By a quirk of fate, the inspiration for Drew's woodwork came not from the Appalachians but from Rudolf Kohler, the 86-year-old Swiss cooper who built the milking bucket that hangs on the wall. The Langsners stumbled upon Kohler in 1972 while hiking in the Alps. A retired farmer and cheesemaker, Kohler has been making farm buckets for over 60 years and coopering fulltime for most of the last 20. Drew spent a total of about 13 weeks with Kohler on two separate occasions learning everything he could about Swiss 'white cooperage,' the making of single-bottom milk pails.

Kohler does much of his coopering at a shaving horse—the workbench of the country woodworker. (Depending on where you live, a shaving horse might be also called a mule, mare, buck, trestle, shave brake, draw brake, draw bench or, in German, *schnitzelbank*.) The tools used by the country craftsman—the ax, froe and drawknife—are ancient and unsophisticated. Although carpenters and furnituremakers have used these tools for centuries, they are much more commonly employed by country craftsmen who prefer unseasoned, or 'green,' wood over dry for almost every purpose and who have developed a sophisticated system of wet-to-dry joint technology to accommodate it. Wet wood can be split and shaved much more easily than dry and is readily available in most rural communities. If an errant split or slice ruins the work—well, there's plenty more where that came from.

Green wood is heavy, so it's more efficient to carry a few tools and a relatively lightweight shaving horse into the woods than it is to tote stacks of freshly cut timber to the shop. Where so much wood is wasted in the process, it makes sense to split the wood where it falls and leave the limbs, bark, gnarled heartwood and excess water behind. If material must be removed from the woods to be processed further, it is reduced to a size and weight that is more easily lugged on one shoulder.

Like the savvy trapper or hunter, the green woodworker has to think like his prey. From selecting and felling a tree, and splitting and shaping it to the finished product, he cultivates an understanding of wood grain and movement. Shaving away at his horse, there's a lively interaction between body, tool and material. All are in motion, and the movement of one causes an immediate reaction in the others.

There's nothing inherently original about country woodcraft. Indeed, that is part of its appeal. Reudi Kohler does not ask himself as he lifts the drawknife each morning whether he is an artist or a craftsman. The tools he uses and the functional objects he makes are of his father's generation—and his grandfather's before that. For Langsner, his bond with Kohler forged a tangible link with a rapidly receding past. He enthusiastically adopted Kohler's tools and techniques—chief among them, the shaving horse.

Shaving horses serve a great many trades. On them, wheelwrights shape spokes, shinglers shave riven shakes; chairmakers, broommakers, basketmakers, carvers and boatbuilders, as well as coopers, each use their own version. Employing basic leverage, the shaving horse does for these country craftsmen what the finest tail vise and quick-action face vise do for their urbanized counterparts.

Kohler's is a dumbhead shaving horse. I've seen one other style, the English bodger's bench discussed later in this chapter, and an endless variety of customized models. But among country woodworkers in many different countries, the dumbhead horse is so common that it is considered the prototypical shaving horse. As shown in the drawing at the bottom of the

facing page, it is a four-legged bench, positioned at a comfortable sitting height, about 18 in. off the ground. An angled platform, or bridge, is erected on one end, to which work is clamped by means of a pivoting arm that fits in mortises chopped through the bridge and bench. A head is attached to the top of the arm and a foot treadle is fitted to the bottom. The craftsman sits on the bench facing the bridge and places his work beneath the head. By applying pressure against the treadle with his foot, the arm of the horse pivots on a pin and closes tightly at the head to hold the work.

The system is ingenious in its simplicity. As you pull the drawknife or spokeshave toward you with your arms, your feet naturally push away—like a rower in a boat. The harder you pull, the harder you push, and the more firmly the head comes down on the work.

Any commercial workbench may be adapted to suit a cabinetmaker's needs, but the country craftsman doesn't have the luxury of buying a shaving horse from his local tool dealer. The shaving horse is truly a 'folk' tool, made by the person who will use it to suit himself; I've never even heard of one for sale. Accordingly, it should be tailored to fit your own body and work. There is no one right way to make it.

The first horse Langsner made when he returned to California from Switzerland was, understandably, a copy of Kohler's. He used what materials were available—local applewood for the head and a scrounged piece of pallet oak for the treadle. When he established his Blue Ridge Mountain homestead two years later, Langsner again used local wood—this time, oak for the head, arm and treadle assembly and pine for the body. He figures that yellow pine, which he considers a neglected 'hardwood,' also would do nicely for the whole thing. Even the legs could be shaved out of yellow-pine limbs.

Today, more than a dozen shaving horses are stabled in the Langsner barn, along with a Massey Ferguson tractor, a disc harrow and assorted farm equipment. Drew built all of the

Drew Langsner's dumbhead shaving horse was adapted from Reudi Kohler's original. Langsner's horse has a straight pivot arm that attaches to the head with a through tenon. The legs are socketed directly into the bench.

horses, along with a bunch of simple worktables, mainly for the students who attend his workshops. In the process, he took the opportunity to experiment with different types of horses and to fine-tune his original. Although he began by learning how to make small Swiss milk pails, Langsner has clocked at least as many hours shaving ladderback chairs, hay forks and shingles, along with the countless other woodworking chores required around the homestead. Only one horse, a close adaptation of Kohler's original model, is set up all year round in the small shop area in the barn. Langsner calls it the 'champ.'

Copying Kohler, Langsner through-mortised the legs of his first horse into a batten, then housed the batten in a sliding dovetail in the underside of the bench, as shown in the drawing at right. Langsner speculates that the batten construction is a descendant of the two-board chair tradition in Switzerland and Germany, where they are used frequently on tables, chairs or almost anything. Certainly, before the era of modern glues and store-bought hardware, the sliding batten offered an accessible and attractive method of restraining wood movement and connecting the assorted members of a piece of furniture. On Langsner's horse, however, the battens have come loose in their slots and he has had to reinforce them with carriage bolts. "Reudi Kohler's bench doesn't have bolts," Langsner says with a smile. On subsequent horses, he has eliminated the dovetailed bat-ten altogether and simply secured the four legs directly in the bench with wedged, through-tenons, as shown in the drawing below.

If built of unseasoned wood, the legs should be drier than the bench so that they will not become loose as they season. Legs are usually splayed in two directions for stability, although the legs on some horses splay slightly in only one direction and are occasionally reinforced with a stretcher.

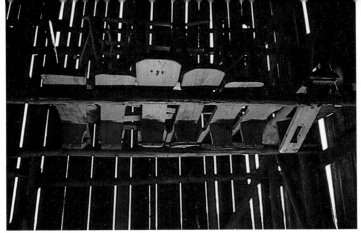

Langsner uses these horses for his summer workshops. They spend the winter tethered to the rafters of his old tobacco barn.

Batten leg assembly

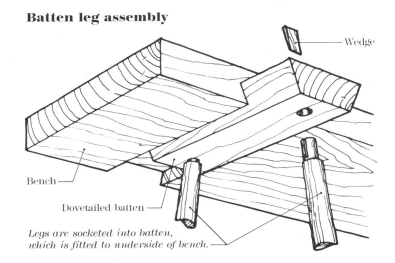

Wedge

Bench

Dovetailed batten

Legs are socketed into batten, which is fitted to underside of bench.

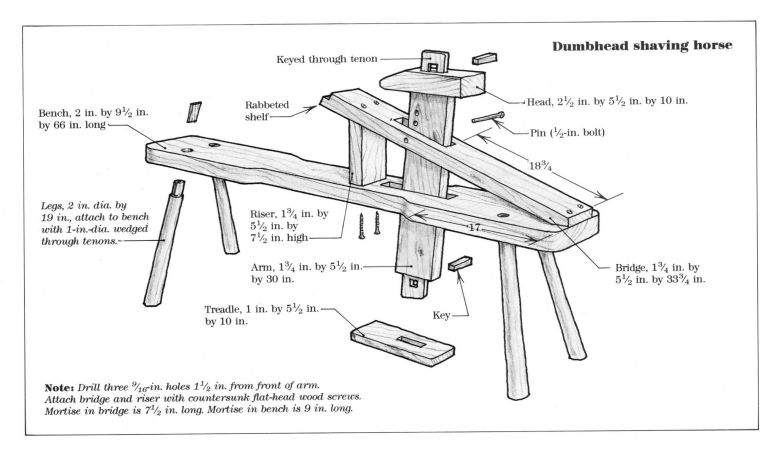

Dumbhead shaving horse

Keyed through tenon

Rabbeted shelf

Bench, 2 in. by 9½ in. by 66 in. long

Head, 2½ in. by 5½ in. by 10 in.

Pin (½-in. bolt)

18¾

Legs, 2 in. dia. by 19 in., attach to bench with 1-in.-dia. wedged through tenons.

Riser, 1¾ in. by 5½ in. by 7½ in. high

17

Bridge, 1¾ in. by 5½ in. by 33¾ in.

Arm, 1¾ in. by 5½ in. by 30 in.

Key

Treadle, 1 in. by 5½ in. by 10 in.

Note: *Drill three ⁹⁄₁₆-in. holes 1½ in. from front of arm. Attach bridge and riser with countersunk flat-head wood screws. Mortise in bridge is 7½ in. long. Mortise in bench is 9 in. long.*

Dumbhead variations

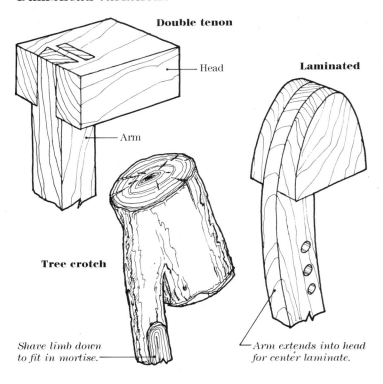

Double tenon

Head

Arm

Laminated

Tree crotch

Shave limb down to fit in mortise.

Arm extends into head for center laminate.

To shave a short stave, Langsner supports one end of the work on a rabbeted shelf on the end of the bridge and holds it in place with a wooden breast bib.

All of Langsner's horses are roughly the same length as Kohler's—slightly more than 5 ft.—which is the minimum length Langsner figures is comfortable for preparing the longest lengths of chair stock. If you're working a lot of long, thin pieces, such as is common in basketmaking or hoopmaking, Langsner suggests a longer bridge. This will provide support over a larger area and keep the wood from flexing. At the very least, you need about 3 in. to 4 in. of solid worksurface in front of the pressure point on the bridge. If necessary, you can slip a plank of wood beneath the work to extend the working platform.

The angle and height of the bridge are not critical—most craftsmen can get used to a certain amount of variation—but it should not be so high that you will have trouble completing a long stroke with a drawknife. One good way to figure the angle, suggests Langsner, is to build the bench first and sit on it. Then prop the bridge in position and adjust its height up or down until the shaving feels comfortable.

The straight arm of the 'champ' attaches to the oak head and treadle with a wedged through tenon. (Langsner's original dumbhead was of a laminated construction, one of several variations shown at left). The wedged joint is very strong, and is easily knocked down for transportation or snugged up if it gets loose. The dumbhead jaw should jut forward enough to get a good grip on the work. The jaw on Langsner's horse has almost 4 in. of bite and is angled to provide a firm, downward clamping action. The treadle extends about 6 in. from the front of the arm at the bottom, making it easy to reach with short legs or when you sit far back on the bench.

The pin that fixes the arm to the bridge is usually positioned near the top of the arm, where it provides a great deal of leverage with very little foot pressure. Langsner bored three adjustment holes near the upper end of the arm; inserting the pin in the appropriate hole, he can clamp anything from the slim tine of a hay fork to a heavy baulk of wood. It's nice to have that flexibility, but most craftsmen find they rarely move the pin. In addition, Langsner places the pin holes $1\frac{1}{2}$ in. from the arm's front edge, rather than on its centerline. This ensures that the head will flop open automatically when pressure is taken off the treadle. It beats pulling the treadle up with your toe—an annoying problem on horses that are not as well balanced.

Speed is critical at the shaving horse, Langsner explains as he straddles his horse and puts it through its paces. "You can work just about as fast as you can move the stick." Two other features of Langsner's horse make it possible for him to move quickly. Because the head is open on both sides of the pivoting arm, long stock can be flipped around in one rapid motion to reverse grain direction or to work on the opposite end. And a small shelf rabbeted into the end of the bridge provides support for one end of short stock held in place by a breast bib hung from the carver's neck, as shown in the photo at left.

In the center of the rabbeted shelf of both Kohler's and Langsner's horses is a 1-in.-wide slot, a vestige from the days before a bench vise was commonly available. The edge of a round bucket bottom could be shaped while being held upright in the slot. At Colonial Williamsburg, the 18th-century interpretive village in Virginia, coopers shave the bottom, or head, of a cask simply by clamping it in the jaw of the horse and slicing a bevel with a drawknife, as shown on p. **177**. These days, however, even Kohler does this job on one of his homemade, German-style cabinetmaker's benches, with the bottom of the bucket firmly locked in a vise.

Times change. There are precious few old-time practitioners left in the vicinity of the Langsners' farm, where stick-frame houses and aluminum trailers (today's covered wagons) are more common than log cabins. In both Europe and North America, shaving horses and many of the tools of the country trade have largely gone the way of the hand-hewn log building. Yet they are still revered by a small knot of contemporary craftsmen who are vigorously reviving country woodcraft. I've met a few who were lucky enough to have been born into a living tradition, but most, like Drew Langsner, have struggled hard to unearth a nearly extinguished history. One fall weekend I attended a workshop in Dover, Delaware, to watch a couple of other country woodworkers in action. The appeal of country woodworking, I discovered, goes well beyond an intellectual fascination with history. For a lot of people, it just feels good.

It was a chilly November morning, and a covey of men stood in a spitting rain ready to learn how to rive thin slivers of wood into sheep hurdles and wrestle a ponderous white-oak stump into stick chairs. As I turned my collar against the weather and looked around at the soggy crew, I couldn't help wondering at the force that inspired them to spend their weekend woodworking in these conditions. One of them brought a small child along, who had the good sense his father lacked, I thought, to get in out of the rain early in the day.

Working in the lee of an open shed, John Alexander, Jr., instructed the huddled crowd in the joys and rigors of the chairmaker's trade. A Baltimore divorce lawyer by vocation, Alexander splits and shaves elegant ladderback chairs out of green wood for fun. He stands about 5-ft. tall and wears baggy corduroy pants, a wool sweater and a rumpled felt hat when he works wood. Kicking off his demonstration, Alexander explains: "If you split fat, you've got more hewing. And hewing is more work. If you hew fat, that means more drawknifing, and drawknifing is more work. If you drawknife fat, that means more spokeshaving..."

Alexander's message is simple. The modern chairmaker handles all ends of the business himself—from the standing tree to the completed chair. He is his own lumberjack, sawyer, joiner and finisher. The idea is to progress incrementally from coarser to finer tools as the work is reduced in size, getting the most out of each tool in turn.

The quickest tools for bulk stock removal are the ax and the froe, which are used along with wedges to fell a tree and cleave its trunk into quarters, eighths, and eventually rough-shaped slats and sticks. (A froe consists of a long blade, sharpened like an ax along one edge, and attached at a right angle to a straight handle. Used with a levering motion, it is a more accurate tool for splitting than an ax.) When a stick is small enough so that it won't hold still on the ground or a stump, it may be held fast in a brake, as Daniel O'Hagan demonstrates nearby on some poles that he is splitting for sheep hurdles, portable fence sections used for herding and enclosure.

O'Hagan is a gentle man of few words—an unlikely foil to the ebullient Alexander. He has been living and working in the country tradition for a long time, and has been an inspiration to Alexander, Drew Langsner and several of the other more recent revivalists I visited. While most of this new breed of country craftsmen focus their attention on chairs, O'Hagan spends much of his time at his secluded log home in central Pennsylvania making simple rural implements like sheep hurdles and hay forks.

Using a froe and a tripod cleaving brake, Daniel O'Hagan splits long strips of wood to make a sheep hurdle. The wood is wedged between two planks nailed to two legs of the tripod.

Cleaving brakes

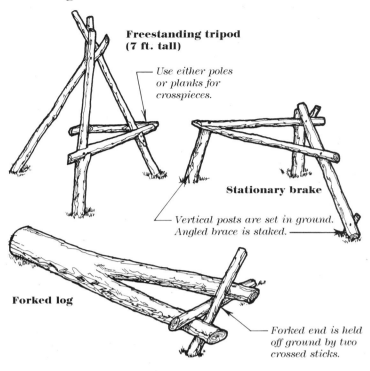

Freestanding tripod (7 ft. tall)

Use either poles or planks for crosspieces.

Stationary brake

Vertical posts are set in ground. Angled brace is staked.

Forked log

Forked end is held off ground by two crossed sticks.

Shaving brakes

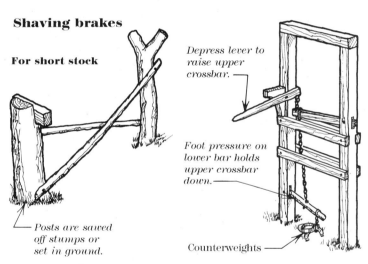

For short stock

Depress lever to raise upper crossbar.

Foot pressure on lower bar holds upper crossbar down.

Counterweights

Posts are sawed off stumps or set in ground.

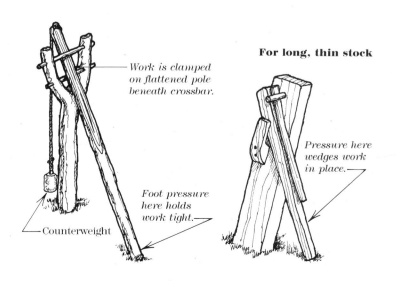

Work is clamped on flattened pole beneath crossbar.

For long, thin stock

Pressure here wedges work in place.

Foot pressure here holds work tight.

Counterweight

The brake is one of the most neglected tools of the country craftsman, but it is more basic, even, than the shaving horse. A brake is used to hold work while it is split or shaved, often in preparation for being finished on the shaving horse. At its most rudimentary, the brake is an informal configuration of interlocking poles into which work is wedged fast. Like the shaving horse, there are a host of variations on the same theme, which may be as casual as a forked log supported by two crossed sticks, or as contrived as a work-clamping guillotine.

O'Hagan uses two different kinds of brakes to make his hurdles—a cleaving brake and a shaving brake. The cleaving brake, as the name suggests, is used to hold wood while it is split with the froe. The stock is slipped between two boards nailed at a slight angle to one another on opposite legs of a large pole tripod. The spacing between the boards permits work of different sizes to be wedged securely. A few of the more common cleaving brakes are shown at left. Cleaving brakes tend to be larger, sometimes quite a bit larger, than shaving brakes, to accommodate heavy timbers. They may rest close to the ground, held in position by their own weight, like the forked log shown in the drawing. Or they may stand as tall as O'Hagan's large tripod. When the sticks have been split to a manageable dimension on a cleaving brake, they may be further reduced using a hewing hatchet; a stump supports the work.

When it comes to splitting, Alexander counsels the audience to follow the grain and allow the nature of the wood to 'reveal itself.' Take chances, he says, by trying to split as close as possible to the finished dimension. Some chairmakers split the wood so close to the mark that they are able to bypass the hewing hatchet, moving directly from the ax and froe to the drawknife and spokeshave for finishing. (The spokeshave is really a smaller, more controllable descendant of the drawknife and a cousin to the plane.) If you succeed, you'll save yourself a pile of work later on. If you fail, and the grain runs out or you hit a knot, you will have gained some useful experience along with some pegs, wedges and kindling.

The shaving brake holds split stock while it is shaped with a drawknife or spokeshave, or drilled and mortised with a twibill—a nearly extinct, double-bladed mortising tool. O'Hagan's shaving brake is a simple affair with two upright posts set into the ground on opposite sides of a board fence, as shown in the photo at left on the facing page. The rough-split stock is slipped below a horizontal board in the fence, placed on top of the post behind the fence and then pulled up and sprung into place on the post in front. The rear post is only a few inches behind the fence, while the front post is a couple of feet away, allowing plenty of room to work the stock. The tops of the posts must be slightly higher than the bottom of the fence rail so the work will be sprung in place with sufficient tension. There are all kinds of shaving brakes—from the pedestrian to the complex—and a few of them are shown at left.

The poles that comprise most shaving brakes are only a few inches in diameter and are often driven into the ground for stability. Shaving brakes are positioned to enable the craftsman to exert pressure on whatever tool he is using—the drawknife, brace or twibill. Outside of holding the work in your hand, the shaving brake is about the quickest way to secure work for rapid shaving. It's also worth noting that the twibill is perfectly suited for use on a shaving brake. The twibill's two blades are used to clean out the wood in a mortise after each end of the slot first has been defined with a bored hole. One knife-like blade is used to slice the edges of the mortise and the other hook-like blade is used to pry out the waste between the holes. In practice, the twibill is pushed or levered, rather than swung. Its obvious alternative—a chisel and mallet—has an immediate disadvantage. The vibration caused by the chopping action would quickly bounce the split stock off the top of the upright post. The steady, downward pressure on a twibill helps hold the stock in place.

It is tempting to write off the brake as a mere rustic ancestor of the shaving horse, which is partly true. In fact, the shaving horse might even be considered a brake with a seat. The photograph on p. 164 depicts what appears to be a transitional form of shaving horse; there, the worker is seated on a separate stump, and the dumbhead is activated by foot pressure. But I think the principle of the brake is much more widely applied than most of us appreciate. Almost 15 years ago, when I learned to build snowshoes in Ontario, I was taught to limber a stick before bending by wedging it between the extended log ends of a nearby cabin. Although I didn't know it at the time, that was the first woodworking brake I intentionally used. I suspect that many of us use them regularly without thinking.

'Appropriate technology' is one piece of modern jargon that could be applied to country woodworkers of every culture and in every age. It means using the right tool for the job. With so much of their lives consumed by the business of living, country craftsmen, perhaps more than others, have had to be truly efficient. Chairmaking is a case in point. We all need chairs, and yet good ones are notoriously elusive. It's easy enough to make a table that will hold dishes for a meal, but not so simple to build a chair that will comfortably hold a diner long enough to enjoy that meal. For this reason, many modern furnituremakers—even those with a shop full of machinery—never attempt to make a chair. But well before the Industrial Revolution, country chairmakers evolved a sophisticated system of joinery that lends itself to production techniques. Traditional chairs like the Windsor and ladderback consist of multiple legs and stretchers, with rails, back slats and spindles that must be made more or less alike. Whether you're making dozens of chairs or only a couple at a time, there are considerable opportunities for streamlining the process by ganging similar tasks.

Until recently, one of the best examples of chairmaking technology could be found among the chairmakers of the High Wycombe district in England. These craftsmen were divided into three distinct categories—bodgers, benchmen, and framers. Of the three, only the bodgers took their tools to the woods, setting up huts and camps in which they split, rived and turned mountains of legs and stretchers to feed the Windsor chair industry. The benchmen and framers worked in town shops, with mostly sawn stock, and were responsible for bottoming (making the seat), bending, assembling and finishing.

The system of busting a tree out on a brake, shaving the parts on the horse and turning them on the lathe is as satisfying as it is elementary—perhaps because it's so sensible. "It's enjoyable," one chairmaker told me, "to see the process from log to rows and rows of chair parts. They've hardly gotten used to being without their bark before they find themselves part of a chair."

To use this simple brake, work is placed on top of the rear post, slipped under the horizontal board and sprung in place on top of the front post. Here, Daniel O'Hagan uses a twibill to clean a mortise.

"Das ist ein schnitzelbank," John Alexander sings, as he shaves away at his English-style bodger's bench.

A modern-day chair bodger, John Alexander has borrowed heavily from old-world tradition. For almost 25 years, when not otherwise occupied with practicing law, Alexander has dedicated his considerable energies to building chairs of green-split wood. Because that tradition has been all but obscured by this continent's 200-year-old love affair with the sawn board, he has found himself embarked on a path of re-invention. With great tenacity, Alexander has dissected chairs and chairmaking, only to discover that the strongest and most efficient techniques are almost always the old ones. (A preliminary report of his conclusions, written almost a decade ago, can be found in his book *Make a Chair from a Tree*.)

Accordingly, when he sought a shaving horse for chairmaking, Alexander patterned the one he built after the so-called bodger's bench illustrated in several English country craft books. The bodger's bench is best suited for the repetitive tasks of the rural chairmaker, who must perform a handful of operations on a large number of sticks of uniform size. Its distinguishing feature is the pivoting yoke, which is used instead of a dumbhead to clamp work to the bridge. Two vertical arms of the yoke flank the bench beam and are joined with three horizontal crossmembers: at the top of Alexander's is a rotating crossbar, at the bottom is a fixed foot bar, and at the middle a wooden screw or metal bolt provides a pivot point (see drawing at top of facing page). The size of the jaw opening is controlled by raising or lowering the adjustable bridge (which is nailed to a pivoting tiller), instead of the yoke itself. Compared with the conventional dumbhead shaving horse, this arrangement has both advantages and limitations.

The rotating crossbar is perhaps the greatest asset of Alexander's bodger's bench. The square bar pivots on its friction-fit, $1\frac{1}{8}$-in.-dia. tenons to seat itself flat on the workpiece. It can provide a different surface on each face, depending upon the requirements of the work. One face may be notched to hold squared sticks with their corner up, another may be flat for slats, while others may have a rubber surface to grip and protect finished work, or even a broken Surform rasp for additional friction. Because the crossbar is held at both ends and is open in the middle, there is plenty of room to center a wide back slat on the bridge—difficult to accomplish on a dumbhead horse.

The jaw opening can be adjusted readily by changing the position of the wooden pin in the series of holes that riddle the tiller. The closer the fit between the crossbar and the bridge, the less you'll have to push the foot bar to close the jaw and the better it will grip. At the bottom of the facing page is a clever variation of Alexander's tiller and bridge, designed by chairmaker Brian Boggs of Berea, Kentucky. Boggs' horse employs a ratchet-adjusted support arm.

Many of the disadvantages of the bodger's bench also relate to the crossbar and yoke. When you encounter a grain reversal in a stick of wood and need to turn it around to shave from the opposite direction, you must pull the entire piece out to turn it. The same is true when you've worked the piece as far as possible and must turn it end-for-end to get at the rest of the stick. On a dumbhead, you can conveniently slip the work out the side of the jaw and flip it around. As Alexander points out, this does not present a serious problem with the chairmaker's normal run of short, straight-grained rungs, but becomes more irritating with longer or less regular stock. In addition, the working surface of Alexander's bridge is only about 2 ft. long. Longer work (back posts, fork handles, hurdle strips, snow-shoes, etc.) would run into the beam of the bench unless it were redesigned with the pivot end of the bridge nearer to the end of the beam. The round foot bar at the bottom of the yoke requires that you sit close to the yoke or stretch farther to operate it, also making it difficult to work long stock.

Although Alexander finds the grip of the yoke more than sufficient for his needs, it offers much less of a mechanical advantage than the typical dumbhead. The pivot-pin placement near the center of the arms provides only a 1:1 relationship, as opposed to the 4:1 advantage of Langsner's horse. And the flat face of the crossbar does not hold work as securely as the corner of the dumbhead's jaw, which digs in for a good grip. (This is a mixed blessing if you're shaving softwood on a hardwood horse, and some dumbheads require a padded jaw to protect the work.)

Alexander's bench has other appealing features to balance these drawbacks. Its double-beam construction, which resembles a traditional lathe bed, is strong and easy to build with a minimum of joinery. When he built it, Alexander didn't have the heavy timber to make a solid bench, but the flexible design that evolved works to his advantage. "I always like to make my shop furniture adjustable in as many ways as possible," Alexander says. So he joined parts with $\frac{7}{8}$-in.-dia. wooden screws and metal bolts instead of wedged or glued pegs and tenons. Parts can be replaced if they wear out, or if their design is to be altered. The whole rig is flexible in operation. The seat can be slid forward or back along the beam, or removed entirely. The yoke can be removed, allowing the tiller to be dropped down for a completely flat worksurface.

In addition, the bench's three-legged design is much more stable on uneven ground—the typical 'shop floor' of the woodland bodger—than four legs. While Alexander's is not the lightest shaving horse I've seen, that hasn't stopped him from lugging it into the woods to gather material for a batch of chairs. If your bench is going to spend a lot of time outdoors and you decide to use wooden screws to assemble it, it's a good idea to drill a hole across the head of each screw so that you can insert a stick to gain extra purchase and leverage. Exposure to the abusive routine of alternating sunshine, rain and shop heat will cause them to swell and shrink dramatically—and to eventually seize up. Alexander also uses tallow, which he renders out of lamb fat, to both preserve and lubricate the threads.

At his suggestion, I straddle the horse and begin shaving a couple of sticks from their rough-hewn state to vaguely square, and even more vaguely round cross section. Drawknifing provides ample opportunity to reflect on our compulsion to turn every irregular, textured surface into a uniform cube or cylinder. Struggling with a piece of hickory that keeps slipping out of the yoke, I am about to write it off to mindless obsession, when Alexander encourages me to apply more pressure.

He takes over to demonstrate. "I play when I shave," he says, depositing a carpet of drawknifed shavings around his feet. "Now and then, when you're going too fast and your knife is too dull, you pull one of these sticks into your stomach...it hurts." Flipping the stick, end-for-end and alternating faces, Alexander works it down evenly, sometimes pausing to eyeball along its length for irregularities. "Halfway from 'thunk' to 'plink,'" he says, snapping the stick on the horse to produce a musical tone. This one's still too thick. "Now I'm starting to get into it," he says. Shavings are flying. "*Das ist ein schnitzelbank*," he sings in accompaniment to the last few '*schnitzes*' on his rung. "Note that I'm using only one foot," he says.

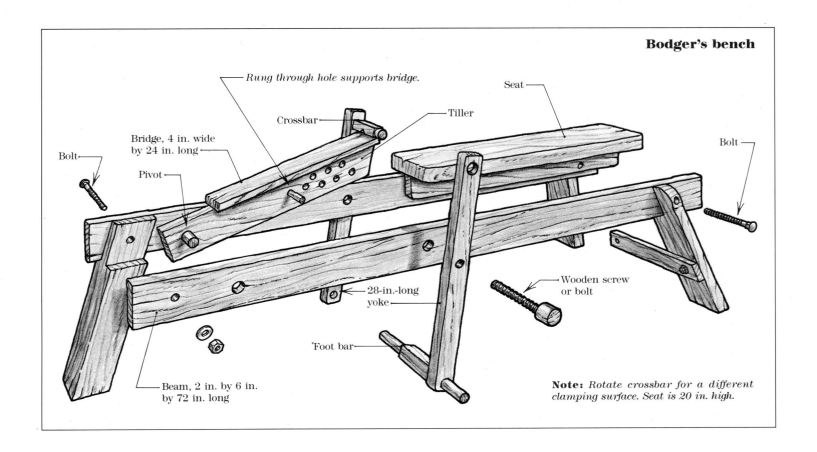

Rung through hole supports bridge.

Crossbar

Tiller

Seat

Bridge, 4 in. wide
by 24 in. long

Bolt

Bolt

Pivot

28-in.-long
yoke

Wooden screw
or bolt

Foot bar

Beam, 2 in. by 6 in.
by 72 in. long

Note: *Rotate crossbar for a different clamping surface. Seat is 20 in. high.*

A ratcheted bridge

Brian Boggs devised this ratchet-adjusted bridge for speed and fine adjustment. These make a big difference to Boggs, of Berea, Kentucky, who spends six hours a day on the horse, shaving the parts for about 100 ladderback chairs each year. The bridge hinges at the far end of the horse and its height is readily adjusted by moving the ratcheted arm that passes through the mortise in the bench. To raise the bridge, lift it up. To lower it, a quick flip on the hardwood key in the seat disengages the ratchet teeth. Let go of the key, and it catches.

Boggs also gets the most out of his yoke by making more frequent adjustments in the bridge's height. (The best grip is obtained—and the least pressure is required—when the yoke is vertical.) Two leather-padded wood blocks on top of the bridge, spaced $\frac{1}{4}$ in. apart, keep round rungs from rolling around when he's shaving their sides. "My first horse was just like Alexander's," Boggs told me, "until I got calluses on my *derriere*." This one has a solid seat, covered with a leather pad.

Ratcheted bodger's bench

Clamp rounded chair parts between padded blocks.

Crossbar

Bridge

Ratchet arm

Key

Pivot

Bench

Yoke

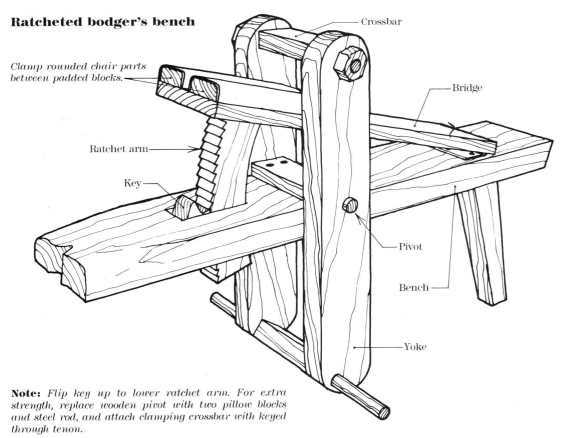

Note: *Flip key up to lower ratchet arm. For extra strength, replace wooden pivot with two pillow blocks and steel rod, and attach clamping crossbar with keyed through tenon.*

In addition to his bodger's bench, Alexander also uses two other low benches for a variety of joint-cutting and assembly operations. The first, shown at left, employs a simple three-peg and wedge holding system. Alexander first discovered it in a contemporary book on rural Appalachian crafts and was delighted to find an almost identical example (shown at the top of the facing page) several years later in M. Hulot's *L'art du tourneur mécanicien*, published in France in 1775. Alexander fitted the legs to the bench with tapered, rectangular tenons, mortised through the seat. Round tenons would work as well if you have access to a conical reamer to taper the holes. The tapered tenons seat themselves more firmly as the bench is used. They can be repeatedly trimmed flush with the top, if necessary, or wedged if they become loose.

Alexander's bench is made out of a heavy slab of white oak, the same height as the seat of his bodger's bench. (If you have room for only one bench, you could incorporate the features of this low bench in the design of the shaving horse itself. If you build separate benches, however, it's a good idea to make them the same height so they can double as sawhorses.) As Alexander says in his book, "Building the bench is a good introduction to wet woodworking in general and to post-and-slab construction in particular. You will learn how to hew, chisel and bore wet wood. Almost no mistake is fatal with this bench. The harder you pound on it, the tighter it will become."

To secure work to the bench for boring or mortising, Alexander places the sticks to be worked between three rectangular pegs driven into round holes drilled in the benchtop. A tapered wedge, knocked between a stick and the single peg, locks them all in place, making it easy to cut matching joints in adjacent legs. Chopping a mortise, Alexander alternately taps the wedge tighter with his hammer and chisels deeper to dig out the slot. He pauses for a moment to explain that the wedge should be placed so that it points in the primary direction in which you are chiseling. This way, it will seat itself more firmly as you chop. A variety of hole spacings in the top of the bench will allow you to clamp one or two chair posts at a time. "The simplicity of this holding system was hard to accept," Alexander says, "until I tried it. It was the last method I tried."

His other low bench, shown in the photos on the facing page, clamps work between two adjustable poppets. Cut nails are driven into the poppets, heated cherry red and bent 90°. Their heads are ground off and sharpened so that work can be chucked between them, as between centers on a lathe, and spun around and shaved on all sides. Although a chair leg can be shaved entirely in this manner, you can't sight along the stick or flip it end-for-end as easily as on a shaving horse. Alexander uses it primarily to trim tenons on the ends of the dried rungs after they've been shaved on the horse. The bench base is a variant on the classic *H*-frame construction, borrowed from the Windsor chair and built to knock down. The end stretchers are cut long enough to spring the legs apart; the tension holds tapered leg tenons securely in their mortises.

Other variations of these workholding devices, shown in the drawing on the facing page, offer even more flexible solutions to the problem of how to hold green wood. The pipe-clamp poppets and the holdfast yoke can be made quickly and attached to any worksurface or removed when they are not needed. The benchdog poppets are used by the wheelwrights at Colonial Williamsburg to hold spokes for shaving, and were designed to operate in the dogholes and tail vise of a standard workbench.

Alexander's low bench holds a chair leg securely between the three pegs with a single long wedge. It's easy to maintain pressure against the spoon bit by straddling the 20-in.-high bench and leaning over the work.

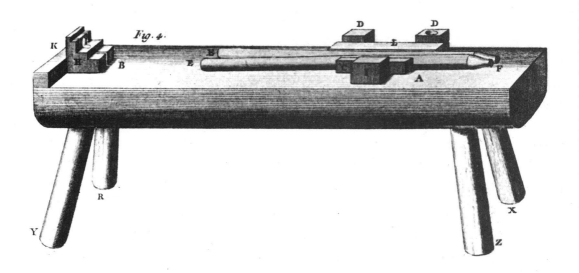

This engraving, from M. Hulot's L'art du tourneur mécanicien (1775), shows the shaving horse/assembly bench used to rough-shape stock in a woodturner's shop. At the end of the bench (about 12 in. wide by 5 ft. long by 16 in. high), a small stepped and notched block supports one end of the work being shaved. A worker holds the stock against the block with pressure from an oak 'bellyguard.' He may sit on the horse or stand to work long stock. At the right end of the bench are the pegs and wedge used to hold chair parts for drilling or notching. Hulot's three square pegs, or bench stops, are made of ash, tenoned into the bench. A single wedge, placed on the side with the single stop, is all that's required. A shallow hole on top of stop (D) holds tallow, which is used to cool the bit of the brace when it becomes hot.

Alexander uses this low bench mainly to trim the tenons on the ends of shaved chair rungs. A bent and sharpened nail at the top of each poppet holds the rung in place. The poppets are adjusted with a wooden screw.

Holdfasts and poppets

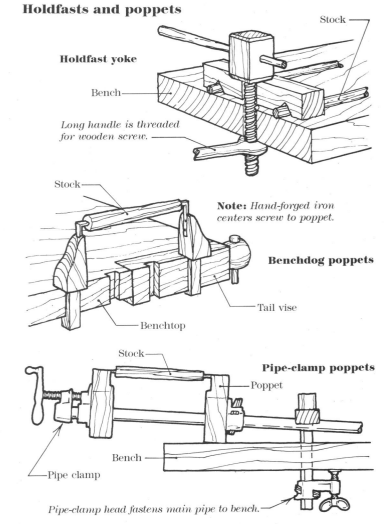

Holdfast yoke

Bench

Long handle is threaded for wooden screw.

Stock

Stock

Note: *Hand-forged iron centers screw to poppet.*

Benchdog poppets

Tail vise

Benchtop

Stock

Pipe-clamp poppets

Poppet

Bench

Pipe clamp

Pipe-clamp head fastens main pipe to bench.

John Alexander is an amateur chairmaker in the truest and best sense of the word. He makes chairs for love, not money. Being a lawyer buys him the time to be a more thoughtful and a more playful, if not a better, chairmaker than the rigors of production would allow. "Not only do I make the thinnest slats," he says, "I also make the thinnest bark seats. That's one reason why I could never go into production." After visiting several other country chairmakers and watching them in action, I found that most prefer the dumbhead horse over the bodger's bench. But in every case, they have adapted the form to suit themselves.

I also discovered in my travels that not all traditional chairmakers work on shaving horses. Some, like Michael Dunbar of Portsmouth, New Hampshire, prefer not to. "I'm like most other people," Dunbar told me. "I've got a shortage of space, and shave horses take up a lot of space." Dunbar prefers to build his Windsor chairs on a traditional workbench (see p. 15). He turns the legs on a lathe and uses a bench-mounted machinist's vise to shave the spindles. "When I'm standing up I can really whale on a piece and I get a much longer reach," he claims. "An awful lot of what I do is hastened by the fact that I can put my whole torso into it." Among chairmakers, Dunbar's speed is legendary.

Having spent most of my time examining shaving horses used by chairmakers, I never fully appreciated the need for a tight horse until I watched Jim Pettengell at work. A full-time cooper, Pettengell regularly shaves bone-dry, white-oak barrel staves with a drawknife and hollow knife (the curved drawknife used to shape the inside of each stave). That's a whole different story from the easy slicing cuts of the greenstick chairmaker. "You can get away with a wobbly horse on green wood," Pettengell explains, "but there's too much energy being dissipated for dry oak. Dry, white oak is tough stuff."

Coopering in general is not something to be embarked upon lightly. As Kenneth Kilby says in his book *The Cooper and His Trade*, "There are no amateur barrelmakers. Coopering is a skill acquired through years of sweating, muscle-aching, back-breaking labour...." Or as one cooper told me, "It's not every jack leg who can put twelve pieces together and have them hold water." If there are no amateur coopers, there surely can't be many professionals.

Jim Pettengell belongs to an endangered species of professionally trained coopers. He learned the trade in England in a traditional apprenticeship as a piecework cooper for Whitbread & Company, where he built casks for Whitbread's ale and Johnny Walker scotch. What did eight years at Whitbread's do

Master cooper Jim Pettengell, of Colonial Williamsburg, works his way through a pile of white-oak barrel staves at this reproduction 17-century shaving horse. The square bridge of the horse allows him to exert pressure on the work from above.

Pettengell saves time by leaning over the head to shave the far end of a stave (at left, facing page), instead of reclamping it. The deep notch just below the back of the head is a sure sign he's been doing it for years. He clamps the head of a small pail in the horse to shave the bevel (at center, facing page). The front of the bridge is worn round from years of 'heading piggins.' The block hook (at right, facing page) is the traditional shaving brake of the piecework cooper. Wedging a stave under the edge of the hook and against his stomach, the cooper can work more quickly and with greater leverage than he can when seated at the horse.

for Pettengell? He points to a 63-gal. barrel sitting in the corner and explains that he learned to crank one out in five or six hours—"not long at all," he confesses modestly. But piecework is demanding in any business, and the life of a brewery cooper is no exception. Like the modern professional athelete, the piecework cooper's productive lifetime is short. "The oldest guy I knew was in his sixties," Pettengell told me. And he was rare. "The rest were thirty-five or forty, and looking around for something else."

There are many operations performed by the traditional cooper that require a number of specialized tools, but the shaving horse is used mainly for shaping the outsides and insides of the staves ('backing' and 'hollowing') before they can be 'raised up' into casks or buckets. It also is commonly used to hold the heads while they are shaved for a tight fit. At the coopers' shop in Colonial Williamsburg, Virginia, where Pettengell now interprets coopering, there are four shaving horses and no two are alike. Although cooper's horses may be generally tighter than the green-stick chairmaker's horses, they are also built out of available materials and are every bit as idiosyncratic. At Whitbread's, Pettengell was one of 15 or 20 coopers in the shop, all of them trading time on the same two or three horses. Under such circumstances, a cooper might never make his own shaving horse. He simply would adapt to whatever was there. Pettengell figures the ideal horse is "what you get used to."

Pettengell's preferred shaving horse was built at Williamsburg about 20 years ago. It is a reproduction of an early seagoing dumbhead horse, which resides in the Mariner's Museum in nearby Newport News, Virginia. For centuries, military and merchant fleets all over the world depended upon wooden casks for storing everything from whale oil to fresh water. Coopers were required regularly to repair casks or dismantle them into bundles of staves, or 'shooks,' for easier storage. The cooper was a fixture on board ships of the Royal Navy as recently as the First World War.

Perhaps the most striking thing about Pettengell's horse is its square bridge. This low, flat surface is an asset for excavating chunks of dry oak. By not having to pull the knife up to his chest, he maintains leverage and control throughout each stroke. The square bridge also makes it easier for him to shave stock on the other side of the head, as shown at left below. Rather than taking the time to reposition the work in the jaw, Pettengell simply leans forward over the head to reach the rest of the stave. Fifteen feet away, however, another Williamsburg cooper, Lew LeCompte, is perfectly happy using a dumbhead with the more common angled bridge.

For a better grip on slippery oak staves, the edge of the jaw on Pettengell's dumbhead is bolstered with a thin, screwed-on metal strip. (Since this dents the staves of cedar casks, however, it is padded with a piece of scrap if softwood is being worked.) The arm and head are balanced so that it opens automatically when he releases pressure on the foot, just enough to accept another stave.

All the horses used by the Williamsburg coopers have a shallow shelf at the end of the bridge like the one Drew Langsner found on Reudi Kohler's horse. Pettengell and LeCompte agree, however, that this is probably a relatively recent accommodation for inferior stock. Given the proper material, they argue, there would be no grain reversal and no need to hold short sticks on the shelf. There is no slot in the middle of Pettengell's shelf, as there is on Kohler's and Langsner's horses. Instead, Pettengell simply grips the head in the jaw of the horse to shave the bevel, as shown at center below.

Like chairmaking, coopering is a production craft. A large number of similar parts must be split and then shaved at the same time. Also like a chairmaker, the cooper uses a form of shaving brake, called the 'block hook' to hold work for rapid hollowing. (It is also referred to as a 'knee block' for the additional support provided by the cooper's knee.) To use the block hook, one end of a stave is hooked beneath the serrated lip of the hook, as shown at right below, leaving the full length of the work unobstructed. It is much quicker in operation than the shaving horse, and much more tiring because of the greater demands placed on the cooper's stomach, legs and back. According to Pettengell, a young pieceworker is more likely to use the block. But once he's past his prime and is a little more interested in retiring than setting records, he will content himself with the shaving horse.

The weight and shape of the dumbhead on this old horse, owned by Bill Phillips of New Tripoli, Pennsylvania, allows it to fall open automatically.

Whether built by a chairmaker, a cooper or some other breed of country craftsman, all the shaving horses I've seen have one thing in common: a rich diversity. Some are planed smooth and square like Drew Langsner's horse, others are full of bark and twist and have been hewn just enough to clear a splinter-free seat. Some are meticulously joined with glue and tapered tenons, while others are scabbed together with nails and baling wire.

Usually, but not always, such variety is grounded in function. The back of the dumbhead on Lew LeCompte's horse at Williamsburg, for example, is curved, which provides a handy bending form for slim hoops. The horse with the keyed trunnion at right was probably designed to combine the stability of four legs with the flexibility of a three-point stance on uneven ground.

In the case of Bill Phillips's pointed dumbhead above and the unorthodox bridge on the Shaker horse at the top of the facing page, however, the function is more obscure.

Every now and then, as in the old German horse at the bottom of the facing page (a part of the Greber photo collection at the Winterthur Museum), the maker's purpose is abundantly clear, though unrelated to function. When you spend all day at the shaving horse, he must have reasoned, it might as well be delightful.

Keyed trunnion

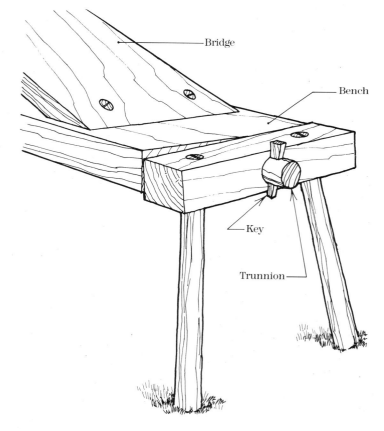

Note: *Leg assembly pivots on trunnion for a solid footing on uneven ground.*

The purpose of a shaving-horse design may be obvious or obscure. In the horse at left, from Hancock Shaker Village, the bridge is joined to the bench by a curious barrel-shaped construction. It appears to be adjustable, but is not. The business end of the bridge has been extended with an extra board, which provides better support for thin stock. The rear leg assembly is attached to the seat with a bolt and wing nut. By contrast, the design of the animated German shaving horse, below, speaks for itself.

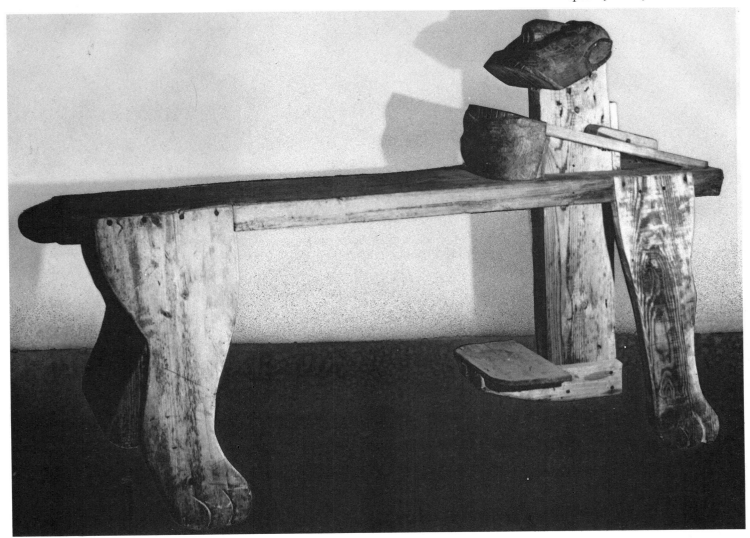

Boatbuilder Alfred Sovik of Lysekloster, Norway, hews a garboard strake on two low horses. Photo by Ted Spiegel.

Boatbuilding

I t's been said that a housebuilder works to the nearest $\frac{1}{8}$ in., a furnituremaker to the nearest $\frac{1}{64}$ in. and a boatbuilder to the nearest boat. While not altogether serious, these comments reveal a fundamental fact about boatbuilding. Most boatwork is conducted right on the boat—laying the keel, hanging the garboard, spiling and planking the hull, and so on. In such a situation, precise measurements have only the barest utility. As John Gardner told me, "All boatbuilding is more or less approximation. The final arbiter is the eye—if it looks right, it is right."

Gardner, who is Associate Curator of Small Craft at Mystic Seaport in Connecticut, has been building, teaching and writing about boats for more than 40 years. He explained that a boat has to be both tight and hydrodynamically efficient, or what boatbuilders call 'fair.' It is much more important that adjacent parts fit tightly for a waterproof seam than that they replicate whatever numerical configuration may have been assigned by a naval architect or designer. The success with which the builder's hand shapes the joint and his eye judges the lines is what spells the difference between a 'hole in the water lined with wood' and a truly successful watercraft.

This chapter is about the benches and workholding devices common to builders of wooden boats. There are, of course, lots of other kinds of boats—steel, fiberglass, ferro-cement, etc. I have focused on wooden craft partly because this is a book about woodworking, but also because wooden boats are at the heart of the tradition. From the earliest fishing boat to the most recent America's Cup challenger, the qualities that make a boat float and move haven't changed all that much.

These qualities can best be assessed in three dimensions, which is why, before cutting the first plank, boatbuilders often shape small half-hull models. They work out the lines on the model, sometimes scaling from it to make full-size drawings—a process called lofting. I also believe that, like the luthier, the boatbuilder thrives on the most tangible, tactile contact with his craft. Both build objects that are at once sculptural and utterly utilitarian in ways that a chair or table could never be. A boat that leaks or won't cut water efficiently won't be a boat for long—just as a guitar that won't play the right notes is not really a musical instrument.

It is not surprising that, given their need for flexibility, boatbuilders have little use for the classic cabinetmaker's bench. The cabinetmaker's bench, essentially a jig for making boards flat, square and regular, is ideal for working relatively short stock that needs to be hand-planed to thickness or fitted with intricate joinery. By contrast, the boatbuilder works long, spindly planks, which are often curved or tapered and are joined simply along their length with wooden or metal fasteners and butted or scarfed at the ends. Most boatwork (save perhaps some cabin fitments and other occasional small-scale work) is neither flat nor square. Regularity is judged by eye and is most effectively achieved on the boat itself. For these tasks, the conventional workbench would be more hindrance than help.

This is not to say that boatbuilders do not use workbenches—only that the forms they have developed are uniquely suited to their work and tools. By and large, these are small, flexible arrangements, built of whatever materials happen to be at hand and heavily dependent upon C-clamps, wooden hand screws or pegs and wedges for their workholding apparatus. The cabinetmaker's bench stands at the center of his universe—all stock is planed, chiseled and sawn while being held in its vises, and work is often assembled, glued and finished on its top. In the boatbuilder's shop, the boat commands the center, with workbenches arranged around it.

Along with the variation among boat designs and the concommitant flexibility of the working method, tradition and speed play a pivotal role in the way the boatbuilder views his work and his bench. Until relatively recently, the watercraft of most cultures were built by farmers or fishermen, not specialists. These boatbuilders-of-necessity were not a literate people. Their traditional skills were passed along by example, without written record, and as British naval historian Basil Greenhill points out in his book *Archaeology of the Boat,* they were "therefore strongly protected against hasty innovation." Unlike the workbenches that evolved in the Guild cabinet shops of medieval Europe, part-time boatbuilders would have pressed into workbench service whatever they happened to have or could quickly cobble together to do the job. They were not cabinetmakers, so there was no instinct or professional responsibility compelling them to build cabinet-like workbenches. A boat's graceful form has always been its most highly prized decoration.

Underlying this pragmatic tradition is the requirement for speed—the ability to make a boat quickly and efficiently. Although pleasure craft have surely existed for as long as there has been a privileged class, until recently the overwhelming majority of boats have been built to work—either to fish, or to fight, to trade or to colonize (and often to engage in several of these activities at once). While the furnituremaker's client could always wait another week for that bed or desk, the ascendancy or the very survival of the boatbuilder's own culture depended upon the rapid completion of his efforts. By relying on hand and eye and only minimally on measurements, by maintaining a supple working method and a strictly functional workshop, the boatbuilder was also able to make speed.

During our discussion, John Gardner pointed out with a waggish grin, "What goes on in boatshops today is a far cry from what went on in boatshops when boats were being built." Although Gardner intentionally overstated his case, the point is well taken. More than 1,000 years of tradition threaten to be lost in the several generations since the advent of power—both in the motive force of the vessels built and, perhaps more fundamentally, in the technology used to construct them. The earliest boats were built of green timber, split and hewn to shape. Modern wooden boats are more often built of air-dried wood, milled into boards and either bent dry into position or made supple through steaming. The preparation of stock with today's machinery requires willing muscle, rather than skilled effort applied to traditional hand tools like the ax and adze, which are used on the material where it lies. Likewise, traditional workbench requirements differ significantly from those of the modern boatbuilder. Nowhere is this more evident than in the Scandinavian boatbuilding tradition, firmly entrenched since the days of the Viking longships.

Paul Schweiss, a Washington State boatbuilder, apprenticed in the town of Bjorkedalen, about halfway up the Norwegian coast near Aalesund. In the small farming valley of about 200 people where Schweiss lived for a year, practically every farm had a boatshop. During the winter months, the farmers moved indoors to build boats for the fishermen down the coast—much the same as their Norse ancestors had done many centuries before. In earlier times (before the advent of large urban centers), the work of these farmer/boatbuilders was supplemented by itinerant boatbuilders, the original 'journeymen,' who traveled the countryside with their tools on their backs. The farmers provided a boat shed (a simple post-and-beam shelter) and building wood. In addition, it was the farmer's responsibility to have on hand a couple of planking benches, the Norwegian boatbuilder's principal workbench.

As the German, Scandinavian and English cabinetmaker's workbench evolved to facilitate the use of the hand plane, and the shaving horse to accommodate the drawknife and spokeshave, the Norwegian planking bench served as an adjunct to the hewing ax, the principal boatbuilding tool of antiquity. The requirements of the clinker, or lapstrake, construction favored by Norse builders are rather specific. Planks, or strakes, are long and curved and must be hand-fitted along their edges to their neighbors. At the ends they are fitted to heavier timbers at the stem and stern, and in the bottom of the boat to the

Paul Schweiss and master builder Dave Foster fit the garboard of a Norwegian Sognabat at the Apprenticeshop of the Maine Maritime Museum. Photo by Steve McAllister.

keel. To keep from having to run around the workshop between the boat and the bench, builders did much of the work—the scribing and fitting—right on the hull. In the Bayeux Tapestry, pictured below, 11th-century craftsmen are engaged in this characteristic approach to the work. Where a workbench is required, the most efficient method is to pull the bench up to the boat and to move it around as the building progresses.

The planking bench that evolved is a simple workholding device that has remained essentially unchanged for at least a thousand years. In its form it is closely related to the benches used by early Roman woodworkers. The bench consists of a long plank and four legs attached to the plank with through-wedged tenons. The builder's own body is the clamp or vise; he sits or kneels in various positions on the stock to hold it while he works. When necessary, he can easily use the bench itself as a step stool to secure shores to the ceiling of the shop, or to sit on while fitting or riveting planks.

Schweiss' own bench, built when he returned to his home on the coast of Washington, is made from a 10-ft.-long piece of Douglas fir 2x12. From a distance, it could be a long picnic bench removed from the local park. But close up, it looks a well-worn veteran of the workplace. Schweiss speaks wistfully about the benches he saw in the old Norwegian boatshops—benches made of pine, finished with a linseed-oil bath, and permeated with all the good smells and the tarry patina that come from years in the trade.

Kneeling on one knee on top of a plank, Schweiss can comfortably hew the edge to a scribed line (or perhaps saw it first with a frame saw). With the hewn strake still overhanging the bench, he can then finish-plane the edge by kneeling on the plank, turning the plane on edge and running it down the length of the board. This looks uncomfortable to me, but Schweiss says "you just get sore in a different part of your back."

Standing the plank on edge and straddling it lightly between his legs, Schweiss can also hew a scarf joint on the end. Because his eye is directly over the board, Schweiss explains, he can easily see the lines on both sides of the joint. The plank can be laid flat on the bench (and sat upon by the builder) while it is planed smooth.

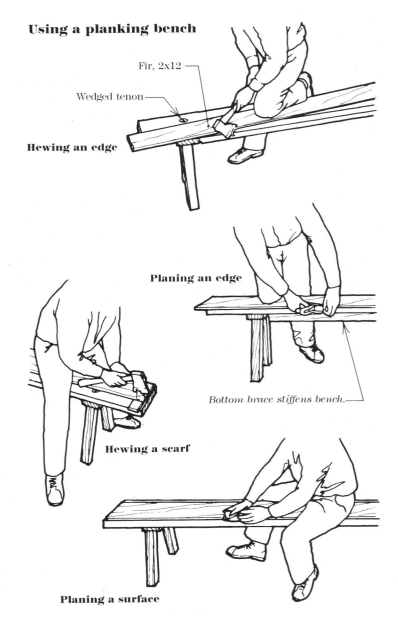

Using a planking bench

Fir, 2x12

Wedged tenon

Hewing an edge

Planing an edge

Bottom brace stiffens bench.

Hewing a scarf

Planing a surface

The Bayeux Tapestry is a remarkable record of the events leading up to the Norman conquest of England. It begins in 1064 with the departure of Harold from England to Normandy, and ends two years later with his death at the Battle of Hastings. In this scene, the fleet is readied for the invasion. Carpenters work directly on the assembled boats with hewing ax, adze and brace.

The bench is typically knee-height, about 22 in. to 24 in. high, to allow the builder to kneel comfortably, or to steady a thin strake on edge between the buttocks—even a 12-in.-wide plank could be straddled like this. But the most important feature is the builder's body. "It's all done with body weight," says Schweiss. "No clamps or vises at all." This obviously allows for quick work and a maximum of flexibility, as the builder can effectively scoot along a spindly plank, hewing, planing or sawing to a line.

If a plank is really big, the builder can prop it up on a hewing horse. Schweiss's hewing horse is 12 in. by 12 in., about 36 in. long, made from a salvaged burnt timber. But the actual dimensions are unimportant. Much more critical are the height and mass—it stands about knee-height on a trestle base and is as heavy as the large chunk of unsplit firewood it is. Four upright 'horns' protrude above the body of the horse between which the work may be wedged. These horns can be either plumb or splayed outward, according to the maker's preference.

The hewing horse is best at holding large timbers such as the keel or the curved stem and stern: one end can be wedged on the horse while the other end rests on the floor or another horse. Likewise, the frames, or ribs, are best secured on the horse for careful ax work. These must be shaped with a complex changing bevel to fit the stepped insides of the lapstrake hull. All such unwieldy operations would be difficult to perform on the longer, flat surface of the planking bench.

These are the benches on which Paul Schweiss learned to build Bjorkedals—a traditional lapstrake fjord rowing and sailing boat—and small, double-ended power boats called Snekka, which are used mostly for day fishing. On p. 180, Alfred Sovik is shown using a horse similar to Schweiss's to perform one of the most impressive tasks of the Norwegian boatbuilder—hewing the garboard strake (the strake adjacent to the keel). Sovik is considered by many to have been the consummate craftsman of the Oselvar, a graceful lapstrake craft similar to the Bjorkedal and a direct descendant of the Viking Gokstad ship. (When Sovik died recently, his workshop—including his tools, benches and a partially constructed boat—was moved to Hordamuseet, a museum of traditional crafts in Fana, Norway.) About 14 ft. to 18 ft. long, the traditional Oselvar has only three planks on a side, overlapping and joined along their edges with rivets. The planks are fitted to each other and to the keel and stem first, before the frames are installed.

Like his Norse ancestors, Sovik used an ax to hew the convolutions of the garboard from a solid board, a practice requiring great skill. By wedging the plank securely on edge between the horns on two adjacent horses he was able to free both hands for control of the tool. Each of Sovik's horses had only two horns, placed near one end and oriented on the centerline. The remainder of the horse was left free for other ax work. I never had the good fortune to meet Sovik or see one of his boats, but I couldn't help but be impressed by one of his garboard shavings, which was carried across the Atlantic by an admirer. The single ax-hewn shaving was 5 ft. long.

A somewhat more common style of horse—not quite as stocky as the others—is shown in the photo on the facing page. Nils Drange, of Drange, Norway (30 km from Bergen), keeps two such horses busy, moving back and forth between them and the boat he is working on. Drange builds a lovely, if more contemporary, Oselvar. As a major concession to efficiency, Drange steams his garboard rather than hews it. For this oper-

Hewing horse

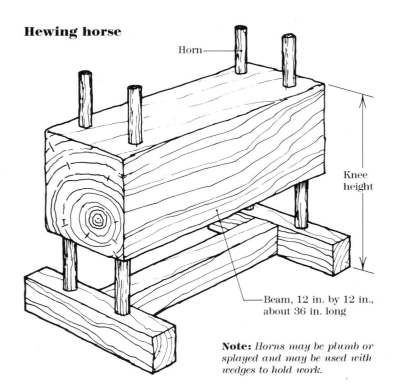

Horn

Knee height

Beam, 12 in. by 12 in., about 36 in. long

Note: *Horns may be plumb or splayed and may be used with wedges to hold work.*

ation, the horses serve simply as low set-up benches and need not be as massive as Sovik's. Like Sovik's horses, Drange's have only two horns, which he uses to help steady wide planks. With the work wedged between the horns, he can support it with one hand or his body if necessary. For Drange, the horses combine the functions of a planking bench and a hewing horse. Using his two horses to support a narrow plank, he can comfortably trim the edge or hew a scarf.

Even in a working boatshop of 20th-century Norway, like Nils Drange's, it is not unusual to find these same medieval workbenches right next to the jointer and bandsaw. And in Norway, Schweiss points out, quality control takes place in much the same way as it does around the world: visiting farmers stop by the shop for coffee and a chat and need a comfortable place to sit down.

Several years ago, Schweiss visited The Apprenticeshop, a wooden-boatbuilding school in Rockport, Maine, to instruct students in the ways of the Norwegian planking bench. The bench they built is a modification of Schweiss's original Norwegian model, designed with a somewhat sturdier leg construction, as shown on the facing page. When I visited the shop recently, the bench was being used as a utility stool and set-up table. According to instructor Dave Foster, most of the students prefer the shop's conventional wall-mounted benches with wooden leg vises to the Norwegian planking bench. Their preference is natural, considering that most North American woodworkers grow up using a vise to hold their work. But at least as important is the fact that machine tools and the hand plane have elbowed the hewing ax out of the modern boatbuilder's toolbox—even Schweiss is more likely to be fitting woodwork to a 46-ft. plastic racing boat than hand-hewing a 14-ft. Snekka. Hewing both is labor intensive and wastes a lot of wood—when Alfred Sovik hewed a garboard for one of his Oselvar boats, more wood was reduced to shavings than remained in the finished board. Hewing also requires a facility with hand tools that most of us no longer have the patience to acquire.

A fortified planking bench (benchtop removed)

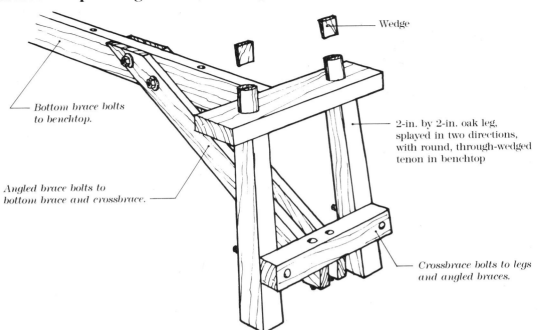

Wedge

Bottom brace bolts
to benchtop.

2-in. by 2-in. oak leg,
splayed in two directions,
with round, through-wedged
tenon in benchtop

Angled brace bolts to
bottom brace and crossbrace.

Crossbrace bolts to legs
and angled braces.

The horse is such a simple workholding device that it relies upon good body contact. Nils Drange (top right) supports a board by straddling it to hew a scarf. With the craft flanked by horses (above), Drange springs the fitted planks into place. Photos by Bertil Quirin.

Just down the coast from the Apprenticeshop, in Camden, Maine, Sam Manning has devised a basic boat workbench that builds on the attributes of the Norwegian planking and hewing benches but can be expanded to perform many other functions. "Every time I have a [traditional] workbench," Manning says, "it gets used as a table." So he set out to devise an utterly flexible, modular bench system.

Manning explains that he learned boatbuilding in the field, making 'real boats,' not yachts, and so gained an appreciation for simple, local woods instead of exotics. "Being a child of the Depression, I don't trust things in stores," he adds. Like the freelance journeymen of Norway, Manning had to travel to the work and make do with available materials. The result is the 6-ft.-long plank bench shown below, which is built of materials commonly found in most lumberyards or scrap piles. A removable 'anvil' top is held to the bench, or horse, at both ends with C-clamps, or with a single chain binder in the middle to relieve the clamps for other work. The chain binder (either chain or rope, with lever and blocking) is also the best non-slip clamp for holding logwood to the anvil while it is hewn.

To tighten the chain binder (Manning calls it his "chain crusher"), Manning slips a stout piece of wood between the chain and the bench. He lifts the long end of the stick and wedges it up. Then, using the wedge as a fulcrum, he pushes down on the stick to raise the short end of the stick on the other side of the chain to slip another wedge beneath it. This process is repeated as often as necessary until the chain is tight. Manning keeps a box of wedges of different sizes handy to his work area for this purpose. The same method will effectively turn almost anything into an on-site workbench; a fallen tree or a heavier log can be used as an anchor for work in the field.

Sam Manning's bench and anvil are great for hewing, planing and just about any boatshop task. The bench is sturdy and portable and was made completely out of materials Manning had on hand. Photo by Roger Barnes.

Bench and anvil

Note: *Drawings here and on facing page are based on originals by Sam Manning.*

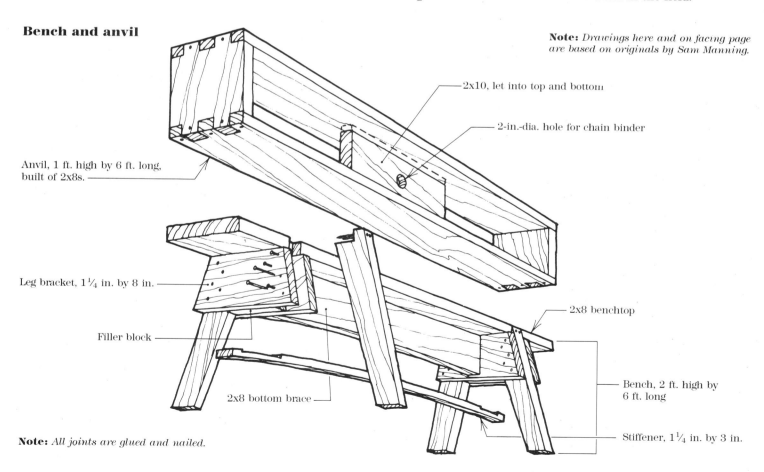

Anvil, 1 ft. high by 6 ft. long, built of 2x8s.

2x10, let into top and bottom

2-in.-dia. hole for chain binder

Leg bracket, 1¼ in. by 8 in.

Filler block

2x8 benchtop

2x8 bottom brace

Bench, 2 ft. high by 6 ft. long

Stiffener, 1¼ in. by 3 in.

Note: *All joints are glued and nailed.*

Jorgenson wooden handscrew clamps, attached with *C*-clamps or other handscrews, provide vises at both ends of the anvil. Virtually any flat surface can be converted into a workbench by arranging a combination of handscrews and *C*-clamps, as shown below. At 2 ft. high, the heavily built horse is the right height for both crosscut and rip sawing. The horse and anvil combination provides modules of staging at 1-ft., 2-ft. and 3-ft. heights. The anvil can also be turned on its side atop the horse to make a first-rate saw-sharpening vise. Simply clamp the sawblade flat to it with handscrews. A single bench with an anvil provides clamping and pounding surface for large work, such as a long plank, panel or door, the other end of which can rest on the ground or be supported if necessary. A panel or door can be clamped on edge alongside the bench for edge-jointing or the setting of hinge mortises.

Add another 6-ft. bench, with or without its anvil, and the two can be used as parallel sawhorses to hold a wide panel at a comfortable working height. The two benches can also be placed end-to-end to support a very long plank or timber, with four places to mount clamp vises along the way.

The main attraction of Manning's bench is its flexibility. It is solidly built, has the weight to stay where it's put, and yet is fully mobile and easily stowed. It has served Manning well for nearly 30 years of doing all kinds of handwork—from house construction to boatbuilding to on-board boat repair. He uses it for hewing, whipsawing, planing, assembling, fastening and finishing, and stands on it to reach an eave board or to install a sheetrock ceiling. With holes mortised into the top of the bench, he can also wedge a plank upright to hew its edge. The bench can be set up on the polished floor of a city apartment as quickly as below deck on a vessel under construction. A further feature is that, at the dimensions shown in the drawings, the legs are the right distance apart for Manning to brace his feet against when standing on a slippery surface or for positioning between frames in a vessel. If carefully built to the sizes of timber shown and fastened with pre-bored nails, the bench will not creep or wiggle. Manning builds his benches of softwood lumber, both for economy and because hardwood is a little too slippery for his liking.

Using the bench and anvil

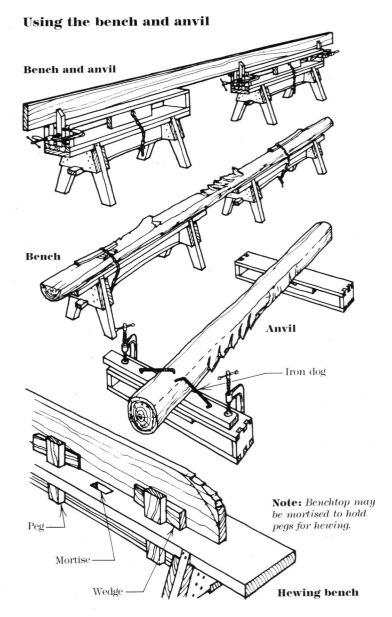

Bench and anvil

Bench

Anvil

Iron dog

Note: *Benchtop may be mortised to hold pegs for hewing.*

Peg

Mortise

Wedge

Hewing bench

Handscrew vises

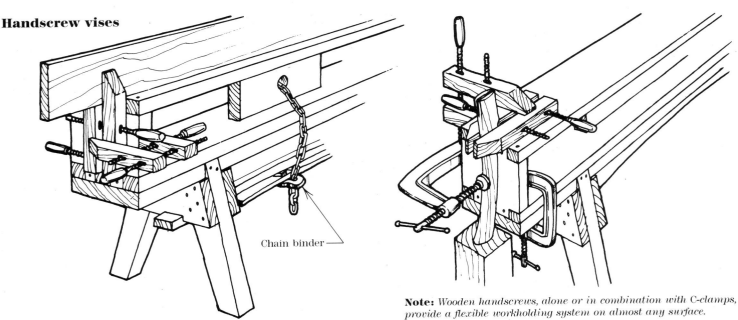

Chain binder

Note: *Wooden handscrews, alone or in combination with C-clamps, provide a flexible workholding system on almost any surface.*

The benches at the Rockport Apprenticeshop are built of standard dimensional lumber, braced with angled brackets to the wall studs of the shop. A vertical leg vise handles most of the standard workholding operations. Everything else is done with clamps or holdfasts, or on the boat itself.

In researching this chapter, I visited quite a few contemporary boatbuilders. The typical boatshop, I discovered, is appropriately long and narrow, with benches built into the walls on both sides of the boat—which sits in the middle of the workspace. Once building has begun, no matter how large the space is, it is always slightly smaller than you'd like it to be. Even if a cabinetmaker's bench were an asset to the boatbuilder, it is doubtful that he could fit such a bulky, freestanding object in the shop without constantly having to move it out of the way. The only times I saw a freestanding bench were in a school that had an unusually large shop and in a boatpattern shop. I suspect that, in common practice, a freestanding bench would eventually find itself supporting one end of a scaffold or buried beneath the hopeless household clutter that overflows so many boat benches—boatshops tend to be living environments, which is evidenced by the junk on the bench.

Benchtop cleats and clamps

Note: *Bracket has three slots to hold wider boards.*

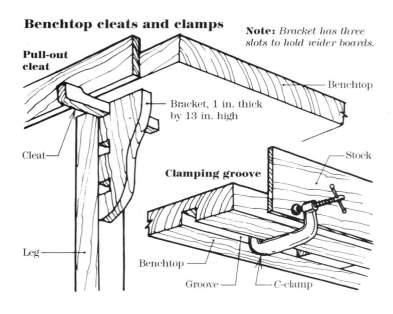

Pull-out cleat

Benchtop

Bracket, 1 in. thick by 13 in. high

Cleat

Leg

Clamping groove

Stock

Benchtop

Groove   C-clamp

In the modern boatshop, stock is horsed through machines before being brought to the bench for final edge treatment and joint preparation. The bench is long—at least as long as the longest plank that will be worked—and as narrow as 14 in. or 16 in. wide. The narrow top is perfectly adequate for supporting narrow planks; were it any wider, it would be even more likely to harbor boatshop flotsam. The longest bench I encountered was a 90-ft.-long sparmaking bench built into the shop wall. It was a bench without a top, consisting instead of simple triangulated 2x4 frames on 3-ft. centers, each supporting a permanent cradle. The cradles, or stations, were carefully leveled so that a long, hollow spar could be glued and clamped accurately on them.

Each bench usually has a single, wooden leg vise or metal woodworking vise mounted on one corner. The vise holds one end of a plank while the edge is planed or shaved to a curve. The other end may be supported by pull-out cleats attached to bench legs spaced every few feet, as shown in the drawing at left. Alternatively, a channel can be cut into the underside of the front edge of the benchtop, so that clamps can be inserted to fasten a plank more securely to the bench. This channel should be blocked every couple of feet so that the edge does not become too fragile. The boatbuilder can also work a plank on top of the bench by drilling or mortising the front board of the bench to accept pairs of pegs at intervals along the benchtop, as shown in the drawing of the hewing bench on p. 187. The plank can then be inserted and wedged upright between the pegs as on the lower bench.

Most of the long benches I've seen are made of standard dimensional lumber (2x6 or 2x8 tops with 2x4 bracing below), but the front board of the top may be made of maple or oak (a couple of inches thick) to provide better support for the pegs. Clearly, the traditional benchdog system would be unable to cope with long, curved planks, which would spring up with the slightest application of clamping pressure on the ends.

Workshop flotsam is as common to most boatbuilding benches as is a leg vise. Photo by Peter Maltbie.

The face of a plank is usually worked by simply clamping the board to the benchtop, near the front edge. Conventional hold-downs are too shallow in the throat to be effective for holding wide planks, so most boatbuilders either hold the work with one hand while they plane with the other, or they use a few deep-throated clamps or wooden handscrews. I have also heard of hold-downs made by boatbuilders out of iron re-bar, quickly heated and bent to the required size and shape. (These would resemble the Roubo holdfasts described in Chapter 2, and could be installed as required in a hole drilled in the benchtop.) Simon Watts, a boatbuilder from California and Nova Scotia, employs a more elaborate hold-down system, shown at right, which is really a homemade version of the Jorgenson hold-down shown on p. 111. In any event, the front edge of the benchtop should overhang any apron that supports it by several inches to allow room for attaching clamps. On one boatshop bench that had not accounted for this, the apron had been hastily hacked away to allow clamps to be inserted.

The leg vise is commonly used because it's cheap and easy to make; the only hardware required is the screw. Most boatshops can afford to have as many as they need and the vise can be made as tall or as short as the bench requires. (Boat benches come in all different heights, depending upon the type of work and the height of the boatbuilder. Benches used for layout tend to be lower than benches used primarily for hand-planing.) To provide a parallel clamping surface, the leg vise has a horizontal beam at the bottom of the leg, which is mortised through the stationary leg of the bench. The beam is drilled with a row of holes and can be pinned in position so that the jaws of the vise are more or less parallel when clamped on a piece of work. Metal machinist's vises, like the Versa Vise, are also popular because they hold curved work above the benchtop, allowing it to be shaped on more than one face. This is especially useful for drawknifing or spokeshaving a round object such as an oar or a tiller handle. (The standard leg vise is described in greater detail on pp. 124–125.)

Once the plank stock for a boat is prepared at the bench, it is ready to be installed. From that point on, much of the work takes place directly on the vessel. There are all kinds of basic wedges, shores, setts and levers employed by boatbuilders in the planking of a hull.

During my talk with Sam Manning, he pointed out that threaded screw devices, such as C-clamps and handscrews, are convenient but relatively modern inventions. Woodworkers have been using wedges and rope to hold their work for as long as they have been working wood. Today's mechanics, he complained, tend to begin in the gadgetry because it is available and it is quicker to employ for special purposes. The simpler stuff gets learned in the field over a long period of time. As we attempt to offset our lack of experience by purchasing more and more sophisticated and expensive tools, we run the risk of losing the ability to improvise—the boatbuilder's stock in trade.

Deep-throated hold-down

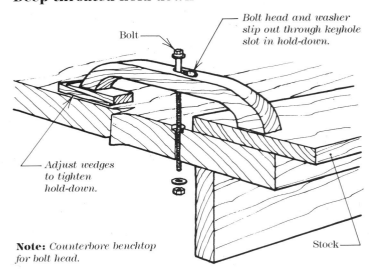

Bolt head and washer slip out through keyhole slot in hold-down.

Bolt

Adjust wedges to tighten hold-down.

Note: *Counterbore benchtop for bolt head.*

Stock

Richard Schneider's workbench and shop face south and the Olympic Mountains. The instruments have to be protected from direct sunlight, but the view does wonders for the maker's spirit.

Lutherie

As a species, luthiers tend to work in tight quarters. Unlike the typical cabinetmaker or furnituremaker, the luthier's attention is lavished on a small musical instrument that requires an inordinate amount of finicky detail work and receives an extraordinary finish. It also must be built with a higher degree of precision than is usually called for in furnituremaking. An out-of-square carcase or a swollen door are problems you might live with or repair without trauma. A misplaced guitar fret or a warped neck, on the other hand, makes an instrument difficult or impossible to play and may be hard to fix. Accordingly, most luthiers I've known measure their materials with a micrometer, check their work with a magnifying glass, and are scrupulous about humidity. It stands to reason that the luthier's workspace is often small, allowing him to burrow in and focus his energy at the bench.

I was interested to see just how the requirements of the luthier are manifested in his workbench. Among the woodworking trades, lutherie is certainly one of the most specific. And within lutherie, the guitarmaking tradition is unique in tracing its roots to a single part of the world. (Although luthiers make a variety of stringed instruments, including those in the violin and lute families, in this chapter I'm applying the term to guitarmakers only.) Unlike early furniture or watercraft, which evolved in the Mediterranean as well as in Northern Europe and the Orient, the guitar appears to have been 'born' in Spain. Antecedents exist in various forms and cultures, but were it not for the influence of Spanish instrument makers and musicians in the 18th and 19th centuries, the modern guitar would, in all likelihood, not exist. So when I set out to visit luthiers' workshops, I expected to find some highly evolved workholding systems, tailor-made for the guitar, that had come out of that Iberian tradition.

Curiously, I was both right and wrong. The highly evolved system amounts to an assortment of jigs and clamps, which are often specific to a particular instrument. But the luthier's workbench itself is often generic. Instead of tailoring the traditional tail-vise and benchdog system to the guitar, luthiers tend to ignore it. At the C.F. Martin Company in Nazareth, Pennsylvania, for years the makers of the most highly regarded production-model steel-string guitars, craftsmen still work at cabinetmaker's benches, most of which date from the late 19th or early 20th century. But many of the benches have been decommissioned, and on those that have not been stripped down, the vises are used not to hold the work, but to hold the specialized vises and fixtures that in turn hold the work.

Still, whether in a factory the size of Martin's, which has about 80 workers in guitar production, or in a one-man operation, the workbench is the focal point of the luthier's shop. The top and base must perform many of the same basic functions as the cabinetmaker's bench. Most luthiers spend all day bent over the bench, either standing, or sitting on a high stool, so it's important that the top be the right height.

The commercially made benches at the C.F. Martin Company rarely are used conventionally. More often they hold specialized jigs, such as the guitar-neck fixture above.

On the ingenious bench at bottom left, designed by guitar repairman Dan Erlewine of Athens, Ohio, the adjustable top enables the luthier to clamp the work at the most convenient height and angle. The benchtop is fully reversible—on one side is smooth, laminated maple, on the other a *T*-slotted surface that accepts a variety of jigs and clamps. The top pivots to 45° and 90° on a central bolt, and locks in place with four pins and a hand knob at each end. Erlewine's bench is 20 in. wide by 38 in. high by 56 in. long, and weighs about 150 lb. It sells for about $600 and is available from Stewart-MacDonald Mfg. Co. Erlewine designed his 'Luthier's Workstation' to enable him to set a guitar action with the instrument in the playing position, but the result has many other applications for all kinds of woodworking.

Although the luthier doesn't pound on the workbench with the same fervor as the cabinetmaker or remove large amounts of stock with a hand plane, the base must be stable enough to resist movement. Their tight quarters—sometimes a garage behind the house, or a spare room inside—lead many luthiers to turn the underside of their bench into a rabbit warren of drawers and shelves to store tools, clamps and materials within arm's reach. These provide additional mass and rigidity to the bench structure. It's not uncommon—particularly in a small shop—to attach the bench to the wall.

I was also surprised to find that, although musical instruments have multiple layers of finish and are buffed to a mirror sheen, the typical luthier's workbench is made of dimensioned lumber and plywood, nailed together and fitted with drawers cannibalized from an old desk. I suspect that this is partly due to the luthier's skills and temperament, which are attuned to craftsmanship on a scale different from that of the cabinetmaker or furnituremaker. In that, the luthier is not unlike the boatbuilder. Although the luthier's work is comparatively small and fine, both are often neophytes when it comes to furniture—and the workbench is essentially a piece of furniture. Their benches tend to be rough-hewn.

While they fall at opposite ends of the woodworking spectrum in terms of sheer scale, lutherie and boatbuilding share a common working method. In both, much of the joinery and shaping work is performed directly on the guitar or the boat. The cabinetmaker and furnituremaker on the other hand strive to complete as much of the work as possible at the workbench before the piece is assembled. There is more control at the workbench and it is easier to standardize parts and cut joints before the members have been shaped. The luthier and boatbuilder often have the opposite inclination. At the bench they do only the work that is absolutely necessary before grafting the part onto the object they are building and then continuing to work. Boats and guitars are curvilinear and usually one-of-a-kind, and it is next to impossible to trim out all the parts at the bench and then assemble them, as you would a piece of furniture. It is also easier to shape parts after they have been incorporated into a structure. But I've a hunch there is another, equally significant rationale, which relates the luthier and boatbuilder more closely to the sculptor than to the joiner. I believe that they identify so closely with the finished form that they cannot bear to work apart from it for very long.

The angle of the top on Dan Erlewine's 'Lutherie's Workstation' can be changed without the work having to be unclamped. Photo by Tom Erlewine.

Richard Schneider, a luthier for almost 25 years, certainly shares that intense connection with the guitar. Although he began making guitars in a traditional, classical form, Schneider now builds according to the Kasha method, an exacting, innovative approach to guitar design originated by Dr. Michael Kasha of Tallahassee, Florida. Schneider's guitars are a vast departure from the 'Model-T' developed by Antonio de Torres Jurado, a Spanish luthier known simply as Torres. During the 19th century, Torres made many fundamental contributions to the shape and dimensions of the classical guitar. Chief among them was his fan bracing system, which enabled the soundboard to vibrate more efficiently, and produce a clearer tone with more power and sustain than its predecessors. Today, more than 100 years later, Torres' work continues to define the standard for classical guitars. Working closely with Michael Kasha, Schneider employs a radically different bracing structure for the soundboard as well as the back, and has altered virtually every other aspect of the guitar's design, from the butt block to the peghead.

The result is a remarkable instrument that stands at the top of its class. Schneider's guitars take several times longer to build than traditional models, and cost three times as much, so he makes very few. He employs a variety of unusual gluing jigs and clamping devices in the process, and his workbench is uniquely designed to facilitate guitar work.

I visited Schneider in the house and shop built by his wife, Martha, on the Olympic Peninsula of Washington State, where they recently relocated from Michigan. It is perhaps to be expected that Schneider's workbench and shop, like his instruments, are unconventional in lutherie circles. The shop is a large, airy, 25-ft.-square workspace. A full wall of south-facing windows flood the shop with light and provide a panoramic view. It was the mountains that seduced Richard and Martha to the Northwest, so despite the problems posed by direct sunlight, Richard situated his workbench to make the most of the view. (He plans to build a movable screen to diffuse the direct rays of the winter sun on the bench.) When I arrived, Schneider confided that he had been consumed by his move to the Northwest and had been away from the bench too long. It was with some anxiety that he was about to immerse himself in the building of a new guitar in a new shop.

At first glance, Schneider's bench is an anomaly among guitarmakers. It is shaped with a cutaway section at one end of the top that provides two protruding 'wings' for working on at least two instruments at one time. An extra-wide leg vise mounted on one wing meets most general workholding requirements. The bench is carefully constructed of hard maple and finished like a piece of furniture, but its design is simple. In fact, the bench is actually an upscale copy of the one used by Schneider's first teacher and mentor, Juan Pimentel of Mexico City, and the one Pimentel's father worked on before him.

Schneider first visited Pimentel in the early '60s as an aspiring guitarist looking for an instrument. Two years later in 1963, aware that he would never become a performer, Schneider quit his job in a Detroit car dealership and drove back to Mexico determined to apprentice himself to the master. Unfortunately, he neglected to warn Pimentel. When Schneider finally showed up—he spent six weeks in Mexico honing his Spanish and mustering his courage—Pimentel said no. As Schneider retells the story, he was crushed. He leaned back against a wall and went limp, sliding down to the floor. By the time he hit the ground, tears were pouring down his face. Pimentel was so startled and moved that he consoled Schneider by saying, "Stick around, we'll see what happens." Schneider returned each day for the next few days, playing and singing Mexican songs and watching Pimentel work. After three days, Schneider finally asked Pimentel to help him select the tools he'd need to set up his own small shop nearby. Pimentel responded: "What do you need tools for? A lot of tools here." When he closed the door to the shop that night, Juan Pimentel turned to Schneider and said, "The Pimentels—they are your friends. You can start a guitar on Monday."

Richard Schneider worked in Pimentel's shop for close to two years, sharing the bench with at least two other workers. Pimentel's shop was of the old-world tradition. Four or five men were crammed into a small, dingy workspace permeated with the mingled odors of wood shavings, melted hide glue, sweat and shellac. The bench was hard to find beneath the assorted detritus of lutherie—mounds of clamps and tools, encrusted glue and old Pepsi bottles.

When Pimentel eventually kicked him out of the nest, Schneider returned to Detroit and set up his own shop in a backyard garage. He built his first bench in three days in his old high-school woodshop, acquainting himself with the marvels of tablesaw and jointer. In its design, this bench was closely modeled on Pimentel's. The work surface consisted of two 5-ft.-long pieces of 2x10 pine, which protruded about 6 in. beyond one end of the 1x6 tool well that lay between them. The base was of green, lumberyard 4x4s. When he finished the bench, Schneider initiated the top by gouging it with a chisel. He had grown up in awe of his father's cherry workbench, which he was never allowed to work on for fear he would damage the top. His own shop had no room for a prima-donna bench.

A young Schneider at work under the watchful eye of the master, Juan Pimentel. Schneider estimates that Pimentel has built about 5,000 guitars in his cramped Mexico City workshop, working alone or with a few assistants.

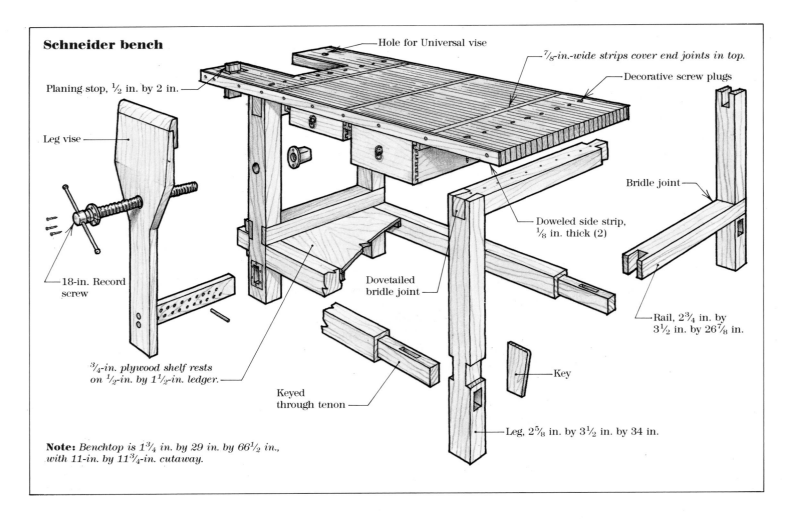

Schneider bench

Hole for Universal vise

⅞-in.-wide strips cover end joints in top.

Decorative screw plugs

Planing stop, ½ in. by 2 in.

Leg vise

Bridle joint

18-in. Record screw

Doweled side strip, ⅛ in. thick (2)

Dovetailed bridle joint

Rail, 2¾ in. by 3½ in. by 26⅞ in.

¾-in. plywood shelf rests on ½-in. by 1½-in. ledger.

Keyed through tenon

Key

Leg, 2⅝ in. by 3½ in. by 34 in.

Note: *Benchtop is 1¾ in. by 29 in. by 66½ in., with 11-in. by 11¾-in. cutaway.*

Schneider worked on that bench for ten years until shortly after he moved to Kalamazoo to begin working for the Gibson Guitar Company. There he built his current workbench, shown above, incorporating several changes in its design, but still basing it on Pimentel's original. He extended the length of the top by 6½ in. to accommodate a Gerstner toolbox on one end for small instruments. He increased the depth of the cutaway at the end to 11 in. and added 3 in. to the bench width. This allowed him much more flexibility, including room to stand in the cutaway and maneuver around guitars clamped up on the wings. The well on the old bench had become an annoyance—guitars fell into it, tools got lost in it and it was hard to keep clean. He eliminated it on the new bench, making a single-thickness top, and upgraded the bench material throughout to hard maple. The jaws of the leg vise were widened to 12 in. at the top—almost as wide as a saw-filing vise—to provide a firmer grip on guitar necks and sides. The mortise in the top for the friction-fit planing stop, and the one in the leg for the horizontal adjustment beam near the bottom of the vise, were inlaid with brass to minimize wear.

The bench construction was a team effort. A friend at Gibson laminated the top out of ¹⁵⁄₁₆-in.-thick maple strips, scraps left over from Les Paul-model, solid-body electric guitars. Another friend back in Detroit sawed out the cutaway in three slices on a large tablesaw and two students assisted in the rest of the joinery. The seams in the top (created where some short laminates were glued end-to-end) are covered with three thin strips of maple that are inlaid across the benchtop and are glued

down with epoxy. I wondered how these stand up to the movement of the top, but Schneider assured me that in more than ten years they haven't broken loose. The top is screwed to the upper rails of the base (also with no allowance for movement), and the screw heads are countersunk and covered with decorative plugs. Although it's always a good idea to allow for movement across a wide benchtop, Schneider avoided serious problems by gluing the top out of narrow, well-seasoned laminates, and maintaining a stable humidity in the stop.

The base is designed to knock down—keyed through tenons fix the stretchers to the legs, and dovetailed bridle joints connect the top rails to the legs. Schneider's first bench had been so lightweight that when he leaned back to wrap a guitar binding with rope the bench would slide across the floor. To counter that tendency, he kept a box of weights on the bottom shelf. These still live on a shelf below the current bench, and are used occasionally in glue-ups, although the bench's increased weight and rigidity make them redundant. For further stability, Schneider plans to pin the feet to the floor once he's absolutely certain about the bench's placement.

Beneath the top are two drawers, which pull out on wooden slides from either side of the bench. The drawers hold chisels and the tissue-thin newspaper Schneider uses for clamping pads, but they are often blocked shut when work is clamped to the top. Schneider admits that their function is more emotional than practical. He put them there, he says, "because Juan had a drawer there and I'm used to seeing a drawer there and if there isn't a drawer there I feel awkward."

The key to Schneider's use of the bench is his workboards. These are pieces of ¾-in. plywood, about 18 in. by 26 in., upon which a guitar or part of a guitar may be clamped while it is worked. (Schneider covers his boards with heavy paper to keep them from getting fouled with glue.) The board in turn is clamped to the bench, and is usually cantilevered off the sides or the wings at the cutaway end. For extra support, Schneider sometimes installs a *pata*, or foot, between the floor and the outboard end of the workboard. Cheap and easy to make, the workboards provide enormous versatility in the use of the bench—they become the worksurface, and the bench holds tools, clamps or glue.

Because the boards protrude from the bench, it is easy to gain working access from three sides. With a workboard clamped across the two wings, Schneider can even insert clamps from the fourth side of the board. With his stool pulled right up to the board, Schneider's legs fit comfortably beneath, rather than bump into the understructure of the bench. When work is complete, the board can be removed—guitar and all—and carried elsewhere for another operation or set aside out of harm's way. "As Pimentel taught me," Schneider says, "everything is portable—put it away." The workboards allow Schneider to juggle several different instruments or projects on the bench at the same time, as shown below. Soundboard braces may be fitted on one, while bindings are clamped to another and fretwork takes place on a third. Workboards are also useful when Schneider conducts a workshop and, as in the old days in Mexico City, three or four students hover around the bench.

Schneider fills in around his guitarmaking with what he calls 'micro-woodworking'—recombinant slices of natural and dyed woods laminated into belt buckles and turned bracelets.

The patterns developed in these playful exercises often inspire future soundhole rosettes. The workboards provide a convenient way to keep these projects moving alongside, but not in the way of, the guitars. All the materials for a run of belt buckles, for example, can be organized on a single board and moved from the saw to the bench to the buffing wheel.

Schneider's two more conventional workbench features—the bench stop and the leg vise—are used somewhat less frequently now than in his early work. The stop is used only occasionally for hand-planing boards, an operation that has by and large been superceded by the jointer. To plane small stock, Schneider still uses a cypress board (Pimentel had a cypress board), roughly 2 in. by 4 in., as a buffer between the work and the bench. The board is planed perfectly flat and clamped on the bench against the stop so that purfling or a fingerboard can be scraped or planed on top of it. Because the cypress is soft and unfinished, it is easily flattened, is not slippery, and protects both the bench and the work.

When Schneider worked with Pimentel, and later in his first shop in Detroit, guitar necks were planed flat by hand with the part held in the leg vise. Although the leg vise continues to provide a variety of workholding functions in the course of a day, neck-planing and other such operations either are performed on the jointer or have been taken over by one of at least nine different vises Schneider uses instead. The most frequently used vise in the shop is the Universal vise (see p. 149). It mounts quickly in a hole drilled in one wing of the bench and is used for everything from burnishing a scraper to virtually all the neck work that isn't done on the jointer. It rotates 360° and its swiveling jaws accommodate non-parallel stock. At about 8 in. above the benchtop, the jaws also bring the work closer to eye level for detailed fitting without bending over.

Schneider uses a flat, unfinished cypress board propped against the bench stop to scrape purfling.

Schneider's workboards provide access to three sides of the work and can practically double the working size of the benchtop. When one project is glued up on a board, it can be set aside to cure and be replaced with another.

Besides the Universal, Schneider uses machinist, drill-press, clock-movement and pin vises at the bench. He carries the smallest of these vises with him on the road if he has to perform emergency hotel-room surgery on a concert guitar. In addition, there are a host of other clamps and fixtures around the shop, adopted over years of guitarmaking—from the U-shaped *soquettes* that support the underside of the neck on the bench, to the printer's quoins that Schneider uses to clamp work in jigs. There are drawers full of aluminum C-clamps, which Schneider had cast in Mexico City and has used for decades.

Schneider learned through bitter experience that the bench-top should not be too smooth. During a recent refinishing repair on one of his own instruments, the guitar squirted off the smooth bench and onto the floor, causing excessive damage to the neck and soundboard. Schneider was able to effect an almost invisible structural repair, but when I arrived at the shop, his apprentice, Mark Wescott, had just finished resurfacing the benchtop with 80-grit sandpaper and a tung-oil/polyurethane finish. After trying 100-grit, Wescott found that the top was too slippery, and went back to rough it up—one more reason why most luthiers don't bother with a fine bench in the first place.

The top of the soquette *(right) is shaped to match the back of a guitar neck. With a pad of newsprint and a single C-clamp, the soquette securely holds an instrument several inches above the benchtop.*

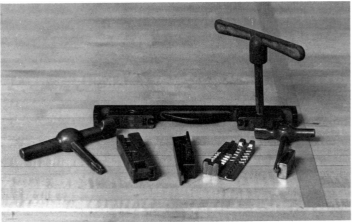

Schneider uses printer's quoins to tighten work in jigs and fixtures. The two wedges are ratcheted against each other with the key, top left. At bottom are two different styles of quoins.

A few hours' drive south of Schneider's workshop, in the shadow of Mt. Ranier, another luthier, Mark Stanley, builds guitars in what amounts to a spare bedroom in his bungalow. Almost a quarter of the shop is workbench. When he built the bench shown at the top of the facing page about ten years ago, Stanley wanted to be able to work from all sides of a clamped-up guitar. The top he designed took on the stylized shape of a guitar. The 42-in.-long tongue, which sticks out on the left end of the bench, steps down from the top's full 35-in. width to 17 in. and then 6 in. Stanley can clamp an instrument to this protrusion and work all around it without removing the clamps. A chunk of a giant oar can be used—like Schneider's *pata*—to support the end of the tongue. A ½-in.-dia. dowel in its end locates in one of several holes drilled in the underside of the tongue. In its construction, Stanley's bench is also much more typical of the trade than Richard Schneider's. The bench underframe is 6x6 and 4x4 Douglas-fir legs joined to 2x6 rails with lag screws and 20d nails, reinforced on the ends with ¾-in. plywood. Between the weight of the frame and the four large drawers filled with tools, the bench stays put.

What he couldn't keep from moving was the top. Stanley made it out of 5/4 Hawaiian apitong, which had been stored in a boatyard shed at an ambient relative humidity of about 60% to 70%. The price was right, but that was about all. Stanley built the bench without allowing the wood to adjust to the drier conditions inside his house—mistake number one. He glued and nailed the apitong to a sheet of plywood—mistake number two. The plywood allowed moisture to escape only through the ends and top side. When it was done, he installed the bench in front of a south-facing picture window—mistake number three. In a day, the top warped like the brim of a Stetson. "It didn't take long once the sun hit it," Stanley recalls.

Apitong is a handsome red wood resembling padauk, but when I arrived the top was covered with plywood to provide a reasonably flat working surface. Over the years, Stanley planed a more or less flat surface on the sprung edges of the apitong. He stopped short of leveling it, which would have drawn it out to a feather edge.

Bench height is important and luthier's benches seem to be higher than most—probably to accommodate the detailed fitting that has to be done. Stanley's bench is 39 in. high, or about 7 in. below his elbow, which he figures is just about right for working on 6-in.-high guitars. Either you adapt your bench to your own height or plan on adapting your body, Stanley explains, and he points to abundant examples of old, stooped craftsmen who have worked all their lives on workbenches that are too low.

Despite its shortcomings, Mark Stanley's bench still works, and he's learned some valuable lessons in the bargain. The top on his next bench will be a sandwich of plywood with a tempered Masonite surface. This will be covered with a sealer coat of thinned varnish, wiped off. He might also make the tongue a little narrower, perhaps 16 in. wide instead of 17 in., to allow his deep-throat clamps to reach the middle from either side. If possible, he will also reorient his shop for a northern exposure, eschewing the mountain view and harsh sunlight for a softer, more even illumination.

A clever feature Stanley incorporated into his bench is a pipe-clamp fixture designed by another luthier, Duane Waterman, of Colorado Springs, Colorado. Waterman's 'guitar-body vise' is comprised of standard pipe fittings available in most

Mark Stanley built his benchtop with a 42-in.-long tongue, which leaves lots of room to work around the narrow neck of a guitar.

This simple pipe-clamp fixture holds an instrument securely on edge in almost any position.

Guitar-body vise

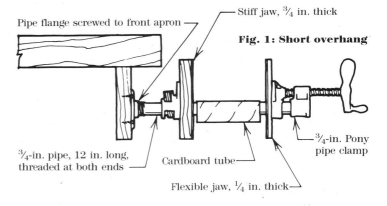

Pipe flange screwed to front apron — Stiff jaw, ¾ in. thick

Fig. 1: Short overhang

¾-in. pipe, 12 in. long, threaded at both ends — Cardboard tube — ¾-in. Pony pipe clamp

Flexible jaw, ¼ in. thick

Note: *Flexible jaw fits curved guitar back. Both jaws should be covered with felt to protect instrument.*

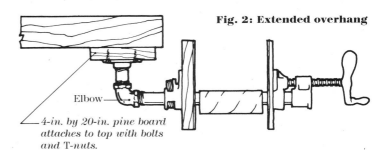

Fig. 2: Extended overhang

Elbow

4-in. by 20-in. pine board attaches to top with bolts and T-nuts.

Note: *Longer pipes may be used for larger-capacity clamping. For extra support, screw flanges to inside of rear apron and pass pipes through tight-fitting holes bored in front apron.*

hardware stores. Two threaded pipe flanges are screwed to the apron of the bench, just under the overhang of the top. (The drawing also shows a version for benches with large overhangs.) Lengths of ¾-in. or ½-in. pipe, 12 in. long, are screwed into the flanges and Pony pipe-clamp fixtures mounted on the pipe. Waterman makes two jaws out of plywood, particleboard or pine. He faces the jaws with felt and slips cardboard toilet-paper tubes over the pipes to protect his instruments. One jaw is stiffened to clamp flat against the top of a guitar, the other is thinner and able to flex slightly to match the contour of a back. (The instrument in the photo above is reversed.) The jaws are slotted to slip over the pipes and hold a guitar securely on edge, as shown above. It's possible to crush a guitar by tightening the clamps too much, though. (Paul Jacobson, of Kansas City, Kansas, adapted Waterman's design for an enlarged clamping capacity, as noted at the bottom of the drawing at right. By attaching larger pipes to the inside of the rear apron, he is able to clamp a guitar body with its top or back facing up.)

Ervin Somogyi is another West Coast guitarmaker who works in a cramped shop that is typical of most of the luthiers' workspaces I've seen. Somogyi recently relocated his business in a 14-ft. by 19-ft. workshop attached to his house. "I have about half the space I had before," he told me, "so I have been forced to be very inventive in my use of it." Like a man confined to a one-room cell, Somogyi has spent a lot of time contemplating the four walls—and everything between. Somogyi's is a mind that wanders while he sands; the result is an unusually lucid perspective on the workshop environment and the place of the workbench within it.

I dropped in on Somogyi in his Oakland, California, shop one warm, summery day in February. With two fans going and the windows open, the room felt air-conditioned. The workshop takes up one corner of the ground floor of his house, which is shoehorned between two others on a hillside. The shop itself is compact, but otherwise rather pedestrian at first glance. It has a kind of organic demeanor—like that of the rambling house and the surrounding garden—which easily might be considered haphazard. Actually, it was carefully considered, reflecting Somogyi's 14 years of experience in his previous workshop. The first shop was already in existence when Somogyi moved in, so he had to adapt it gradually to his needs. Most of the tools and fixtures that evolved in the old shop were moved to the new shop, but this time around, Somogyi is trying to design his workspace around his needs right from the start.

The workbench was built in the old shop and modified for the new space. "You kind of make modifications and grow into it," Somogyi says. It's a fir-4x4 and plywood job—a handyman classic—with six drawers transplanted from an old oak desk. Underfoot is a large chunk of closed-cell foam, picked up at a flea market for $2. The bench design is deceptively simple. Just about every cubic inch above and below the top is used. The large open space between the drawers holds hand tools within easy reach. A slim plywood drawer slides out beneath the benchtop and holds precision measuring instruments and small files. The bench is backed with a 6-in.-high carpeted ledge to protect guitars from damage. Above it, a small tool rack (also carpeted) holds the most frequently used tools at an accessible angle and casts less of a shadow on the bench than would a vertical rack. The bench itself is 37½ in. high, 30 in. wide and 10 ft. long (including a short extension on one end that serves as Somogyi's office). It is built against the wall and has a Versa Vise on the right-hand corner.

Next to clamps, which are the luthier's most frequently used accessories, Somogyi likes to use the small benchtop jig shown at the top of the facing page. Similar to Richard Schneider's workboards, this jig allows Somogyi the flexibility to clamp work to the surface and move it into position or store it out of the way. The 18-in. by 24-in. surface is made of 1⅛-in.-thick flooring plywood screwed to four 5-in. legs (high enough to fit over a guitar body for temporary storage). The jig is lightweight, easy to store and, best of all, extremely versatile. Upright, with all four legs on the benchtop, it serves as a mini-workbench to lift detailed work up to a comfortable working height. Or, one or more legs can be hooked over the edge of the bench to provide a stable, sloped worksurface. When Somogyi leans against the board, holding it firmly in position, he can work from several different directions and angles. There is a hole drilled in the surface in which he can mount a Dremel router for cutting soundholes or rosette channels. He has two such jigs, but could make another in a few minutes if he had to.

Ervin Somogyi fits a lot into his garage-size workshop. His benchtop jig holds a guitar top or back in a variety of positions. Here, he hooks one leg of the jig over the edge of the bench to get a better angle on the work.

"The bench is where I spend eighty percent of my time," Somogyi says, "but it is the relationship of the bench with respect to other things in the shop that really makes it work." Behind the bench, only 4 ft. away in the middle of the room, is a 2-ft. by 4½-ft. island workstation. It, too, is cobbled up of scavenged 2x4s, plywood and broadloom. It's a multipurpose set-up table—useful for student work, glue-ups or any other task that the main bench is too busy to provide. There's also room for carpeted storage of guitars on two shelves in the base. Around the workshop perimeter floor-to-ceiling shelves cover the walls wherever there are no windows.

While others speak in terms of workspace, Ervin Somogyi speaks about 'systems.' "Workbenches aren't set up in a vacuum," he says. Airflow, lighting and noise reduction are as important to him as the workholding devices. These are the factors that are often forgotten, or left to chance and whatever remains in the checking account after the shop is built. But for Somogyi it's all connected, and his solutions are not expensive. "Most of the really good ideas are low tech," he says, and most of his equipment was picked up secondhand. Dust control, for example, is achieved mainly by regulating the airflow through the shop with a combination of household and exhaust fans and open windows.

After air quality, good lighting is the next critical component in Ervin Somogyi's workshop system. "I like light," Somogyi says. "I like to be able to see what I'm doing." He has located his main bench under two north-facing windows for a soft, shadowless light, which is ideal for accurate grain and color matching. But unless your workshop has the luxury of a full wall of windows, like Richard Schneider's shop, it's likely that natural light will account for only a portion of the workshop's illumination, even during the day. For background and spotlighting, Somogyi uses a combination of full-spectrum fluorescents and incandescents. These offer a true color rendition, and Somogyi believes that the full-spectrum lights (GE 2000 multi-spectrum bulbs) are also healthier and less fatiguing. When the shop is complete, the fixtures will be recessed flush with the ceiling to eliminate shadows completely. The smaller, flexible incandescents help to diffuse the flickering of the fluorescents. This combination of indirect natural light, overhead full-spectrum fluorescents and spot incandescents minimizes eye fatigue, shadow and glare.

The last major system in Somogyi's workshop is noise reduction. While lutherie doesn't often call for the high-pitched banshee of the router or marathons of machine work, there is enough noise to be concerned about—especially if your shop is attached to your house. For his own safety, Somogyi wears ear protectors when operating power tools. To protect his family (and neighbors) from shop racket, he soundproofed the walls and ceiling. Resilient channel iron holds ⅝-in. sheetrock away from the wall studs and a layer of Owens Corning's 'Sound Attenuating Blanket' between the studs and the sheetrock further dampens vibration.

In a sense, Somogyi's workbench is his shop—a most important quality, I think, for such a tight space. From the benchtop jigs to the shop fans, the whole place works together. His low-tech solutions are simple and cost next to nothing. Somogyi demonstrates that, by capitalizing on good planning and the hawkeye of the scavenger, it's possible to assemble a healthy, efficient workshop environment in a space not much bigger than a one-car garage.

Benchtop jig

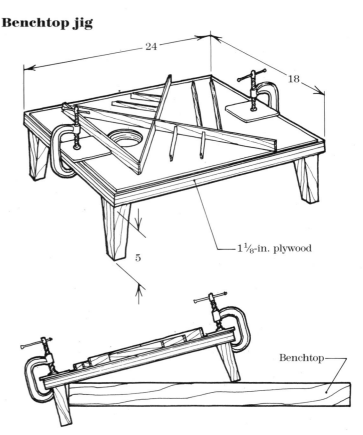

Note: *Jig may be used with all four feet on benchtop, or with any one or two legs hooked over edge.*

Somogyi shop layout

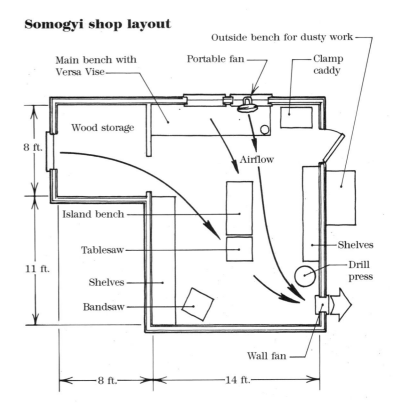

Furnituremaker C. Stuart Welch mounted a laminated maple top on the base of a Ritter dental chair that he scavenged for $50. Welch, of Marshall, California, got the idea for this hydraulic carving bench from Wendell Castle. It adjusts in height from 17 in. to 35 in. to provide a carving surface right where it's needed and can be rolled around the shop, or even taken outdoors.

Carving

For all its endless variations, the workbench is really quite simple. Strip away the nifty accoutrements and fancy joinery—the tool trays, cabinets, dovetails, through-wedged mortise-and-tenons, quick-action screws, and so on—and you're left with a rudimentary holding device and a worksurface. If pressed, I think I could define a workbench by two basic essentials: it must hold the work securely and it must hold it in the most accessible position. Everything else is enhancement.

As other chapters have illustrated, each branch of woodworking has its own special requirements in these respects. The size of the bench, its workholding mechanism and its height are all a function of the shape of the stock being worked and the tools that are used. For many of us, this involves rectilinear material, shaped in one or two planes using straight-edged tools like the saw or hand plane or chisel. Others—notably green woodworkers, boatbuilders and luthiers—have special requirements based on their need to work wood in three dimensions or to handle curved stock. To this list I would add woodcarvers.

Carvers, of course, share the two universal workholding requirements, security and accessibility. But, whether cranking out cornice moldings by the yard or sculpting large figures, the carver routinely faces a more varied range of shapes than does the cabinetmaker. Rectilinear panels or regular moldings may

be trapped between dogs on a bench with the aid of a tail vise. But even then, carvers often develop their own specific workholding methods for handling flat stock, and a few of these will be described later. To secure three-dimensional, round or irregularly shaped objects is the greater challenge and the focus of this chapter. Such items frustrate the typical tail vise or face vise and demand a more flexible arrangement.

What's more, the use of carving tools creates forces of movement and vibration not normally encountered by a workbench. In addition to drawknives, spokeshaves, rasps and rifflers, and an array of power grinding burrs, carvers commonly use gouges to shape their work. The gouges may be pushed by hand or struck with a mallet. Striking turns to pounding for sculptors, who need to waste a lot of wood quickly using large gouges. Such pounding—often from several different directions—puts the most versatile holding device to the test.

For maximum control and to reach all parts of the work, it may be important for the carver to be able to adjust the height, and even the angle, of the work. As always, the nature of the problem dictates the solution. Fine lettering, delicate relief or detailed wildlife carving are often facilitated by working much higher than usual. Large, powerful tool movements are inappropriate, while heightened visibility and control are crucial. Many carvers prefer to do this kind of work sitting down, so they can concentrate comfortably for a long period of time. Heavier sculpture, on the other hand, usually means getting above the work and being able to whack away at it from every angle. When an object has only one flat surface—and some-

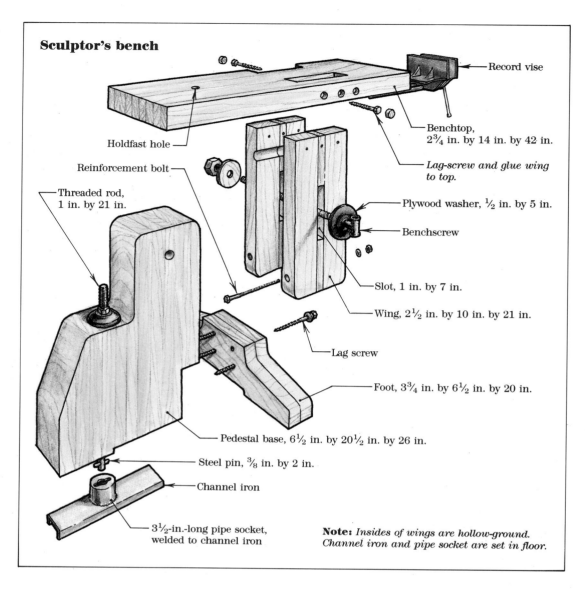

Sculptor's bench

Record vise

Holdfast hole

Benchtop,
2¾ in. by 14 in. by 42 in.

Reinforcement bolt

*Lag-screw and glue wing
to top.*

Threaded rod,
1 in. by 21 in.

Plywood washer, ½ in. by 5 in.

Benchscrew

Slot, 1 in. by 7 in.

Wing, 2½ in. by 10 in. by 21 in.

Lag screw

Foot, 3¾ in. by 6½ in. by 20 in.

Pedestal base, 6½ in. by 20½ in. by 26 in.

Steel pin, ⅜ in. by 2 in.

Channel iron

3½-in.-long pipe socket,
welded to channel iron

Note: *Insides of wings are hollow-ground.
Channel iron and pipe socket are set in floor.*

*The wings are secured with a
benchscrew and the pedestal pivots on a threaded rod. Photos by
Richard Starr.*

times a small one, at that—it helps to be able to fix that surface firmly in some kind of device, and then to be able to adjust the position of the work without having to reset the fixing.

Many woodworkers are furnituremakers, joiners, luthiers or whatever first, and carvers second. Perhaps they carve the occasional linenfold panel for a door, a ball-and-claw foot on a cabriole leg or a figure on a banjo peghead. Such craftsmen tend to work on conventional benches, which they simply adapt to meet their occasional carving needs. Several accessories useful for carving have been described elsewhere: note particularly the Universal and Zyliss vises shown on p. 149, and the several shop-built carving vises described at the end of Chapter 9. In addition, the benchdog poppets on p. 175, and the holdfasts shown in various places throughout the book provide even more options. I will describe a few others later in this chapter. But first I'd like to introduce a couple of benches that were specifically built for woodcarving. These were never intended to serve the average woodworker, but they excel at their specialty.

Pierre Cloutier, a Canadian woodcarver, made the
bench shown here. Cloutier, who died in 1986, was one of the most accomplished professional carvers in a village full of woodcarvers—St.-Jean Port Joli, Quebec, on the south shore of the St. Lawrence River. (Cloutier's sudden death came just as I was preparing to visit him to see his bench first hand. So for much of this account I've relied on Richard Starr, a woodworking teacher and author from Vermont, who visited Cloutier several years ago.)

The essence of Cloutier's workbench is versatility. Simply put, it is a hefty, articulated vise, reproduced on a workbench scale. The bench consists of a top and pedestal base. The 14-in.-wide, 2¾-in.-thick, laminated-maple benchtop is through-mortised and lag-screwed to two slotted wings, which are in turn attached to the heavy-duty maple pedestal with an Acme-threaded benchscrew. Loosening the benchscrew allows the height and angle of the top to be adjusted. The inside faces of the wings are hollowed slightly with a disc grinder, and two ½-in.-plywood washers are inserted beneath the screw's head and the nut to provide a firm grip when the screw is cinched down. The benchtop can be adjusted to the standard horizontal position for carving bas relief, or tilted almost 90° in either direction, as shown at right, for easy access to larger, three-dimensional carvings.

The entire bench can be rotated 360° on a keyed socket mounted in the floor. Cloutier's shop has two such sockets, welded out of 3½-in.-long pieces of pipe and channel iron, cast into the concrete floor. By loosening the nut on top of a 1-in.-dia. threaded rod that passes through the base, the bench may be rotated or lugged to the other socket to take advantage of natural light flooding through a large bank of windows along the west wall. The 20-in.-long foot, which is lag-screwed to the pedestal, provides stability across the width of the bench.

This combination of flexibility and security is an obvious boon to any carver. It is particularly useful for the bas-relief and life-size, three-dimensional carving of the human form that was Cloutier's specialty. Work can be clamped to the benchtop with a carver's screw, holdfasts, benchdogs, or any number of other devices. Cloutier also mounted a quick-action Record vise on one end of the benchtop for handling standard rectilinear stock.

Once a workpiece is attached to the top, it can be positioned at the most convenient angle and height for carving. Photo by Richard Starr.

While Cloutier's bench is ideal for heavy sculpture, Elmer Jumper built several sit-down benches for carving on a much smaller scale. I visited Jumper at his home in Philadelphia, Pennsylvania, where he demonstrated the bench shown in the photos and drawing on this page. This bench was designed to take advantage of the Versa Vise, which swivels on a base flange screwed to a wood block on one end of the bench. (The Versa Vise, which was temporarily out of production, is now available from Gaydash Industries, Inc., while a similar product, manufactured in Taiwan, is offered by Leichtung Inc. See Sources of Supply for their addresses.) Work can be clamped directly in the vise in its upright or side-mounted position, or on a 10-in.-square work platform that Jumper mounts in the vise. The carving is attached to the platform using a holdfast or *C*-clamps, and the whole assembly may be rotated 360° on the Versa Vise base. The backrest adjusts forward and back along the bench, allowing Jumper to sit as close to the work as he wants, and the leg panels, which are hinged to the underside of the bench, angle outward to prevent wobbling.

Another attractive feature of Jumper's bench is its collapsibility. The whole thing can be set up or knocked down in a couple of minutes using just five stove bolts. "You can put it together with a quarter," Jumper explains, and he proceeds to do just that on the floor of his cramped garage. The seat, back, legs and platform are ½-in. plywood, and the undercarriage is made from angle iron manufactured for industrial shelving, cut and welded together, as shown in the drawing. All the nuts are welded to the angle-iron frame, so there are few loose parts to keep track of.

Collapsible carving bench

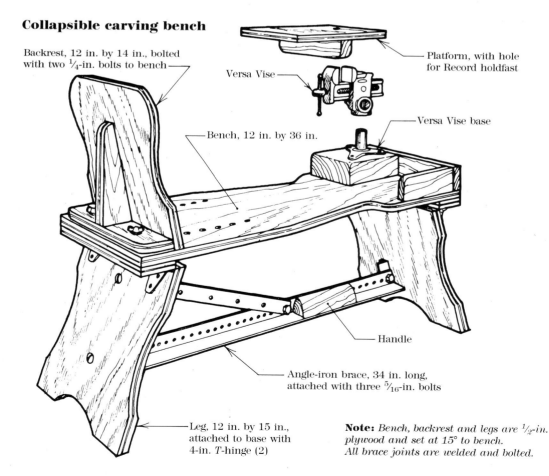

Backrest, 12 in. by 14 in., bolted with two ¼-in. bolts to bench

Platform, with hole for Record holdfast

Versa Vise

Versa Vise base

Bench, 12 in. by 36 in.

Handle

Angle-iron brace, 34 in. long, attached with three 5/16-in. bolts

Leg, 12 in. by 15 in., attached to base with 4-in. T-hinge (2)

Note: *Bench, backrest and legs are ½-in. plywood and set at 15° to bench. All brace joints are welded and bolted.*

Elmer Jumper clamps this 10-in.-square platform in the Versa Vise for a flat worksurface (top). When Jumper hits the road, the whole rig knocks down to a compact, one-hand load (bottom).

Knocked down, the bench is only slightly longer than its 36-in. seat and makes a reasonably compact, 8-in.-thick bundle, including the vise. To position the carrying handle, Jumper folded the frame and moved the wood handle along the beam until it was balanced. Surprisingly rigid and stable in use, it's easily carried in one hand and stowed in a closet or the trunk of a car—real advantages for Jumper, who frequently carts it off to carving shows. One of Jumper's friends, a New York City detective, built a similar horse to work on in his spare time. He stores it in a closet and sets it up in his kitchen on a 10-ft.-square tarp. When he's done carving, he packs the bench away, picks up the corners of the tarp and dumps the shavings out the window. (Another version of a carving horse is described in the sidebar below.)

Jumper has several other carver's benches—including another sit-down horse (built with knock-down keyed tenons in the base) and a Danish-made Lervad workbench, which he keeps in a small shop in his house. But for the most part he prefers portable, flexible devices like the carver's lap bench shown at right and the ingenious ball-and-socket bench shown on the following page. The latter was inspired by a photograph of a similar gadget made by Texas carvers Bill Bryant and Ben Hargis, which was published in the bimonthly newsletter of The National Woodcarvers' Association. Jumper uses the bench primarily for duck carving, where the ability to swivel the bird into almost any position really pays off.

Jumper's portable lap bench travels anywhere. Two blocks of wood wedge a relief carving in the cork-lined tray, and a piece of rubber floor tread glued to the bottom keeps the bench from sliding off his lap.

A two-headed horse

To get a bench that would work for both carving and shaving diamond willow and sumac walking sticks, E.D. Lyman of Lincoln, Nebraska, bolted this small carving bench to the back of a conventional dumbhead shaving horse (see p. 167). The carving bench is based on an ancient Chinese design (and is closely related to the Japanese foot vise on p. 6). The clamping mechanism is simple. A piece of rope runs from one end of the foot pedal to the benchtop, loops around the work on both sides, then returns to the other end of the pedal. To use it, Lyman turns around in the sliding seat on the horse, slips the work beneath the rope loops and steps on the pedal. Two notched 2x4 risers hold the work off the benchtop, making it easier to clamp irregularly shaped objects and provide more room for carving. The risers can be changed for different work. The stock is held securely, but can be repositioned in an instant.

Two-headed horse

Dumbhead horse

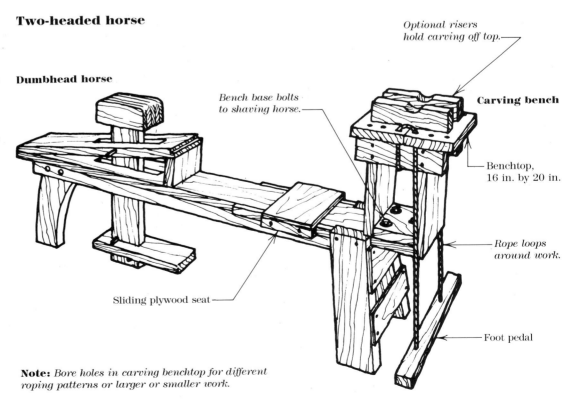

Optional risers hold carving off top.

Carving bench

Bench base bolts to shaving horse.

Benchtop, 16 in. by 20 in.

Rope loops around work.

Sliding plywood seat

Foot pedal

Note: *Bore holes in carving benchtop for different roping patterns or larger or smaller work.*

A stand-up bench

Smaller is often better when it comes to carving benches. A small bench allows the best access to the work and still provides a variety of clamping surfaces. R. Bruce Hoadley of Amherst, Massachusetts, built this small stand-up bench for detailed carving and general woodworking. Hoadley (who is 6 ft. tall) finds the 42-in.-high benchtop ideal.

The key to Hoadley's bench is flexible clamping. The slot between the two halves of the hardwood top leaves room for attaching clamps or carving screws. The top can be drilled anywhere for attaching a favorite vise or holdfast. A 1x6 apron attached to one side of the top presents a vertical surface for clamping flat stock to work an edge. Rows of holes could be bored in the apron for support pegs.

Hoadley has several such benches. On all, the splayed legs of the base are standard 2x4s, nailed together and braced with 1x6 and other dimensional lumber. Hoadley piles rocks or scrap metal on a bottom shelf to keep the bench from moving. On one bench, built behind his summer cottage, Hoadley sheathed the base with plywood and filled it with sand. Then he poured in as much water as the sand would absorb. Watering the base every now and then, Hoadley reports, keeps the bench rock-solid—making it the perfect stand for his backyard carving.

The bench itself is straightforward. An 8½-in.-dia., 17-lb. bowling ball is gripped between the radiused tops of four 2-in. by 3-in. fir legs. Opposite legs are bolted to 1-in. by 3-in. pine stretchers at the bottom and are tightened below the ball with turnbuckles, eyebolts and T-nuts. To augment the ball's weight, Jumper rests one foot on the 15-in.-dia.-plywood platform while he carves, lending the whole unit "190 lb. of stability." He has been able to work a full-size Canada goose with a 2½-in. fishtail gouge without causing the ball to slip. "Of course, I had to go at the turnbuckles a couple of times," he adds.

When Jumper makes a ball-and-socket bench, he bores and taps a hole for a carver's screw in the plastic-resin ball to hold the work. He uses a two-part screw, made for him by Clifford Huston, a fellow Philadelphia woodcarver, but other screws are available from carving-tool suppliers. Huston uses zinc or rust-free cadmium-plated 4-in. by ½-in. 'all-thread' for the sleeve. One end is ground square for a wrench, the other is tapped for a standard ⁵⁄₁₆-in. hanger bolt, one end of which has machine threads, the other wood threads. (Jumper prefers Huston's screw to the commercial variety for the gentler pitch of the wood threads, which are less likely to split the work.) The hanger bolt is screwed into the underside of the work and the whole assembly is tightened by means of an oversize wing nut and a wooden washer. Several of Jumper's friends have built similar benches, using defective bowling balls that he was able to pick up for free from the manufacturer. The dimensions can be varied, but at 5 ft. 8 in. tall, Jumper finds a 30-in. height just right. (Bruce Hoadley, a carver and wood technologist in Massachusetts, has come up with another method for securing three-dimensional carvings, shown at left.)

Stand-up bench

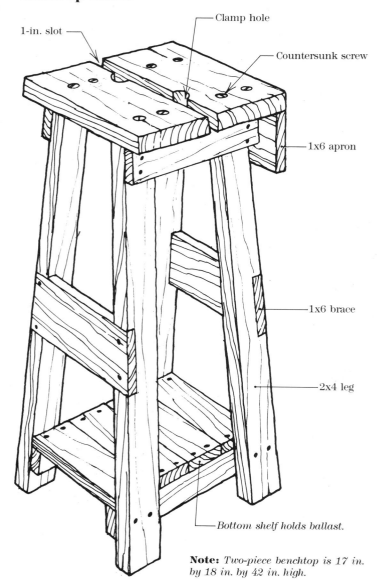

1-in. slot

Clamp hole

Countersunk screw

1x6 apron

1x6 brace

2x4 leg

Bottom shelf holds ballast.

Note: *Two-piece benchtop is 17 in. by 18 in. by 42 in. high.*

The ball-and-socket bench is as flexible as any articulated vise. Jumper stands on the base platform to stabilize the bench.

Cloutier's and Jumper's benches were designed specifically for carving. If you prefer some of the advantages of a conventional bench, but require more flexibility, you may want to modify one to suit your needs. With the addition of specialized clamping devices any work surface can be transformed into a versatile carving station. Indeed, some of the more adaptable carving rigs that follow would work with or without a bench.

Birdcarver William Schauber of Still Pond, Maryland, mounted the simple leg vise shown at right on one corner of a rough-hewn workbench in the shop behind his house. Schauber's vise employs a cam-action arm instead of a carver's screw to keep his prey from squirming. Pressure is released with a quick yank on the lever, which allows the bird to be repositioned. The jaws and lever of the vise are made of oak, joined at the bottom with a pinned, horizontal beam, and at the top with a heavy U-shaped iron strap. An earlier vise had a standard bolt for the pivot pin, but it sheared right off when Schauber cranked down on the lever. Now he uses a hardened bolt. Once a year Schauber heats the cam-end of the handle and the matching area on the back of the vise with a propane torch and rubs paraffin into both surfaces. This stops the squeaking and keeps the vise running smoothly.

Schauber carves geese, ducks, brants and loons in his spare time—about a hundred waterfowl a year. When he's really moving, he can knock out a bird in two hours, aided in part by the speed of his vise. When he's touring the bird-carving circuit, Schauber works on a freestanding vise, a portable model of his bench-mounted version. (He also carries an extra handle, although he's never had to use it.) To keep the portable vise steady, Schauber plants his weight on a heavy steel plate bolted to the bottom of the rear jaw. Either vise can handle up to 8½ in. between jaws—a large enough bite to swallow his biggest bird, an 8-in.-wide goldeneye. The idea for Schauber's vise came from Charlie 'Speed' Joiner, a retired birdcarver who got him started whittling ducks 20 years ago. According to Schauber, "Most decoy makers learned all this stuff from each other," and Charlie probably picked it up somewhere else.

Richard Hicks, of Cedar Crest, New Mexico, copied the leg vise at right from one used in the workshops of Marcus C. Illions, a master turn-of-the-century carousel carver. The vise clamps in the shoulder vise of his cabinetmaker's bench and opens to a maximum of 20 in. It's built beefy to handle the carousel animals that Hicks also carves when he's not building reproduction wooden printing presses.

Hicks's vise is 36 in. high, with 5-in. by 5-in. laminated red-oak jaws, operated by a 2-in.-dia., 36-in.-long standard Acme-threaded rod. At the bottom of the vise is the traditional horizontal beam; a wood pin keeps the jaws parallel. At the head of the screw, Hicks brazed a nut and bored an oversize hole for the steel handle. A free-floating flange beneath the head provides a little more flexibility in clamping irregular shapes than it would if it were attached to the screw. At the other end of the vise screw, Hicks welded the nut to a flange screwed to the rear post.

Hicks uses an assortment of carpet-padded jaw inserts to accommodate everything from seahorses to 4-ft.-long rabbits. The inserts are T-shaped in cross section, with the leg of the T fitting in a long channel built into both jaws of the vise. The inserts are pinned to the jaws and can be turned upside down or replaced to provide another clamping surface. Hicks makes a new insert whenever he needs one and has five sets so far.

The cam-action leg vise takes a firm bite on the tail of a canvasback. In the shop, Schauber keeps the vise bolted to the corner of his bench. On the road, he bolts the back jaw to a post mounted on a steel plate.

Carousel leg vise

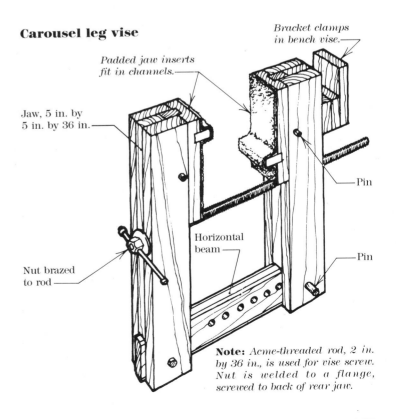

Bracket clamps in bench vise.

Padded jaw inserts fit in channels.

Jaw, 5 in. by 5 in. by 36 in.

Pin

Horizontal beam

Pin

Nut brazed to rod

Note: *Acme-threaded rod, 2 in. by 36 in., is used for vise screw. Nut is welded to a flange, screwed to back of rear jaw.*

Carver Sam Bush of Portland, Oregon, uses an assortment of benches and workholding devices, depending on the kind of work he's doing. For conventional cabinetry, Bush uses his full-size joiner's bench with two vises and a row of dogholes. But when he's carving detailed lettering or other flat relief, Bush prefers to use small cleats to hold the work in position. The cleats can be screwed to any benchtop or workboard. They're cheap and quick to use, and their low profile won't obstruct the work. To make it easier to handle the small nubs of wood, Bush cuts a long strip, drills and countersinks the screw holes before slicing them apart on a bandsaw.

When he's working away from home, Bush clamps the frame shown below to the nearest flat surface. A simple box of screwed-together 2x12s, the frame converts any table into a carving station, or raises the working height of an existing bench. Flat work can be secured to the top with Bush's screw cleats. Larger work can be clamped to its surface or attached with a carver's screw.

Another useful carving device that can be clamped in any standard face vise is shown at bottom. It is used by students who attend woodcarving workshops at The Woodsmith Stone (formerly Rosewood Tool Supply) in Berkeley, California. The top and one end of the box contain a slot large enough to fit a small cam clamp. The jaws of the clamp protrude from the top of the box and can be adjusted to clamp the work against two small pivoting dogs for a firm, three-point connection.

Screwed-down carving cleats grip the corners of a thin board securely, without interfering with the work. Bush raises the worksurface of his carving by cleating it to a heavier board clamped between benchdogs. Photo by Rick Mastelli.

Carving cleat

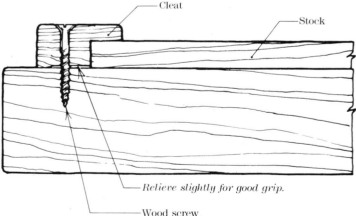

Carving frame

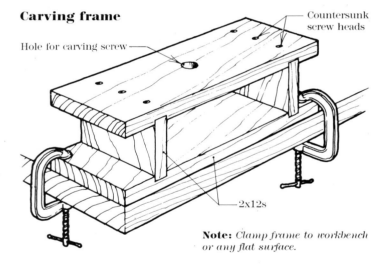

Note: *Clamp frame to workbench or any flat surface.*

Carving box

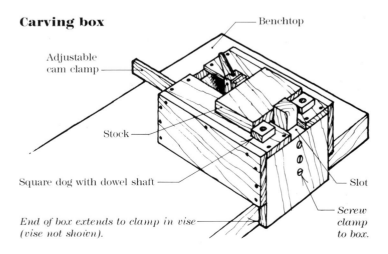

Moving away from the realm of the workbench entirely are the two unusual workholding devices at right. For Michael Fortune, the chair-carving vise at top right is like having his own robot arm. Pressure on one side of the foot pedal engages a piston within the pillar, which is welded to a flange bolted to the floor. The piston locks the ball in the desired position within the socket. To release pressure and change the position of the ball, Fortune simply steps on the opposite side of the pedal. A small harness attaches a large handscrew clamp to the end of the ball. Two pads are pinned to the jaws of the handscrew and contoured to fit the musculature of Fortune's delicate chairs. A backing of no-slip, corrugated rubber floor tread is glued to the matching surfaces of the handscrew and the pads to keep them from squirming around.

Fortune uses the vise primarily for carving and sanding. It resolves one of the biggest problems he faced in working chairs: holding them down. As Fortune says, "You're put into fewer contortions"—the vise does it for you. The piston assembly is manufactured by Interwood Ltd. of London, England. The good news for Fortune, was that it cost him nothing. He rescued it from a dusty corner of the Sheridan School of Crafts and Design in Ontario, where he teaches. Nobody knew where it came from, how it got there or how much it was worth. The bad news is that he'd like another at the same price.

Next to Fortune's vise, the carving box at bottom right is the opposite end of technology. Sculptor Margaret Neerhout of Portland, Oregon, patterned her carpeted wine crate after one made by her former teacher, Joe Goethe of Santa Barbara, California. According to Neerhout, "Sculptors who use clamps are unfailingly lazy in doing and undoing them the dozens of times necessary to avoid the 'this is the front' look." And Neerhout's free-form sculpture is the kind of work that defies holdfasts, carver's screws and vises of all description.

Positioning work in her carving box is as easy as turning it over. (Size the box to suit your work; Neerhout's crate is 8 in. deep by 14 in. wide and 19 in. long.) The deep V-notches in the sides and the 1x2 nailed to the bottom present a variety of corners and planes for propping and wedging. Work can be turned constantly to gain access to the best surface, yet it is held securely enough to be whacked with large chisels and a mallet. To hold it in place, Neerhout tacks strips of wood to her workbench around the base of the box. For a more flexible method of attachment, a piece of plywood could be attached to the bottom, extending beyond the box on all sides to provide a flange for C-clamps.

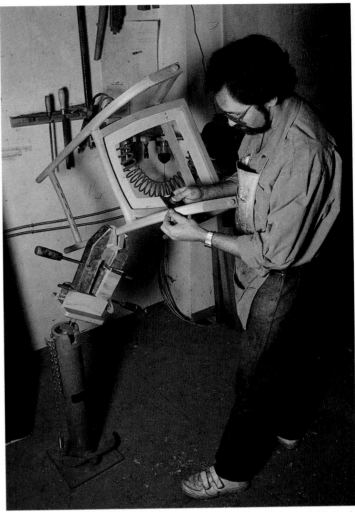

Fortune's chair-carving vise is less precarious than it looks. Although he tries to clamp the piece close to where he will be working, Fortune finds he can range pretty far afield before the leverage becomes excessive. Photo by Jack Ramsdale.

Wine-crate carving box

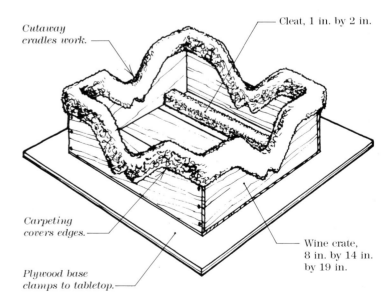

Cleat, 1 in. by 2 in.

Cutaway cradles work.

Carpeting covers edges.

Plywood base clamps to tabletop.

Wine crate, 8 in. by 14 in. by 19 in.

Inventor Ron Hickman, standing on the Mark II Workmate. Photo by Arthur Steel.

The Workmate®

**A**bout 25 years ago, while working on some built-in wardrobes in his new house, Ron Hickman goofed. He had dragged a chair over to serve as both sawhorse and workbench while he cut a panel to fit. Holding the board steady on the chair with one foot and concentrating on his pencil line, Hickman sawed away—into the seat of the chair.

The chair, a Swedish-made Windsor, was one of the few pieces of store-bought furniture Ron and his wife, Helen, owned. He was building most of the rest himself. A few nights later, Hickman, who worked as a designer for Lotus Cars Ltd. in Great Britain, sat down at his drafting table to sketch something that he figured would rescue other household furniture from a similar fate.

But for that serendipitous sawcut, Ron Hickman might still be sketching bumpers and grills. The small, clumsy object he drew that night was the first effort in what was to become a consuming passion to design a better workbench. It was almost seven years before Hickman had turned that first tentative prototype into the bench the world has come to know as the Workmate.

You would have to be a latter-day cave dweller not to have heard of the Black & Decker Workmate—as I write, sales approach 20 million worldwide. Characterized by its spider-like stance, the flagship of the Workmate line perches on stamped steel legs that are adjustable in two heights. Collapsable to a compact armload, it is the ultimate in portability. But the Workmate is probably best known for its integrated top and vise. The two halves of the benchtop function as the wide jaws of a powerful twin-screw vise, which is capable of clamping tapered stock or round pipe as securely as it clamps standard rectangular boards.

You may well ask, as did I, whether a mass-produced, handyman special like the Workmate belongs in a book devoted, by and large, to handmade traditional workbenches. Indeed, the Workmate raises all sorts of difficult questions about the very nature of the workbench.

In the first place, is it possible to *invent* a workbench? The basic elements of the modern cabinetmaker's workbench—tail vise, shoulder vise, benchdogs and holdfasts—have been around for centuries and their origins reach back to prehistory. After seeing hundreds of new and old workbenches, it is obvious to me that the woodworker—and, by extension, the bench—is a most conservative creature. Almost every woodworker I know works on a bench closely modeled after the one on which he or she learned. The notion of inventing a workbench would be in itself absurd, if not offensive, to many of the people whose benches are described in this book.

"If a man write a better book, preach a better sermon, or make a better mousetrap than his neighbor, though he build his house in the woods, the world will make a beaten path to his door."
—Ralph Waldo Emerson

Granting the possibility of inventing a new workbench, is the Workmate a serious tool, or a toy? To many who recall the television commercial depicting the Workmate as the handy helper to a singing, weekend tool jockey, it is difficult to take seriously. Nevertheless, in my travels I discovered that a surprising number of serious woodworkers have used one. The more candid among them admit a grudging admiration for the thing. It is an excellent job-site companion—easy to tote around and good at gripping or supporting a host of different objects. The fact that it is not easily compared with the cabinetmaker's workbench—or any other bench, for that matter—only supports Hickman's assertion that he had created a different animal. (The application of the word *workbench* to the Workmate has proved to be contentious; it's been an asset in court battles but a continuing liability in marketing.) As the ad jingle droned in pop/country-western cant: "It's more than a bench, and more than a vise; it's got both things workin', and that's kinda nice...."

At closer look, I had to agree that Workmate does what it purports to do and does it well. Clearly, a book about the workbench would be deficient if it didn't include this most prolific and portable of benches. I was also intrigued by the idea that it actually might be something new in the hoary world of benches.

Ron Hickman was born in South Africa in 1932, but now resides on Jersey, one of the idyllic Channel Islands, described by Victor Hugo as "bits of France fallen into the sea and picked up by England." Linguistically and politically, the islands retain a curious blend of Anglo-Saxon and Norman cultures. On Jersey, the Hickmans joined a relatively recent invasion, that of a growing number of well-heeled Britons who have fled the motherland in favor of the island's more lenient tax structure.

I visited Hickman in his 'granite wigwam,' the newly built 20,000-sq.-ft. estate he and Helen call home. "Every stone represents another Workmate...or a few of them," Hickman quipped as we drove up to a four-car garage. Built of Jersey granite, the house sits on about four acres of manicured lawn and garden, and looks south on St. Brelade Bay. An expanse of roof stretches to the ground on two sides in long, Welsh-slate tentacles, making the place look vaguely extraterrestrial. Hickman designed the house and has populated it with a host of clever inventions. Flip-up stair treads provide access to plumbing and electrical services; a modified Wedgwood chandelier over the dining table drops a 13-amp toaster socket on a coiled extension cord to the breakfasters below. In the bowels of the house is a fully automated engine room that would be at home in the Sixth Fleet. It's easy to see how the local rumor mill exaggerated the motorized brushes that clean the 1200-sq.-ft. expanse of pool-roof glass into a houseful of self-cleaning windows.

It's a remarkable, but somehow unpretentious castle—a commodious extension of the inventor's lab—and Hickman makes an engaging guide. At the end of my tour we settle down to talk in his top-floor studio, surrounded by the Workmate prototypes, as well as patent infringers and models of other recent inventions that have had more checkered careers.

Imagine for a moment that you have been challenged to design a new workbench. It has to be portable, compact and lightweight, yet strong and rigid, and it must offer a secure workholding system. Imagine further that it has to adjust in height to function equally well as either a sawhorse or a workbench and that its vise must be able to grip irregularly shaped objects as effectively as square stock. Now try to imagine that it has to be manufactured and sold for under a hundred dollars.

Fortunately for Hickman, the problem didn't present itself in those terms. When he had the good fortune to saw into the seat of his Windsor chair, he was a weekend hobbyist fixing up his house. Had he been a professional woodworker in need of a bench, he probably would not have arrived at one of the most critical aspects of the Workmate's design—its size. It was not a traditional workbench Hickman sought, but a minimal structure, something chair-size and portable, that would serve as an enhanced sawhorse. A professional designing a bench would likely have started with a bulky cabinetmaker's bench and dismembered it in stages to reach his goal. Repeatedly, as the Workmate evolved, Hickman would navigate his way around an impasse by dint of his commitment to his original concept. "I was," he says, "an absolute worshipper of lightweight."

The very first bench, shown at left in the photo at right, was conceived as a compromise between the flat top and ample work surface of a bona fide workbench and the low height and portability of a sawhorse. Hickman hadn't the room for a full-size, built-in bench and reasoned that 4-ft. by 8-ft. sheet materials would be more easily held and worked on a low, freestanding bench. And, it would always be easier to prop up a low bench than to shorten a tall one. He built the bench 3½ in. higher than the seat of the chair and twice as long as the chair was wide.

When he added the top, Hickman took another critical step. A piece of thick plywood big enough for the top wasn't readily available so he scrounged two separate chocks of hardwood left around his house by the builders. A 3¾-in.-thick by 4¾-in.-wide beam became the primary worksurface in front and a 1-in. by 5¾-in. board provided extra support in back. Hickman left a 1¼-in. gap between the boards for a sawing slot, handily solving the problem he'd encountered with the chair. Work could be firmly supported on both sides of a cut. The slot was also useful when boring holes and knocking out nails.

Hickman wanted a vise on the bench and he automatically selected a British standard, Record's #52½. Its quick-release capability allowed him to rapidly clear the vise's guide rods and screw from beneath the sawing slot whenever they were in the way. Still, Hickman knew that if he was going to do any planing, he had to devise some way to keep the whole business from sliding around. The solution, he figured, was a ¼-in. plywood platform upon which he could stand while working. The platform was bolted to the Z-braced leg frame, which he had welded by Lotus mechanics out of 1-in.-dia. steel tubing.

With his new bench, Hickman was amazed at the ease with which he could handle large sheets of material. The bench's small size was a real advantage. And being freestanding and easy to maneuver, it could support large pieces while allowing him access to the work from all sides.

But he began to identify serious limitations in the vise. Being right-handed, he wanted it at the bench's right corner for crosscutting and at the left corner for planing. The temporary solution was simply to keep the vise mounted on the left for planing, and to approach the bench from the back for sawing. The vise was also heavy—its 33 lb. were 2 lb. more than the rest of the bench—and, as Hickman notes, "it was a pig of a thing to store." But for the next six years he used the bench without alteration. Despite its crudeness, he continued to find it useful, lugging it about to build two conservatories, a garage, a carport and to work on projects in every room of the house.

The first three Workmate prototypes. Model 1, at left, has two stationary beams bolted to a welded tube frame. The beams on model 2, at right, are smaller, spaced farther apart and unobstructed at both ends to leave room for wedging stock in the slot. It may be knocked down and stored, or transported on its lumbercore base. Model 3, the Minibench, at center, is the first collapsible prototype. The beams themselves don't move, but a separate clamping bar in the slot is operated by two screws from the back of the bench. A Record vise lends the primary clamping source on all three models.

By 1967 he was tired of designing cars and the tedium of manufacture. He had joined Lotus nine years earlier as a self-taught designer, had been principal designer on the Lotus Elan, Elan Plus 2 and Europa sports cars and, ultimately, Design Director. (Among other innovations Hickman introduced at Lotus were the pop-up headlamps and built-in plastic bumpers used on the Elan. It's interesting to note that the Lotus Elite, on which Hickman had been project engineer, was the world's first do-it-yourself automobile, offered in kit form.) Over the years at Lotus, and before that, at Ford, he had acquired extensive production and design skills, and now he decided to apply them to his own inventions.

He cashed in his 5% Lotus shareholdings for £23,000, and cast about for new projects. Of 24 different ideas—from the fledgling workbench to a jam-proof dispenser for artificial sweeteners—Hickman decided to "do something commercial with this baby here," he says, affectionately kicking his earliest model. It seemed to have all kinds of practical applications. But the cost and the weight would be crucial, he reasoned, because anybody could make one for nothing out of scrap, as he himself had done.

In 1968 the workbench design began to move forward. Hickman's second model took a step toward collapsability and rationalized parts. It was a bolt-together structure with identical metal end frames and wooden top bars. The two-piece top was just over 12 in. wide—two 4¾-in.-fir bars separated by a 2⅝-in. slot—and 27 in. long. The bench stood 21 in. high on a ¾-in.-thick lumbercore-plywood (blockboard) base, several of which could be economically cut out of a 4x8 sheet. Like the first model, it was stabilized laterally by a transverse metal tube, which in combination with the broad base kept the structure amazingly rigid. In another stroke of good fortune, Hickman removed the two ½-in.-thick end battens that had connected the beams in the top of the original bench and found that he could now wedge boards in the enlarged sawing slot. For the first time, long pieces could be run right through the open-ended slot and off either end of the bench.

Hickman replaced the Record #52½ vise with the much smaller Record #55, which lacked the quick-action feature. After removing umpteen bolts and screws, he could stack the whole business on top of the lumbercore base for storage or shipping.

But there were problems. Hickman still wasn't satisfied with the left-hand mounted vise and he considered drilling an additional set of holes at the right corner of the top to provide an option for left- or right-handed workers. The vise alone would have added about £6 to the price of the bench. Could it sell for £13 mail order? Hickman had his doubts. Without the quick-action feature, the guide bars were an annoying obstruction beneath the sawing slot, and although it was much lighter to carry, the bench was not exactly streamlined. While it was beginning to look intentional, it was obviously still homemade.

If the bench was to have a future, Hickman would have to get serious about portability and storage. It had to fold, Hickman felt, to the size and weight of that classic of portability, the suitcase. It had to store in a closet or fit in the trunk of a car and you had to be able to lift it without getting a hernia. Hickman began to think about folding mechanisms.

Hickman focused his attention on the base and, for the time being, left the top and vise intact. Within four months, he had mocked up his third prototype, shown in the center above. Mechanisms for ironing boards and folding chairs were just too flimsy for a workbench, so for his folding undercarriage Hickman adapted what was to him a familiar design—the classic 'double wishbone' that had been the basis for independent front suspension in automobiles since the 1930s. The base had four feet for stability, the rear left foot doubling as a height adjuster and a set of rollers when the bench was collapsed. Although the bench still wasn't exactly svelte—it was somewhere between carry-on luggage and a small steamer trunk—Hickman added a fold-away carrying handle to remind people that it was meant to be carried, in case they didn't notice.

At this point, Hickman filed his first patent application. He claims he never doubted that he had a salable product, "I just had to get the formula right." The myriad bolts had been greatly reduced in number, and the bench now collapsed easily by unscrewing a large knob on either side of the base. When folded, it stood up like a suitcase and could be rolled or carried and stored in a small space. As a workbench, it was exceptionally strong and rigid for its weight. But as a workbench, everyone he showed it to thought it was too small. Rather than duck the issue, Hickman instinctively tried to make it an asset. He called it the 'Minibench,' after Britain's fascination at the time (mini-cars, mini-skirts), and cast the name into the aluminum base. He continued to feel that if he could narrow the price gap between the bench and a pistol-grip electric drill (coincidentally, Black & Decker's original innovation), people would eventually judge it on its merits, not against traditional benches. After all, he thought, the public didn't know it couldn't live without refrigerators until it had them.

The Minibench was beginning to look marketable. Hickman dreamed of selling the bench to every do-it-yourself (DIY) enthusiast in England, so he turned to a giant in the DIY industry, Black & Decker, Ltd. (U.K.). Upon the advice of a marketing friend, he decided to jazz up the bench to show Black & Decker how it could be adapted for use with all the tools in their power hand-tool line. He began to develop a set of accessories that would turn the bench into a stand for their horizontal drill holder, drill press, lathe and tablesaw attachment. All were designed to clamp in the sawing slot.

It was while figuring out how to fasten these and other accessories in the slot that Hickman received what was perhaps the only bona fide flash of inspiration in the entire course of the Workmate's evolution. He observed, for the first time, how similar the jaws of the Record vise were to the two beams in the benchtop. "If I can just move one of the beams," he thought, "I can hold the tool anywhere along the top." With his neurons sizzling, Hickman realized, "If I can hold the tool, I can hold the work as well."

But how to do it? Taking his cues from the Record vise, Hickman was temporarily thwarted. First, a mechanism like the Record, with a central screw and flanking guide rods, would obstruct the sawing slot, which he now planned to use for clamping. And since even the precision-engineered Record vise skews out-of-square when an object is clamped off-center, common sense told him not to even attempt it on a 26-in.-wide jaw made mainly of wood. He concluded that two widely spaced screws were the answer, but he hadn't a hope of keeping the jaws parallel without guide bars. Hickman realized that the ability to hold tapered stock would be an asset, but he nearly abandoned the idea because of the resistance he anticipated in the marketplace. The parallel-jaw vise was an icon he was reluctant to tackle.

Twin screws posed another problem. "If it required two hands to wind the screws in simultaneously," Hickman reasoned, "how would the user hold the work while clamping it in the bench—with his teeth?" Hickman considered, but quickly rejected, the notion of linking the two screws with a bicycle chain—it was too complicated and the tolerance on stretching chain would be a problem.

The moving jaw presented another dilemma. Traditionally, it is the front jaw of a vise that is moved to apply pressure against a stationary rear jaw. But Hickman's Minibench would need solid support near both ends of the beams if the top was to function as a work surface as well as a vise. These supports were certain 'knuckle-busters' if vise handles were mounted on a movable front beam. How would people react to a moving rear jaw, he wondered?

For the moment, Hickman sidestepped these questions. He enlarged the sawing slot between the two fixed beams and added a third moving bar made of $1\frac{3}{8}$-in.-thick fir between them. When fitted with two screws, which could be operated from the back of the bench, the bar functioned as a secondary vise (he remained reluctant to part with the Record) and as a

clamping unit for power tools and accessories. When you wanted to clamp with the bar, you simply reached across to the other side of the bench. The idea of using half the top for a moving jaw was still appealing, but Hickman worried that tradition ran strongly against working on a moving vise jaw (he had been around a cabinetmaker's bench enough to know it is taboo to work on top of the tail vise). Plus, there was something inherently ludicrous, he thought, in designing a small worksurface that grew even smaller as the vise was used. His compromise allowed him to keep the two outside beams stationary—only moving the clamping bar between them. Wooden guides were screwed to the underside of the bar to keep the jaw vertical when viewed in cross section. To cope with its inevitable, non-parallel action, the guides were allowed to swivel, and Hickman made a pivoting joint at the ends of both screws.

In practice, the new bar-vise worked pretty well. It enabled either square or tapered stock to be clamped between the beams of the top much more securely than in most single-screw vises. Initial concerns proved groundless. The non-parallel action was not a problem, and the floating jaw could be tightened easily with one hand holding the stock while the other cranked the handles, one at a time. But it had its limitations: the fir clamping bar held work only up to about 2 in. thick, and even when reinforced with a steel plate on its underside, it flexed when an object was clamped in the middle. And the Record's guide rods still got in the way of the slot.

Having proved in principle that the twin-screw vise was going to work, Hickman was finally ready to jettison the vestigial Record. It had grown smaller over the years and, although the Record vise was to British woodworking what chrome bumpers were to the auto industry, he had to admit that the traditional vise had been holding him back. Getting rid of the Record enabled him, in one stroke, to slash the cost and weight of the bench while making it equally suitable for planing and sawing and for right- and left-handed users. His vise would be five times as big and half the price—an unbeatable value, Hickman figured. "I got bolder and bolder by the minute." Reversing what had been an immutable vise principle for centuries, he tackled the knuckle-bashing problem by fixing the front beam of the top and moving the rear one. He went on to eliminate every redundant bolt, screw and washer, swapped the cast portions of the frame for pressed steel and turned the steel 'wishbones' into H-frame aluminum castings, which were to become the hallmark of the early Workmate.

Armed with the Minibench and the drawings of his recent modifications to it, Hickman arrived for his first interview with Black & Decker in December, 1967, at their office in Maidenhead, England. He had beaten the manufacturing cost down to £5, which meant it could be sold for less than twice the price of the cheapest electric drill on the market. To counter any possible objection to its size, he came prepared to describe several different ways it could be altered to gain height. Hickman was confident.

Black & Decker, however, was unconvinced. Why would the do-it-yourselfer want such a big vise? How often do you need to hold tapered work? Could they call it a workbench? Hickman had spent 30 minutes trying to explain the bench—how could they hope to sell a customer in 30 seconds? In hedging his bet, he had confused the issue by showing an out-of-date prototype and by counting on the drawings to demonstrate how the 'new and improved' version would be better. "I made a tactical error," Hickman says now, "by immediately pointing to its limitations, rather than its advantages." He also learned a fundamental lesson about sales—"if you presented the guy a faulty prototype, you were giving him 101 reasons not to do it."

"If they'd offered me £50,000, I would've jumped at it," Hickman told me. "If they'd offered me £100,000, I'd have been over the moon." On his way out, Hickman told them: "One day you're going to come back to me, and you'll have to pay me a lot more." "I wouldn't be the least surprised," one of the senior marketing executives replied.

Undaunted by this initial setback, Hickman took his proposal to Stanley Tools Ltd. "I got a very good reception, had a very good demonstration and received a very good rejection." On April 1, 1968, Stanley wrote Hickman that the potential of the product "could be measured in dozens rather than in hundreds."

Over the next several months, Hickman peddled his bench to every major hand-tool manufacturer in Great Britain who would listen, at the same time that he continued to fine-tune the design. "I was beginning to lower my sights," Hickman admits. "All right, maybe it's not going to be the big winner I thought, but anybody's got to be better than nobody." He approached Record, Spear & Jackson, and Marples, among others, and was rejected by all. Meetings were getting shorter and less encouraging. He was being backed into a corner and the corner was getting smaller. "I knew the British market, and there was nobody left who could market it," he recalls.

Finally, late in 1968, Hickman turned to the only one who believed in the product; he decided to manufacture it himself. His wife, Helen, supported him: "If you're really convinced it's going to go, you have to do it. If you don't...you'll never forgive yourself." They bet all they had on the bench—the windfall of Lotus shares and the title deeds on their house. In 12 days of working flat out, Hickman finished his drawings and made several prototypes (with the cast-aluminum *H*-frame base). He had been thinking about changing the name to break away from the 'too small' reaction and to emphasize its assets. He checked the listings of registered trademarks and cast 'Workmate' into the new model, which he called Mark I.

With the Record vise gone for good, Hickman refined his giant, built-in vise. He replaced the fir in the two top beams with beech. He realized that for the two vise screws to function independently, the nuts that connected them to the movable rear jaw would have to act as pivots, permitting the jaw to angle. Hickman's solution allowed each nut to float in an oversized channel, which also supported the beams and held them flat. Although the steel channels had to be near the ends of the beams for maximum support, the farther apart the screws were located, the more leverage could be developed. This was, theoretically, the weakest link in the Workmate's design and Hickman worried that if someone tried to open one end of the vise while the other was fully closed, the closed screw would act as a fulcrum, developing sufficient force to destroy the top or crush the workpiece. (In reality, this never occurred—people seemed to realize the problem almost instinctively and they relieved the pressure by backing both screws off.)

Hickman had effectively 'de-engineered' the vise. "I took all the critical engineering out of it," he says, pointing for comparison to the rigid precision-machined castings and steel guide rods in the conventional Record vise. The Workmate de-

veloped more effective pressure with less effort (and much less leverage on its tiny handles) than the Record, and would continue to work well even when the threads began to wear. It also was able to grip small stock outboard of the screws. "True angularity [of the jaws] was icing on the cake," Hickman says, referring to the Workmate's capacity to clamp tapered work. "Although it was nothing I set out to design, it is arguably the thing that made it work....In innovation, you find solutions to things you aren't looking for and problems you don't know exist." The jaws were able to contact the whole length of a workpiece, resulting in a better grip with less pressure than any conventional woodworking vise. Hickman remains convinced that the eventual success of the Workmate has a lot to do with the fortuitous versatility of what he likes to call his 'gentle-giant' vise. He admits that while there are many ways to design a portable, folding workbench, without the vise it never would have gotten off the ground.

Hickman's next step was production. He rented a barn at an old brewery, hired two production workers, and lined up a foundry to do low-pressure, gravity-fed aluminum die casting—a much improved method to the tedious sand casting he'd used on the prototypes. (The *H*-frames for the first couple of hundred Mark I Workmates were actually machined on another Workmate by clamping a drill press between the jaws of the top and mounting an extra vise on the front to hold the part being worked.) Hickman called his new company Mate Tools Ltd. "I reckoned I could break even and maybe make some money on 5,000 units per year—if I was lucky," he says. He swallowed hard and printed an optimistic 10,000 brochures.

Hickman hit the exhibition circuit and the Mark I began to attract attention. He was encouraged by the 120 orders taken during his first two-week stint. The shows were a good proving

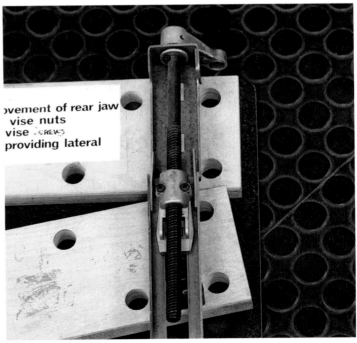

This bottom view of Workmate's vise screw reveals several important features. The nut is attached to the rear jaw by means of a pivot, which allows the jaw to be angled. The slotted channel-iron guide provides lateral restraint and supports the top; the screw is free to angle slightly as required by the configuration of the jaws.

ground for Workmate—customer response was immediate—and Hickman discovered an unanticipated marketing predicament: the vise was invisible. "Nobody could *see* the vise because it didn't *look* like a vise," he recalls. "How is the wood held?" folks wondered. "Why doesn't it fall out?" Demonstrations and good literature were the answer. Hickman hired an acrobat to demonstrate that, as the advertising claimed, "You can do anything on a Workmate." At one show, the actor quickly flipped a Mark I into position, bounced it on the floor and then did a handspring on the top. The bounce jumped the safety catch and the bench collapsed under the actor's weight. He rolled across the floor unhurt and popped up some distance away, exclaiming: "That's the one thing you shouldn't do on a Workmate." Sales climbed to 1,500 units the first year and doubled in each of the next four years. In 1971, David Johnson, editor of Britain's *Do-It-Yourself* magazine, predicted: "...I have no doubt that within just a few years...Workmate will become as invaluable as the electric drill."

In the marketplace, Hickman's worst fears went largely unrealized. Nobody complained about the details of its construction, the moving rear jaw or about having to work on the moving jaw of the vise. Nobody found the bench too narrow. Nobody pinched their fingers in the vise. But Workmate was still fighting a larger battle of identification. It had been conceived as a combination workbench, sawhorse and vise, a concept that was at once its greatest strength and Hickman's greatest challenge. As a workbench people thought it was too small. As a sawhorse it was too heavy and expensive. As a vise it seemed unnecessarily large and complicated. In response, Hickman pointed to the testimony of one journeyman carpenter who responded to the question, "Aren't you bothered by the fact that it's 9 in. lower than a standard workbench?" by saying: "No—it's 23 in. higher than the floor I usually work on." Hickman also argued (largely in vain) that it could be comfort-

ably sat upon in many woodworking operations. Attempting to put the workbench comparisons to rest, Hickman developed the Team-Mate, an attachment that provided a second higher and larger worksurface by (17½ in. wide by 37 in. long by 32 in. high), with a shallow tool tray in the middle and a standard Record vise.

In Mate Tool's third year, with more than £100,000 invested in advertising, tooling costs and inventory, Hickman and his new partner, Derek Bernard approached another watershed. They had adopted a 'marginal costing' method, which got them out in the field and allowed them to boost production but provided slim profit margins. They were flying; what they needed to really gain altitude was to bring the cost down still further through mass-production. At the same time, Team-Mate had been only partly successful and Hickman felt that the time had come to tackle a dual-height Workmate. Hickman went back to the drawing board for nine months to concoct it.

In the midst of the Workmate's redesign, Hickman was approached by three different tool manufacturers, one of which was Black & Decker (U.K.). Despite the success of Mate Tools, Hickman had always believed that only a company like Black & Decker could take it worldwide. Not wanting to repeat his earlier mistake of tipping his hand before he was ready, Hickman tried to put them off. "We have a major new development underway," he told them. "We'll get back in four months." But Black & Decker was impatient. They worried about Hickman's patents, which were all pending, so preliminary talks began. At the outset, Black & Decker offered a ½% royalty—"rubbish," according to Hickman who was asking for 5%. "We just said, 'Cheerio, gentlemen,'" Hickman recalls. Not long afterward, the managing director of the company, Walter Goldsmith, requested another meeting. Hickman consented and six months of serious negotiations began. The deal they finally struck was for a 3% royalty—and Hickman would continue to receive half

Workmate's versatility is difficult to describe. Hickman credits the success of early marketing ventures to good publicity demonstrating a variety of operations, and to photos like these, showing the Mark II Workmate in action.

that amount even after the patents ran out. Black & Decker was so keen to get moving that they stationed two draftsmen in Hickman's dining room, where they worked to his specifications. They adopted what was an unprecedented attitude toward a new product, allowing Hickman virtually total design authority.

"The Mark II Workmate looked like nothing on earth," Hickman says, pleased with his creation. Holes to lighten the bench had been cast into the aluminum base. The large, lumbercore platform was replaced with a single, narrow step in front and, most importantly, the bench was fully adjustable to two heights—a sawhorse height of $23\frac{1}{4}$ in. and a full bench height of $32\frac{1}{4}$ in.—by means of four, flip-out legs. The rear legs hinged toward the back, the front legs toward the sides, giving the Mark II an exceptionally rigid and stable footprint that resisted movement when working either across (crosscutting) or along (planing or ripsawing) the length of the top. Height levelers were added to both back legs. The imported beech used in the top beams, which had become increasingly difficult to procure, was replaced by a $\frac{7}{8}$-in.-thick, 17-ply birch laminate. To both jaws was added a longitudinal V-groove for holding pipes or other round objects, and two rows of round dogholes. The four plastic dogs included with each bench could be swiveled in their holes to provide a whole range of new clamping possibilities. The vise mechanism was left essentially intact except for the screws themselves. These were changed by Black & Decker engineers to a twin-start thread, which eliminated the need for the complicated quick-release devices Hickman had been developing. The twin-start threads worked at twice the speed of the standard Acme thread used on the Minibench, and provided more than adequate clamping pressure.

"Many people to this day say it was the best model," according to Hickman. To my eye, surrounded by a host of more recent Workmates, it is unquestionably the most attractive. Reminiscent of NASA's lunar module, there is something aesthetically appealing in the angular, *I*-beam construction of the aluminum castings, an appeal lost in the plethora of stamped-steel models that followed.

Black & Decker began Mark II production at their Spennymoor plant in County Durham in 1972. Almost overnight, they became Britain's largest consumer of aluminum castings, outside of the automotive industry. The dual-height Mark II hit the market at £24.95, and the following year Black & Decker launched the Workmate in Europe and most of the rest of the world (excluding the Americas, Australia and Japan). In the beginning, their greatest challenge was not how to sell it, but how to keep up with demand. In 1974, construction began on a new plant in Kildare, Ireland, dedicated to Workmate production (a half-million of them a year).

The previous year, shortly after its English launch by Black & Decker, the Mark II Workmate won one of Britain's prestigious Design Council Awards. It was the first time that this annual award had been granted to a tool, and it precipitated one of the toughest debates the Council had ever had. They didn't know what to make of the Workmate. Yes, it works, they agreed. Yes, it sells like hotcakes. But is it good design? Finally, the chairman of the judging panel asked the members if there was anything they could suggest to improve it. They responded in unanimous silence, and the award was presented to Hickman by The Duke of Edinburgh. (Hickman guesses that the award never sold a single Workmate.)

Hickman's drawing of the Mark II Workmate, the bench that finally clinched the deal with Black & Decker (U.K.).

While the Workmate was selling out in Europe, across the Atlantic Black & Decker Inc. (U.S.) continued to reject the product—not just once, but three more times. An informal study seemed to confirm what the U.S. managers suspected—that its success was due to factors peculiar to the European environment: the number of apartment dwellers, the strength of the do-it-yourself industry, etc. Hickman approached Stanley and Sears in the United States and was rejected by both. (This didn't help the Sears case several years later, when Black & Decker and Hickman sued them for infringement of patent with their Work Buddy, the Sears clone of the Workmate. See the sidebar on p. 218.) The Workmate was thought to be irrelevant to the U.S. market. Being a non-electric tool, it was also problematic for Black & Decker, which had built its reputation and marketing around portable power tools.

Years later, Hickman noted the irony of an 'invention-based' company that was so reluctant to stick its neck out with the Workmate. In 1910, S. Duncan Black and Alonzo G. Decker cofounded a small machine shop in Baltimore, Maryland. The first products to carry the Black & Decker trademark were inventions that would have a profound effect on mechanics and craftsmen for the rest of the century—a portable air compressor and a $\frac{1}{2}$-in. pistol-grip electric drill.

Finally in 1973, on the initiative of two of their North American managers, Black & Decker negotiated a license for

In defense of the Workmate

Ron Hickman has devoted enormous energy (and over £1.3 million) first to establish, and then to defend the Workmate from patent infringements. Working closely with his patent agent, Michael Roos, Hickman and Black & Decker have won every legal action they've initiated. The patents—about 20 of them—were filed on virtually every significant aspect of the Workmate. Were it not for the fact that Hickman acquired at the outset several 'foundation' patents, identifying such features as the vise-as-work-surface, the angularity of the jaws, collapsibility, dual-height and the 'double-wishbone' construction, it's unlikely that the Workmate would have survived the years of intense market exposure. It's certain that Hickman (not to mention a host of patent lawyers and Black & Decker employees) would be a lot poorer.

The object of the patent system is to buy breathing space for the inventor. The patent grants him a 17-year head start (in the United States) on any potential competitor who would use his idea to commercial advantage. To effectively prosecute an infringer, however, the inventor must prove first that it is a bona fide infringement, and second that his original idea was indeed original.

Meanwhile, in his defense, the accused infringer will trundle out any number of examples of what is known in patent jargon as 'prior art' (a pre-existing object that is alleged to perform similar functions as the besieged patent). If a piece of prior art is demonstrably and significantly similar to the item in question, the patent is void. Moreover, if any two examples of prior art can be put together to arrive at the invention in question, their combination also constitutes a knock-out punch, unless they have been combined 'with a surprising result.' No matter that the examples of prior art might have existed in different centuries or on opposite sides of the globe. The inventor need not even have been aware of them.

It's well known to furnituremakers how hard it is to protect a new design. Knock-offs are common in the trade and the law is riddled with loopholes. In the case of the Workmate, the infringement itself has been easier to prove than originality. Indeed, what constitutes an *original* workbench—one of the most traditional tools in a traditional craft? It's not an easy question, and patent law comes in many shades of gray.

In its first 15 years, the Workmate has been cloned more than 20 times around the world. These range from the 'Alko' and 'AEG' benches in Germany, which copied the top and vise but ignored the base, to the Japanese 'Kinzo,' which copied the Workmate so closely that Black & Decker was able to re-label and sell the confiscated stock after they'd won an out-of-court settlement.

Only a few clones have been a major commercial threat and have resulted in court battles, the most notable being with Sears, Roebuck and Co. in America. After selling a 'Craftsman Portable Workcenter' (made by Black & Decker) for several years, Sears began to market their own 'Work Buddy.' The top and vise resembled the Workmate closely—according to Hickman, some were even built with Black & Decker vise screws and handles.

Sears contended that the essence of the Workmate was its twin-screw vise, which permitted angularity of the jaws. Citing the 18th-century Dominy benches (see Chapter 1) at Winterthur, Delaware, among other examples of prior art, they tried to establish that this type of vise belonged in the public domain. But despite its superficial similarity, the Dominy vise is not nearly as effective at clamping stock of different sizes and shapes as the Workmate. More to the point, however, was the fact that its vise jaw does not constitute a worksurface.

Sears then introduced as evidence a 1½-ton, cast-iron, Fay & Egan door-gluing machine (patented in 1909), and a tiny hand-held jeweler's

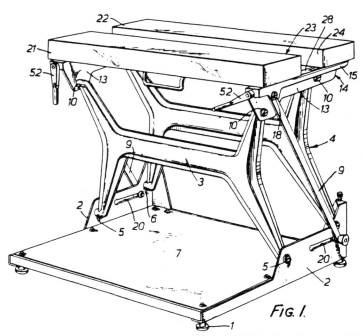

The original British Workmate patent (No. 1267032), filed March 4, 1968.

chuck (patented in 1928), which in combination were said to provide the Workmate's essential components. Hickman likened this comparison to mating a mosquito with an elephant. "It might produce something interesting," he told the jury, "but I very much doubt it would produce a Workmate." In any event, it could have been argued that the result would have been 'surprising.'

With examples of the behemoth and the jeweler's chuck on hand at the Baltimore trial, Sears claimed that the relationship would have been obvious to any person 'skilled in the art' (another ill-defined bit of patent terminology). The jury disagreed, as did the U.S. Appeals Court for the 5th Circuit, which upheld the lower court ruling.

With this kind of scrutiny, it was perhaps inevitable that someone would eventually ask the question: what is a workbench? As it happened, a definition by Sir Patrick Graham in the High Court of Justice ultimately saved the day in Hickman's closest call in the British courts. At the trial, two bookbinding presses were introduced as prior art. It was determined that these presses met all the specifications of the

Workmate's main claim but one. They were not workbenches. In High Court patois, a workbench is "a surface of sufficient area to enable one to carry out manual work upon it, such surface being supported at a convenient height, and sufficiently rigid, having regard to the nature of the relevant work."

In his extensive research, Ron Hickman has identified only two pieces of prior art having the clamping gap, or vise, running down the middle of the worksurface. But one (Gillette of 1916) had no vise screws, only a lever action, while the other (Fleming of 1948) did not use the moving jaws as a part of the worksurface. In both cases, the vise was largely ineffective, leaving the Workmate's claim of originality intact.

Although the court settlements cost Hickman and Black & Decker more than they received in damages, their cumulative effect eventually was to discourage counterfeits. "Some we've killed quickly, some we've killed more slowly; the point is we've killed them all," Hickman says. After surviving nearly a decade of litigation, the wave of infringers has begun to subside, never to return, Hickman hopes.

'the rest of the world' with Hickman, and tooled up their Brockville, Ontario, plant for Workmate production. Black & Decker, Inc. (U.S.) placed 10,000 Workmates in about 50 stores in southern California to test their marketing strategy. People snapped them up and the following spring Black & Decker went national. Once they had decided to jump, they went all the way. According to Jim Martz, Black & Decker's Director of Corporate Market-Product Development in the United States, "We really bet the farm on it." Martz, who was marketing manager on the Workmate production line and took it national, is convinced that television carried the day. "It was the first time we had gone on T.V. betting that something was going to happen." And it did. In a display case in his studio, Ron Hickman has a miniature silver Workmate, made as a surprise by Derek Bernard and given to Hickman in April, 1976, in commemoration of the one millionth Workmate sold. It is inscribed: "To Ron Hickman For Having The Courage of His Convictions." For those convictions, Hickman had also won himself a 3% royalty on all 'rest-of-the-world' sales.

As far as Hickman is concerned, there was no other way to go. "With a really new concept, you need enormous confidence to launch and to advertise...it's not a case where you can put a toe in the water to test the temperature," he says. "You have to get out there and heat up the ocean." He had learned from his very first days flogging the Mark I for Mate Tools that the Workmate's biggest challenge was demonstrating its applications. Hickman agrees with Martz about the impact of T.V. on the Workmate's success; T.V. was used extensively in Great Britain and Holland, where Workmate's 'penetration rate' is strongest, as well as in the United States and Canada. By contrast, sales have been sluggish in the Scandinavian countries, which have no commercial television. And, to date, there has been no serious attempt to sell Workmates in the Third World or the Soviet Union.

To enter Hickman's design studio, you must pass below a single cast-aluminum *H*-frame hung over the door—Workmate's answer to McDonald's 'golden arches.' In the middle of the energy crunch of the 1970s, Black & Decker abandoned the aluminum *H*-frame, arguably Hickman's greatest aesthetic stroke, and replaced it with stamped steel. If the aluminum *H*-frame represents the apex of the Workmate's design, I wondered why it was discarded only a year after being introduced in North America? Among tool and machine aficionados, quality is often synonymous with casting, and the recent Workmates with their garish blue steel frames pale by comparison.

Basic production economics is the answer—it suddenly got a whole lot cheaper to 'bash steel,' as Hickman says, than to cast aluminum, which requires a lot of electricity to produce. And Black & Decker, which assumed control of the Workmate's design and development along with the manufacturing rights, was always looking for ways to trim costs. The switch to steel not only reduced the cost of the bench, but according to several Black & Decker managers I spoke with, also improved it. If you accidentally dropped an aluminum model, or dropped something on it, the frame might snap—especially if it had an undetected flaw in the casting. The stamped-steel Workmate will always bend or buckle first. "Most of the feedback we get is, 'Why did you ever cheapen the Workmate?' It's really a safer, stronger product today," they argue.

Following its North American launch in 1975 and throughout the ten years that followed, all the large Workmates for the U.S. and Canadian markets have been manufactured in Brockville. (Overseas, they have been made in nine different countries.) When I visited the Brockville plant in August, 1985, the Workmate line was being packed up and moved to São Paulo, Brazil. 'Updated positioning,' 'rationalized markets' and 'globalization' are the marketing euphemisms Black & Decker's executives use to describe cutting costs. (Canada is not exactly a low-cost producer of either steel or labor.) If the Brazilian venture is successful, Workmates may eventually be built there on one huge production line, and shipped all over the world with instructions and packaging in different languages and universal symbols.

"I don't like or agree with everything they do," Hickman told me between laps in his Jersey pool. And he continues to feel that, while the aluminum castings may cost a bit more and may have been a little more fragile, "it was a retrograde step to lose that hallmark of the design." But with the signing of each contract, Hickman effectively handed over the marketing and product development to Black & Decker.

The old adage 'don't fix it if it's not broken' doesn't seem to be in the marketing vocabulary. In fact, in their own way, marketers are inveterate tinkerers. At Black & Decker, they asked: why not make it larger, or smaller, with a flip-up top, paint it gray, substitute plastic for metal or particleboard for plywood? And they did. They also reckoned that they could sell five times as many of a cheaper model. The result was a slimmer, fixed-height 'continental' version, which was created by the design chief of the company's Italian division. Soon after, the Americans tried a similar model, and the Germans weighed in with a much bigger one.

While the engineers continue to tinker, the Workmate has reached the point where, as one Black & Decker executive put it, "There's less and less meat on the bone to fine-tune." In the United States alone it's getting hard to tell one generation from another. At last count, there were at least five different models for sale, but none has come close to the phenomenal levels of

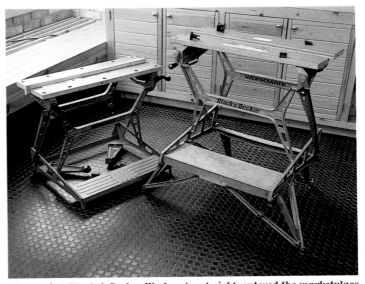

The earliest Black & Decker Workmate, at right, entered the marketplace with a cast-aluminum base. On recent British models, at left, the classic H-frame was retained, but wedded to a stamped-steel base. On the latest North American Workmates the base is made entirely of steel.

the classic dual-height model. As Hickman put it: "It was never as ripe as when I made it…riding a crest of the do-it-yourself tidal wave. It was the classic example of the right thing at the right time."

After the unprecedented success of the North American launch, Workmate sales continued to multiply exponentially through the 1970s. Then, in 1980, Black & Decker raised the price by about 25% in the United States and, as one company executive put it, "The prices were up and down like a toilet seat." At the same time, they cut back on promotion and in 1982 took it off T.V. The results were predictable. Some Black & Decker executives are now beginning to wonder if the Workmate has outlived the natural life cycle of a product. Black & Decker's Martz doesn't think so. "Life cycles are man-made," he told me, and he's convinced you can rejuvenate a product.

Whatever its future, by any accounting, Workmate's success has been extraordinary. Outside the door to Hickman's studio is another trophy. This one is a full-size, chrome-plated Workmate, presented in 1981 to mark the ten millionth Workmate; five years later, that figure has been doubled.

Clearly, the Workmate is a tough act to follow, and I wondered what Ron Hickman had planned for the future. He had made his millions and could luxuriate in a well-deserved retirement. Hickman explained that he had considered about 40 Workmate accessories—only a few of which were ever pursued. "It's the future I came unstuck on," Hickman says.

Before I left, Hickman took me on a tour of his storeroom beneath the house. Located in an old concrete swimming pool, on top of which the new house was erected, Hickman's storeroom is really a morgue. There he keeps a complete collection of all the dead-end accessories and would-be Workmates from around the world—those that never saw the light of day or wilted quickly in the marketplace.

Shelves are jammed with Mitremates, Routermates, Gripmates, Sawmates and Hobbymates, along with Log Grippers and the original chain-driven test-model Workmate. A few accessories enjoyed some limited success: the Handy Jack step stool, which combines a stepladder with a mini-Workmate,

sold 50,000 in England before being rated a flop; the 8-in. Hobby Crafter, a lightweight plastic Workmate on an articulated ball-and-socket joint, the Bench Top Workmate, a drop-leaf Workmate and the Quick Vise, are all still available in the United States. Mixed in among them are the imposters—the world's most complete collection of Workmate clones.

When Hickman moved to Jersey, he set up Tekron International Ltd. as a vehicle for developing future designs. In six years of operation (with 17 employees) Tekron made over £500,000 in licensing a number of the 50 other inventions they had developed. They spent five times that much to do it. Hickman keeps a board labeled: "Some of Tekron's Less Successful Ideas!" upon which he displays a diverse collection of their creations. Among them are an improved claw hammer, a drill guide, The Peeping Tom (an articulated pocket mirror and lamp), The Footprint Potty (an unspillable child's toilet training pot) and an adjustable drafting square. Too large to fit on the board, are The Dizzy Whiz and The Bug, marvelous spinning toys that look to me as good as anything on the market. "These gadgets highlight the fact that it was damn close to being a good idea," Hickman explains, "but we didn't get the product right." Several times during my visit, he made oblique reference to one or two other unspecified projects in the works—hints that I was not invited to pursue.

Will lightning strike twice in the same place? Unlikely. But, as Hickman points out, it's a matter of finding that ephemeral combination of ingredients—the right formula. "I'll never have that amount of energy again," he says, "but you don't have to do it all yourself." He can now afford to hire energy when he needs it.

Despite Workmate's unparalleled success, Hickman expressed a touch of disappointment. "It may seem contradictory," he confessed, "but it is still misunderstood—it's not taken seriously." Reflecting on the day when the critical patents begin to expire, Hickman said earnestly: "I'll be interested to see what free enterprise does." Having spent almost 20 years of his life inventing, refining and defending the Workmate, Ron Hickman is getting ready to sit back and see if the world can build a better portable workbench.

Appendices

Shaker bench

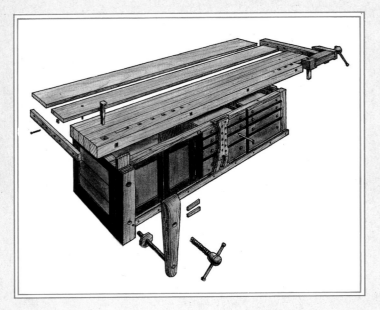

Note: *Many parts of this 19th-century Shaker workbench are not dimensionally regular, and some measurements have been altered slightly for clarity. Moreover, where internal joints and hardware could not be examined, probable dimensions have been recorded.*

This bench was measured and drawn with the permission of Hancock Shaker Village, Inc., Pittsfield, Massachusetts.

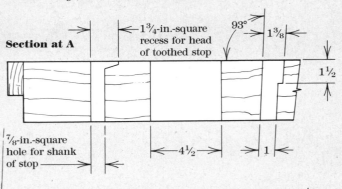

Section at A

1¾-in.-square recess for head of toothed stop

93°   1⅜

1½

⅞-in.-square hole for shank of stop

4½   1

141

Top (vises removed)

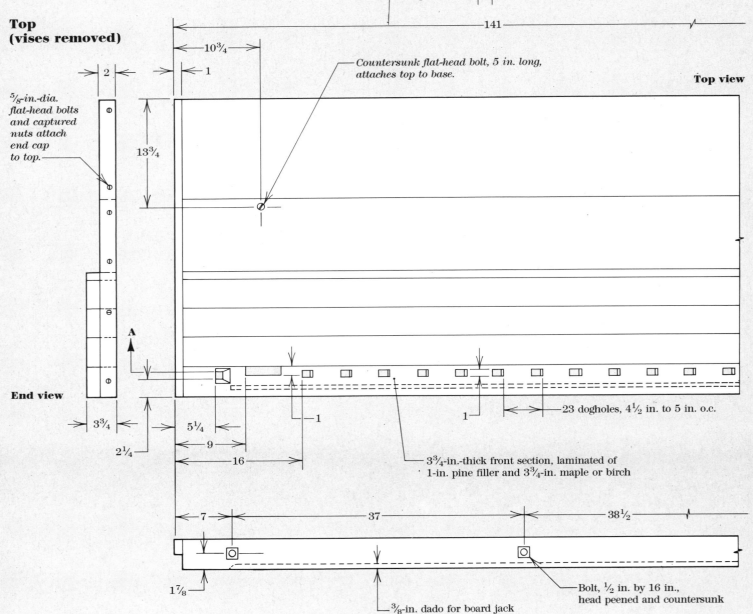

⅝-in.-dia. flat-head bolts and captured nuts attach end cap to top.

10¾   1   2

13¾

Countersunk flat-head bolt, 5 in. long, attaches top to base.

Top view

End view

A

3¾   2¼   5¼   1   1

9

16

23 dogholes, 4½ in. to 5 in. o.c.

3¾-in.-thick front section, laminated of 1-in. pine filler and 3¾-in. maple or birch

7   37   38½

1⅞

Bolt, ½ in. by 16 in., head peened and countersunk

⅜-in. dado for board jack

Right end cap

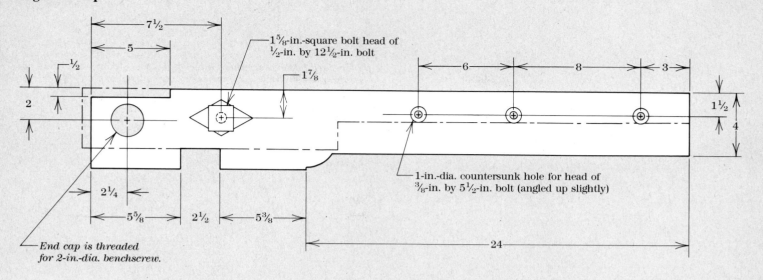

$7\frac{1}{2}$

5

$\frac{1}{2}$

2

$1\frac{5}{8}$-in.-square bolt head of
$\frac{1}{2}$-in. by $12\frac{1}{2}$-in. bolt

$1\frac{7}{8}$

6

8

3

$1\frac{1}{2}$

4

1-in.-dia. countersunk hole for head of
$\frac{3}{8}$-in. by $5\frac{1}{2}$-in. bolt (angled up slightly)

$2\frac{1}{4}$

$5\frac{5}{8}$

$2\frac{1}{2}$

$5\frac{3}{8}$

24

End cap is threaded
for 2-in.-dia. benchscrew.

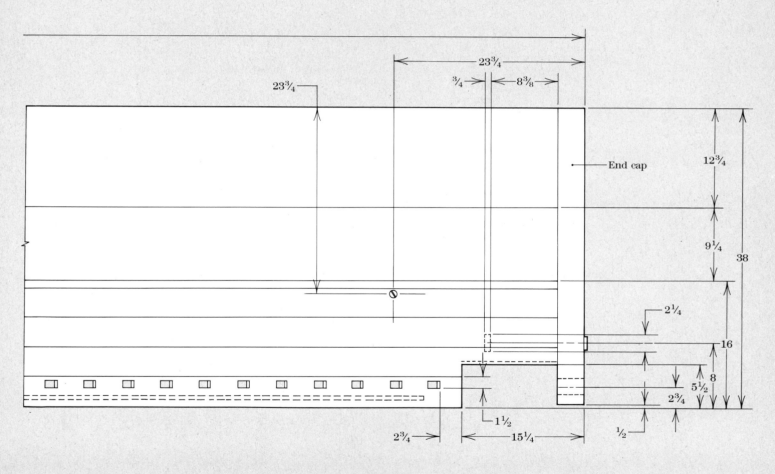

$23\frac{3}{4}$

$23\frac{3}{4}$

$\frac{3}{4}$

$8\frac{3}{8}$

End cap

$12\frac{3}{4}$

$9\frac{1}{4}$

38

$2\frac{1}{4}$

16

8

$5\frac{1}{2}$

$2\frac{3}{4}$

$1\frac{1}{2}$

$2\frac{3}{4}$

$15\frac{1}{4}$

$\frac{1}{2}$

Front view

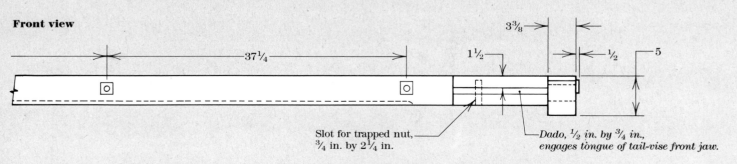

$37\frac{1}{4}$

$3\frac{3}{8}$

$1\frac{1}{2}$

$\frac{1}{2}$

5

Slot for trapped nut,
$\frac{3}{4}$ in. by $2\frac{1}{4}$ in.

Dado, $\frac{1}{2}$ in. by $\frac{3}{4}$ in.,
engages tongue of tail-vise front jaw.

Carcase (doors and drawers removed)

Note: *All blocking, backboards, drawer guides and runners are nailed in place.*

Top view

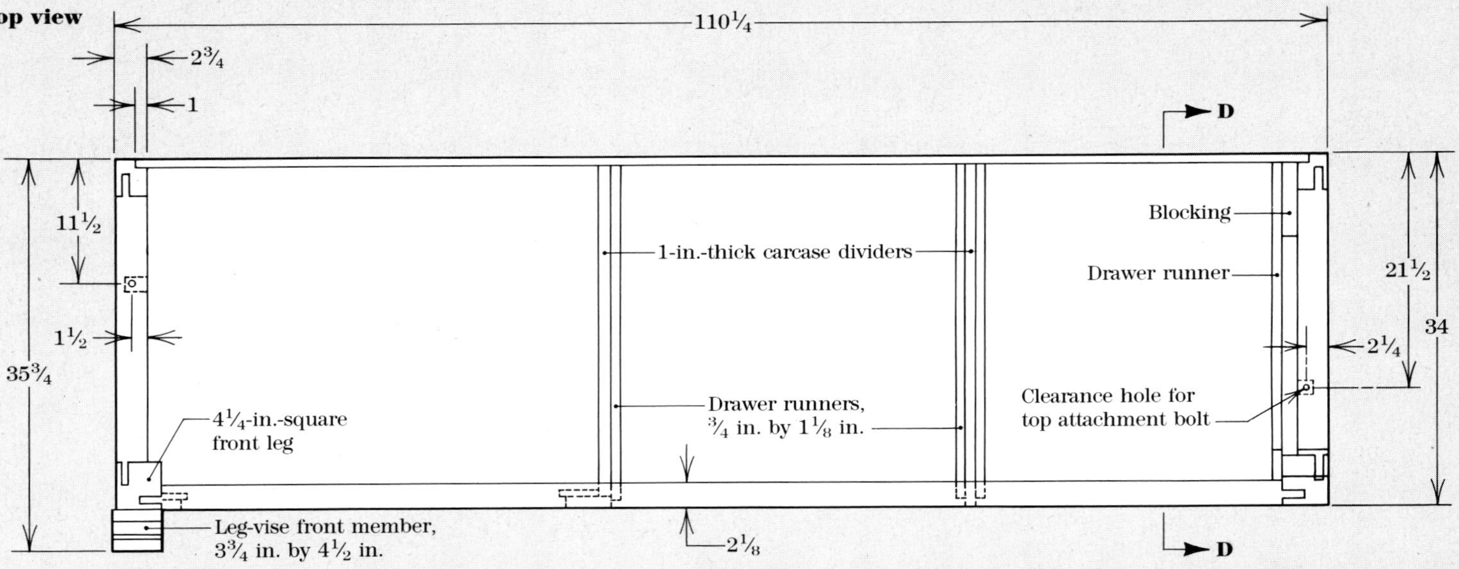

110¼
2¾
1
D
11½
1½
35¾
Blocking
1-in.-thick carcase dividers
Drawer runner
21½
34
2¼
Clearance hole for top attachment bolt
Drawer runners, ¾ in. by 1⅛ in.
4¼-in.-square front leg
Leg-vise front member, 3¾ in. by 4½ in.
2⅛
D

Section at B

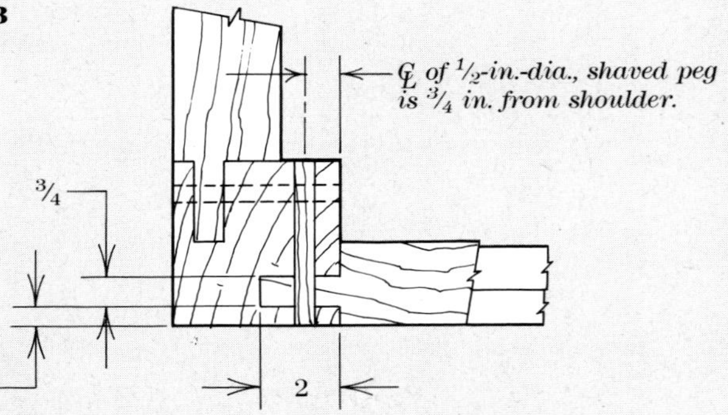

℄ of ½-in.-dia., shaved peg is ¾ in. from shoulder.
¾
½
2

Note: *This detail is typical of frame assemblies throughout. Pegs are at different heights front and end.*

Section at C

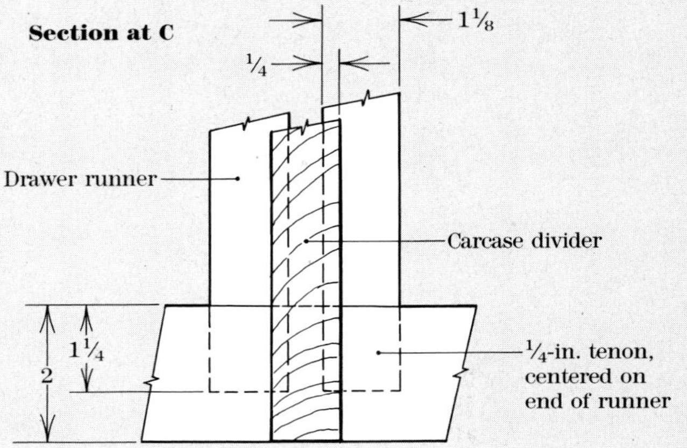

1⅛
¼
Drawer runner
Carcase divider
1¼
2
¼-in. tenon, centered on end of runner

Front view

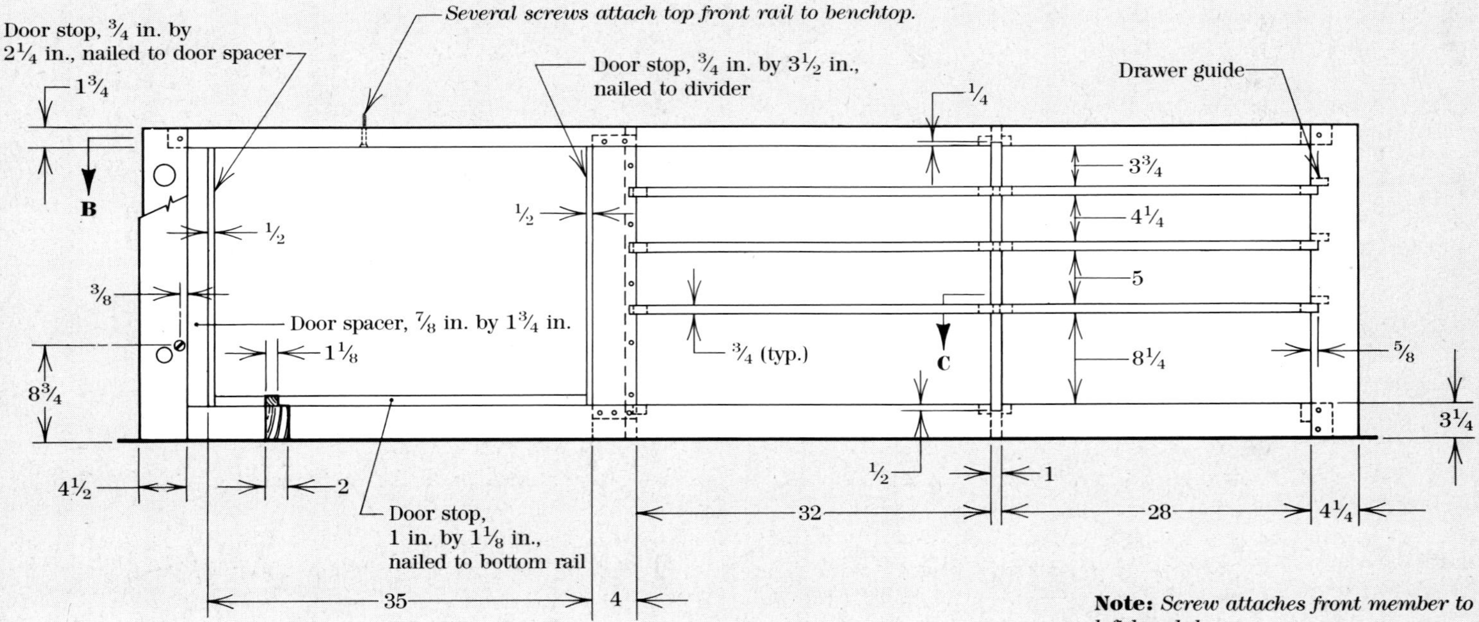

Several screws attach top front rail to benchtop.
Door stop, ¾ in. by 2¼ in., nailed to door spacer
1¾
B
Door stop, ¾ in. by 3½ in., nailed to divider
Drawer guide
¼
3¾
½
4¼
½
5
⅜
Door spacer, ⅞ in. by 1¾ in.
¾ (typ.)
C
8¼
⅝
8¾
1⅛
3¼
4½
2
½
1
32
28
4¼
Door stop, 1 in. by 1⅛ in., nailed to bottom rail
35
4

Note: *Screw attaches front member to left bench leg.*

Section D-D

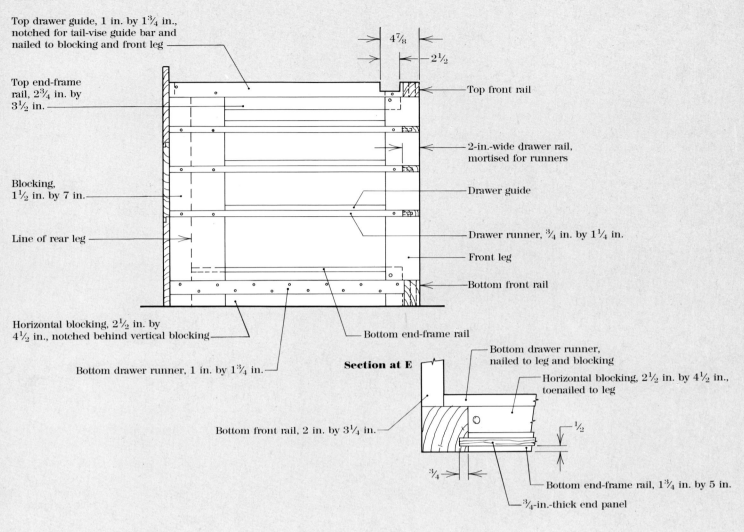

Top drawer guide, 1 in. by 1¾ in., notched for tail-vise guide bar and nailed to blocking and front leg

Top end-frame rail, 2¾ in. by 3½ in.

Blocking, 1½ in. by 7 in.

Line of rear leg

4⁷⁄₈

2½

Top front rail

2-in.-wide drawer rail, mortised for runners

Drawer guide

Drawer runner, ¾ in. by 1¼ in.

Front leg

Bottom front rail

Horizontal blocking, 2½ in. by 4½ in., notched behind vertical blocking

Bottom end-frame rail

Bottom drawer runner, 1 in. by 1¾ in.

Section at E

Bottom drawer runner, nailed to leg and blocking

Horizontal blocking, 2½ in. by 4½ in., toenailed to leg

Bottom front rail, 2 in. by 3¼ in.

½

¾

Bottom end-frame rail, 1¾ in. by 5 in.

¾-in.-thick end panel

Left end view

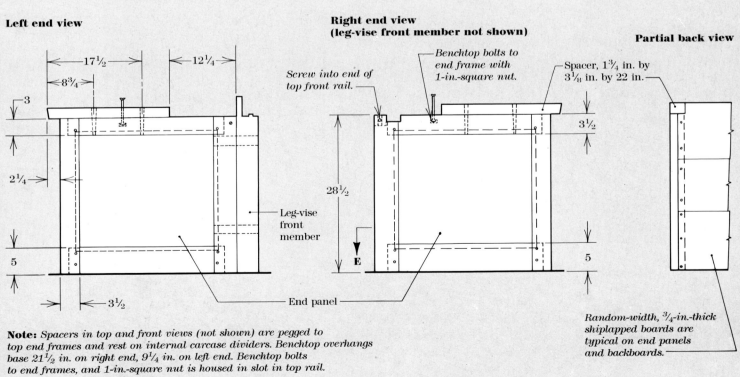

17½

12¼

8¾

3

2¼

5

3½

Leg-vise front member

End panel

Right end view
(leg-vise front member not shown)

Benchtop bolts to end frame with 1-in.-square nut.

Screw into end of top front rail.

28½

E

Partial back view

Spacer, 1¾ in. by 3⅛ in. by 22 in.

3½

5

Random-width, ¾-in.-thick shiplapped boards are typical on end panels and backboards.

Note: *Spacers in top and front views (not shown) are pegged to top end frames and rest on internal carcase dividers. Benchtop overhangs base 21½ in. on right end, 9¼ in. on left end. Benchtop bolts to end frames, and 1-in.-square nut is housed in slot in top rail.*

Drawer

Top view

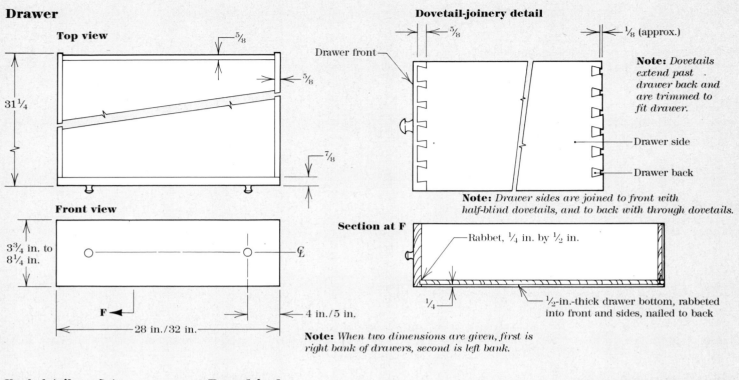

$31\frac{1}{4}$

$\frac{5}{8}$

$\frac{5}{8}$

$\frac{7}{8}$

Front view

$3\frac{3}{4}$ in. to $8\frac{1}{4}$ in.

₵L

F

4 in./5 in.

28 in./32 in.

Dovetail-joinery detail

$\frac{5}{8}$

$\frac{1}{8}$ (approx.)

Drawer front

Note: *Dovetails extend past drawer back and are trimmed to fit drawer.*

Drawer side

Drawer back

Note: *Drawer sides are joined to front with half-blind dovetails, and to back with through dovetails.*

Section at F

Rabbet, $\frac{1}{4}$ in. by $\frac{1}{2}$ in.

$\frac{1}{4}$

$\frac{1}{2}$-in.-thick drawer bottom, rabbeted into front and sides, nailed to back

Note: *When two dimensions are given, first is right bank of drawers, second is left bank.*

Knob detail

1

$1\frac{1}{8}$

$\frac{3}{4}$

$\frac{1}{2}$-in. dowel, turned on end of knob, glued in drawer front

Note: *Benchtop and cabinet front are shown as phantom lines.*

Note: *Wood screws attach board-jack runner to bottom front rail of bench.*

Board jack

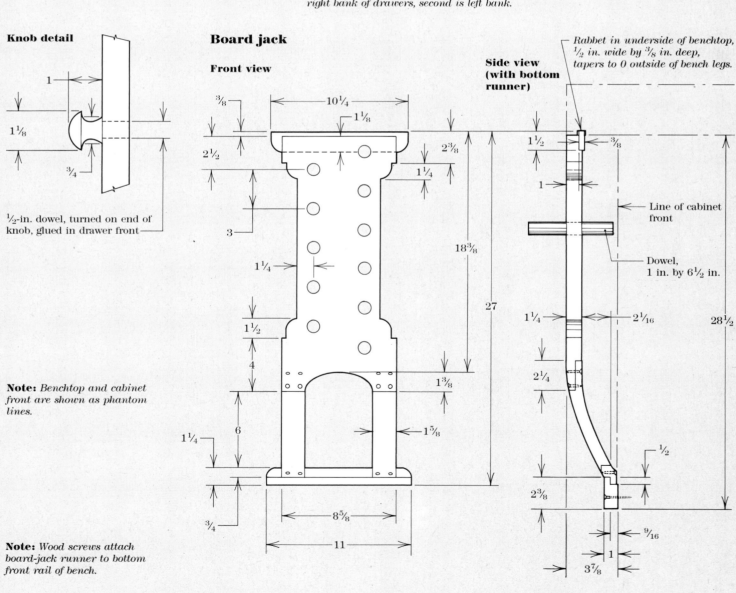

Front view

$\frac{3}{8}$

$10\frac{1}{4}$

$1\frac{1}{8}$

$2\frac{1}{2}$

$2\frac{3}{8}$

$1\frac{1}{4}$

3

$1\frac{1}{4}$

$18\frac{3}{8}$

27

$1\frac{1}{2}$

4

$1\frac{3}{8}$

$1\frac{5}{8}$

6

$1\frac{1}{4}$

$\frac{3}{4}$

$8\frac{5}{8}$

11

Side view (with bottom runner)

Rabbet in underside of benchtop, $\frac{1}{2}$ in. wide by $\frac{3}{8}$ in. deep, tapers to 0 outside of bench legs.

$1\frac{1}{2}$

$\frac{3}{8}$

1

Line of cabinet front

Dowel, 1 in. by $6\frac{1}{2}$ in.

$1\frac{1}{4}$

$2\frac{1}{16}$

$28\frac{1}{2}$

$2\frac{1}{4}$

$\frac{1}{2}$

$2\frac{3}{8}$

$\frac{9}{16}$

1

$3\frac{7}{8}$

Doors

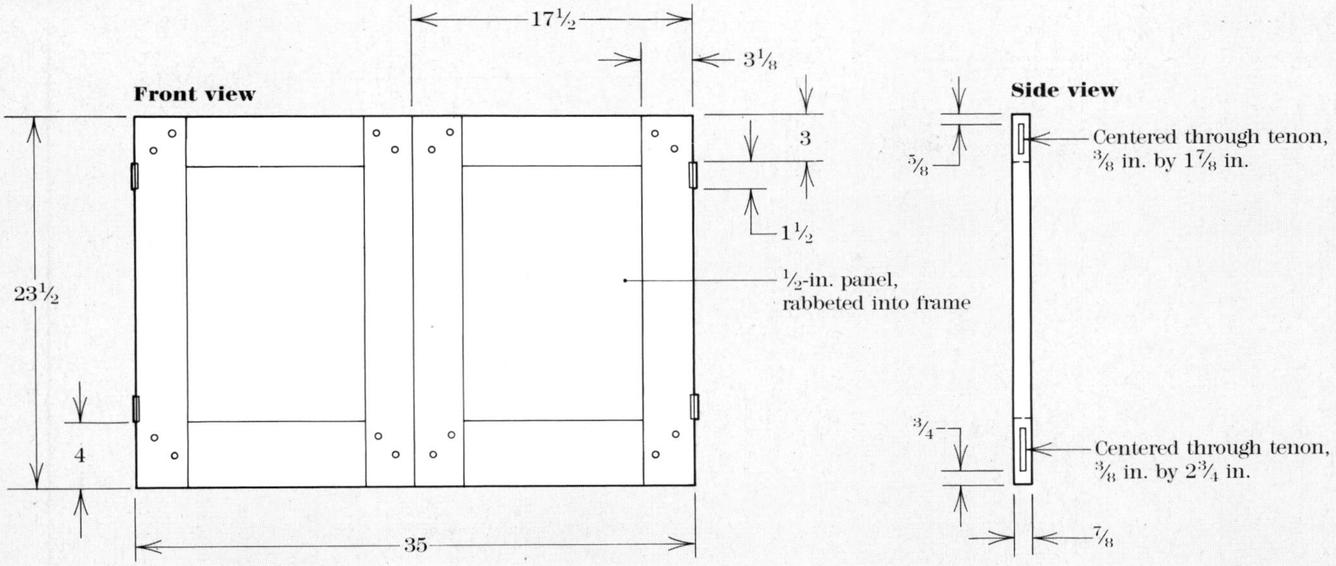

Front view
Side view
17 1/2
3 1/8
3
1 1/2
23 1/2
4
35
1/2-in. panel, rabbeted into frame
5/8
Centered through tenon, 3/8 in. by 1 7/8 in.
3/4
Centered through tenon, 3/8 in. by 2 3/4 in.
7/8

Leg vise

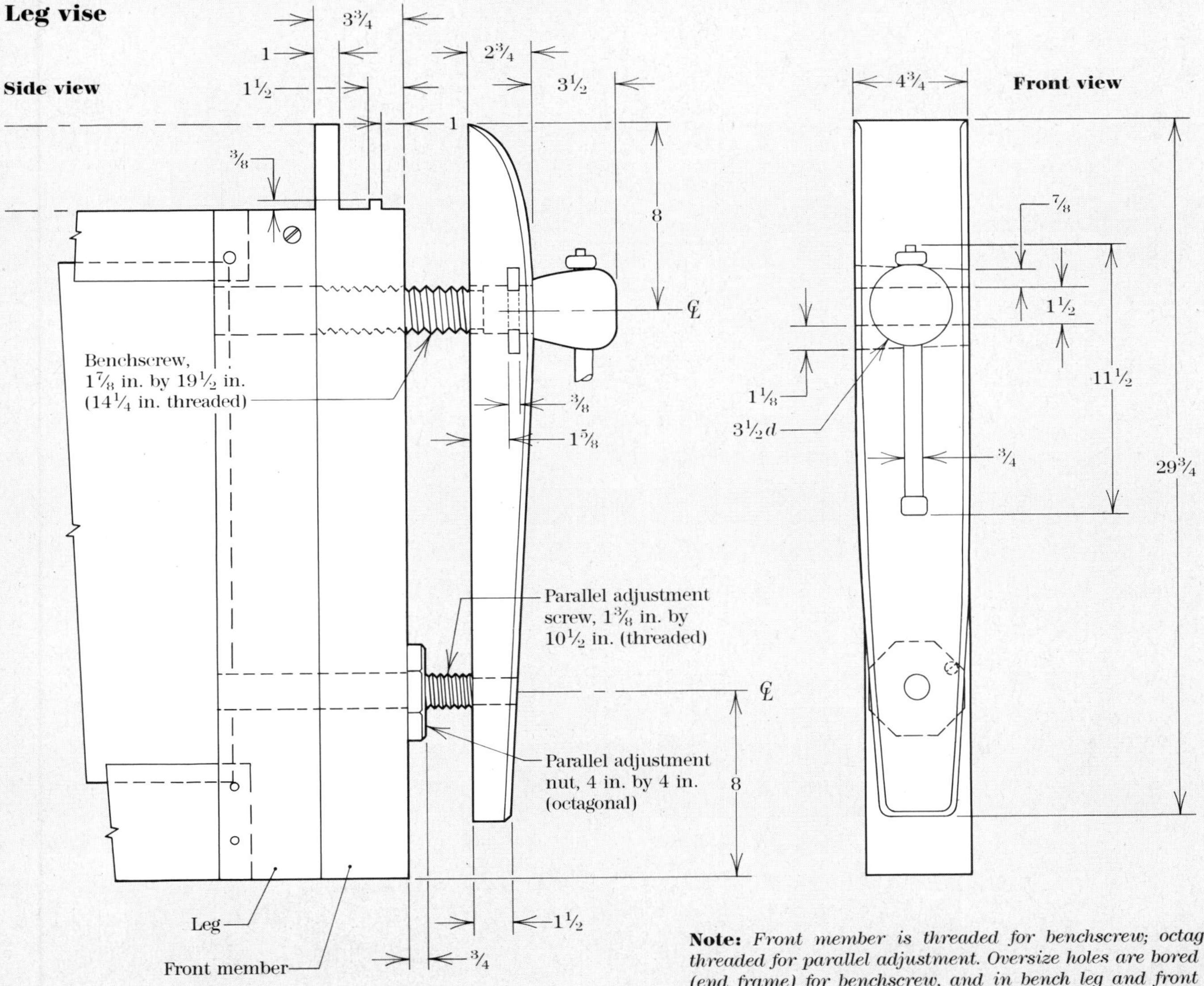

Side view
Front view
3 3/4
1
2 3/4
1 1/2
3 1/2
1
3/8
8
℄
Benchscrew, 1 7/8 in. by 19 1/2 in. (14 1/4 in. threaded)
3/8
1 5/8
Parallel adjustment screw, 1 3/8 in. by 10 1/2 in. (threaded)
℄
Parallel adjustment nut, 4 in. by 4 in. (octagonal)
8
1 1/2
3/4
Leg
Front member
4 3/4
7/8
1 1/2
11 1/2
1 1/8
3 1/2 d
3/4
29 3/4

Note: *Front member is threaded for benchscrew; octagonal nut is threaded for parallel adjustment. Oversize holes are bored in bench leg (end frame) for benchscrew, and in bench leg and front member for parallel-adjustment screw. On original bench, interior end panel was excavated slightly to permit free travel of these screws.*

Tail vise

Top view

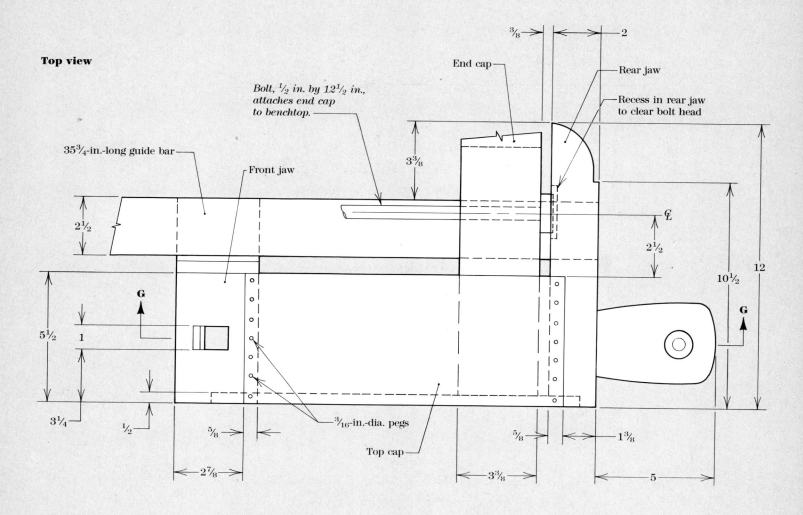

Front view

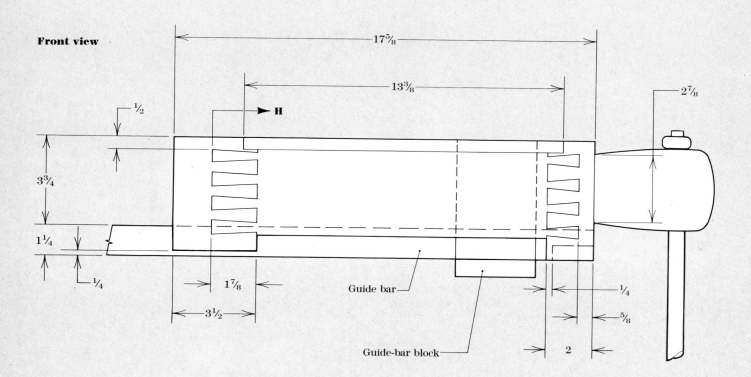

Section G-G

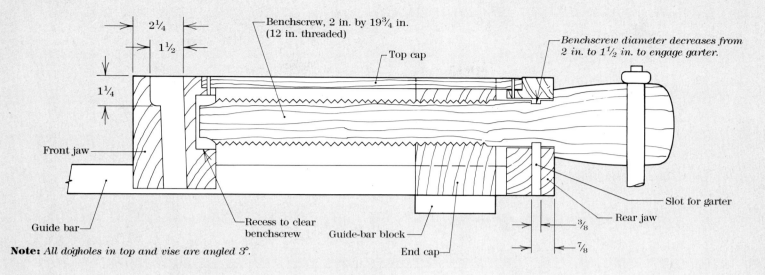

$2\frac{1}{4}$

$1\frac{1}{2}$

$1\frac{1}{4}$

Benchscrew, 2 in. by $19\frac{3}{4}$ in. (12 in. threaded)

Top cap

Benchscrew diameter decreases from 2 in. to $1\frac{1}{2}$ in. to engage garter.

Front jaw

Slot for garter

Guide bar

Rear jaw

Recess to clear benchscrew

Guide-bar block

End cap

$\frac{3}{8}$

$\frac{7}{8}$

Note: *All dogholes in top and vise are angled 3°.*

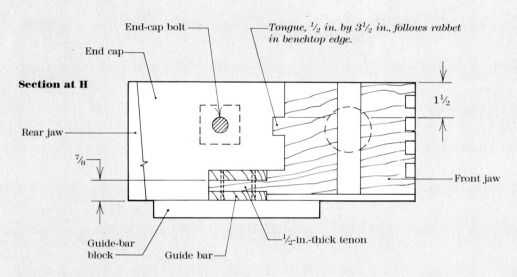

End-cap bolt

End cap

Tongue, $\frac{1}{2}$ in. by $3\frac{1}{2}$ in., follows rabbet in benchtop edge.

Section at H

$1\frac{1}{2}$

Rear jaw

$\frac{7}{8}$

Front jaw

Guide-bar block

Guide bar

$\frac{1}{2}$-in.-thick tenon

Garter

$\frac{3}{16}$

$1\frac{5}{8}$

7

$\frac{3}{8}$

2

End view

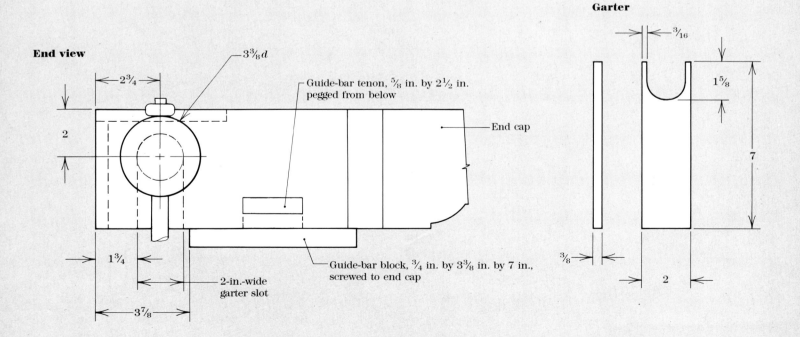

$3\frac{3}{8}d$

Guide-bar tenon, $\frac{5}{8}$ in. by $2\frac{1}{2}$ in. pegged from below

$2\frac{3}{4}$

2

End cap

$1\frac{3}{4}$

2-in.-wide garter slot

Guide-bar block, $\frac{3}{4}$ in. by $3\frac{3}{8}$ in. by 7 in., screwed to end cap

$3\frac{7}{8}$

Klausz bench

Note: *Procure all metal hardware—vise screws, dogs, bolts, etc.—before building your bench. Then make any necessary adjustments to the plans. Make a full-scale drawing of all vises before starting to build.*

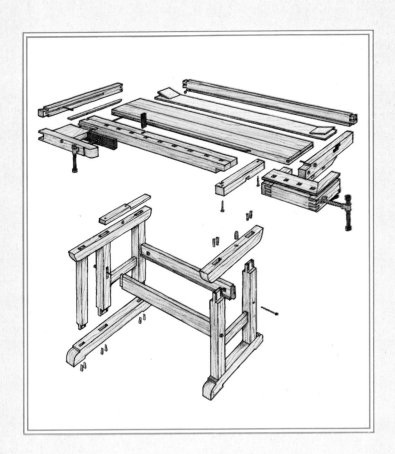

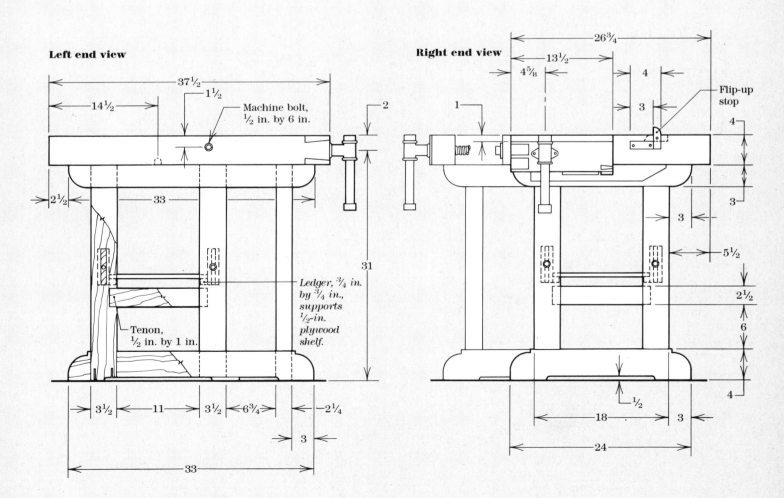

Left end view

$37\frac{1}{2}$

$14\frac{1}{2}$

$1\frac{1}{2}$

Machine bolt, $\frac{1}{2}$ in. by 6 in.

2

$2\frac{1}{2}$

33

31

Ledger, $\frac{3}{4}$ in. by $\frac{3}{4}$ in., supports $\frac{1}{2}$-in. plywood shelf.

Tenon, $\frac{1}{2}$ in. by 1 in.

$3\frac{1}{2}$ 11 $3\frac{1}{2}$ $6\frac{3}{4}$ $2\frac{1}{4}$

3

33

Right end view

$26\frac{3}{4}$

$13\frac{1}{2}$

$4\frac{5}{8}$ 4

3

1

Flip-up stop

4

3

3

$5\frac{1}{2}$

$2\frac{1}{2}$

6

$\frac{1}{2}$

4

18   3

24

Top view

84

3¼ 7¼

1

Plywood ramp,
glued in place

7¼ 3¼

7¼

3⅜

13⅛

5⅜

2⅝

1¼

Hardwood stop,
¼ in. by 2 in.

Clamping board

14½

23

7½

Doghole, ¾ in. by
1⅜ in. at top, ¾ in. by
1⅛ in. at bottom

3¼

12

Threaded rod,
½ in. by 30 in.

17½ 5½

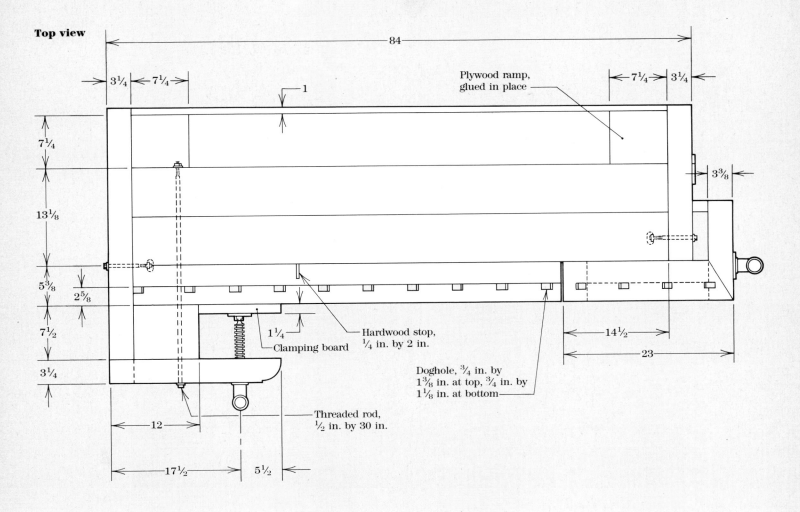

Front view

B

9½

1¼

88° 88° 88°

A

4 3

3⅜

3

Machine bolt,
½ in. by 6 in.

29

5

1½

Tenon, ½ in. by 1 in.

13

Double-wedged through
tenon, ¾ in. by 3½ in.

2¾ 2¾

48½

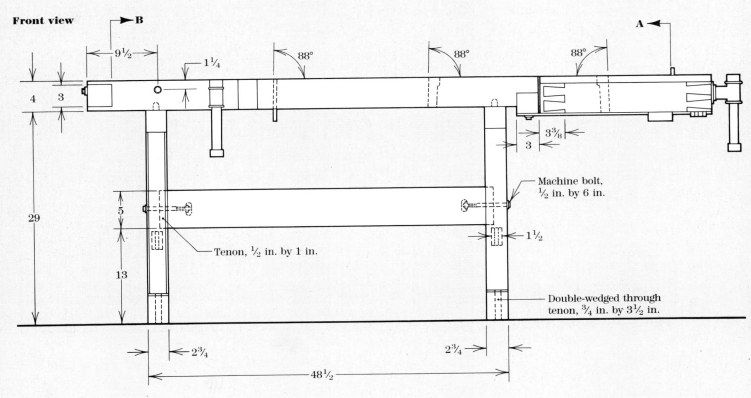

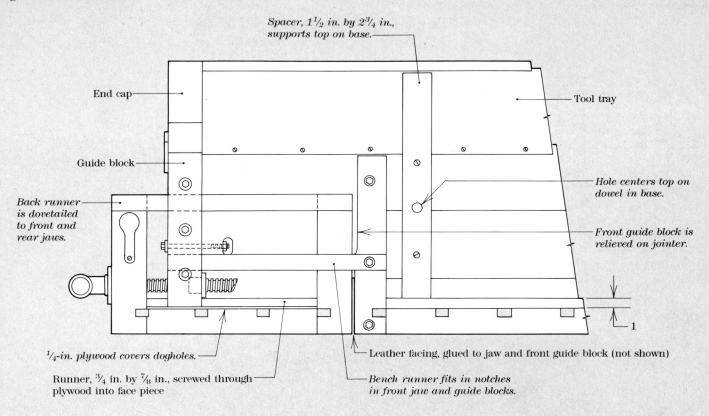

Spacer, $1\frac{1}{2}$ in. by $2\frac{3}{4}$ in., supports top on base.

End cap

Tool tray

Guide block

Hole centers top on dowel in base.

Back runner is dovetailed to front and rear jaws.

Front guide block is relieved on jointer.

$\frac{1}{4}$-in. plywood covers dogholes.

1

Leather facing, glued to jaw and front guide block (not shown)

Runner, $\frac{3}{4}$ in. by $\frac{7}{8}$ in., screwed through plywood into face piece

Bench runner fits in notches in front jaw and guide blocks.

Section at A

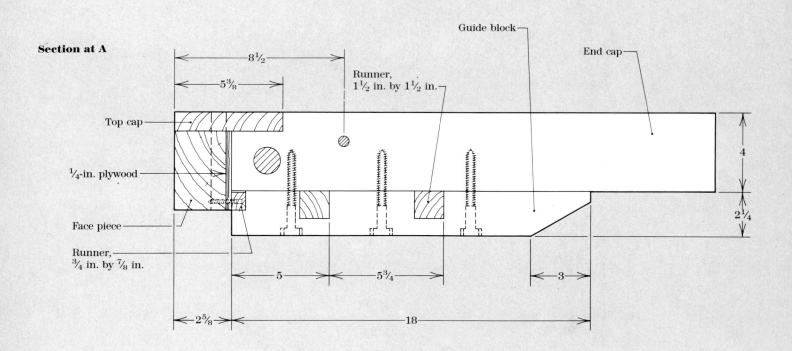

Guide block

End cap

$8\frac{1}{2}$

$5\frac{3}{8}$

Runner, $1\frac{1}{2}$ in. by $1\frac{1}{2}$ in.

Top cap

4

$\frac{1}{4}$-in. plywood

Face piece

$2\frac{1}{4}$

Runner, $\frac{3}{4}$ in. by $\frac{7}{8}$ in.

5

$5\frac{3}{4}$

3

$2\frac{5}{8}$

18

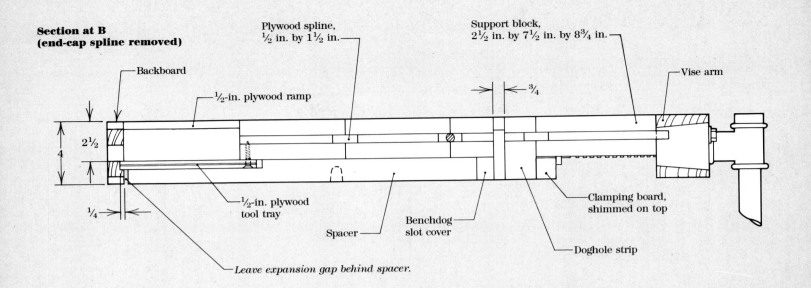

Section at B
(end-cap spline removed)

Plywood spline, ½ in. by 1½ in.

Support block, 2½ in. by 7½ in. by 8¾ in.

Backboard

Vise arm

½-in. plywood ramp

¾

2½

4

½-in. plywood tool tray

¼

Spacer

Benchdog slot cover

Clamping board, shimmed on top

Doghole strip

Leave expansion gap behind spacer.

Bench slave

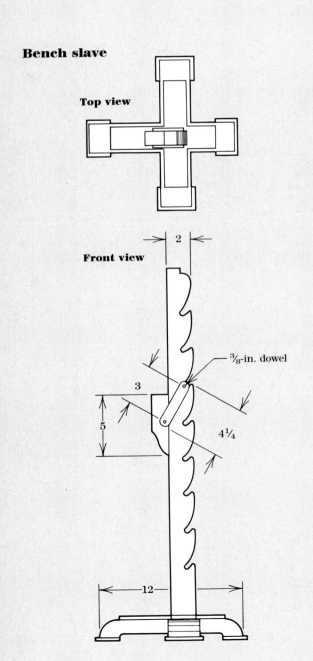

Top view

Front view

2

⅜-in. dowel

3

5

4¼

12

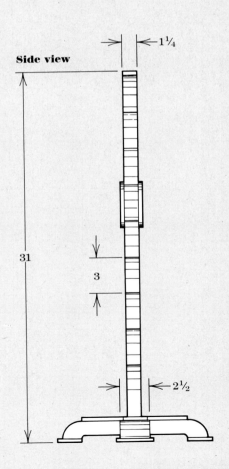

Side view

1¼

31

3

2½

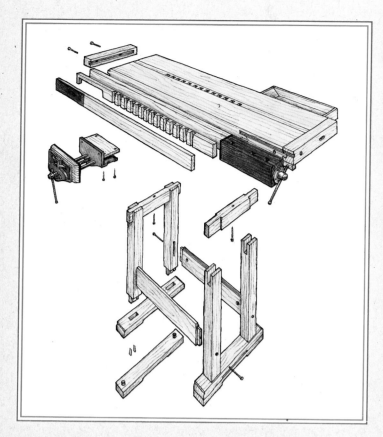

Fortune/Nelson bench

Note: *Michael Fortune's original bench (at left) has been modified in these drawings to accommodate readily available, German-style tail-vise hardware and Tom Nelson's vise-core construction.*

Note: *On pp. 234-235, vises are shown in top view only. Tail-vise hardware is located for maximum capacity.*

Doghole details

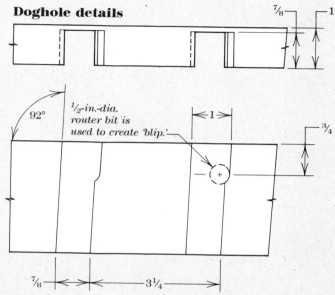

$\frac{7}{8}$ 1

$\frac{1}{2}$-in.-dia. router bit is used to create 'blip.'

92°

←1→

$\frac{3}{4}$

$\frac{7}{8}$ $3\frac{1}{4}$

Base details

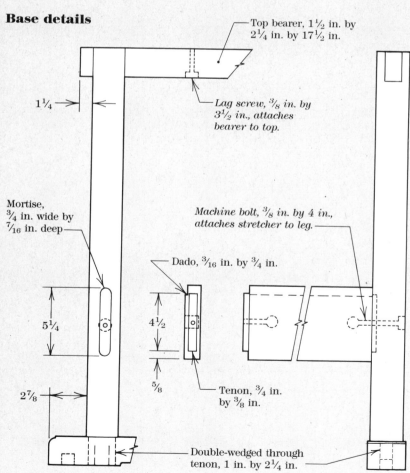

Top bearer, $1\frac{1}{2}$ in. by $2\frac{1}{4}$ in. by $17\frac{1}{2}$ in.

Lag screw, $\frac{3}{8}$ in. by $3\frac{1}{2}$ in., attaches bearer to top.

$1\frac{1}{4}$

Machine bolt, $\frac{3}{8}$ in. by 4 in., attaches stretcher to leg.

Mortise, $\frac{3}{4}$ in. wide by $\frac{7}{16}$ in. deep

Dado, $\frac{3}{16}$ in. by $\frac{3}{4}$ in.

$5\frac{1}{4}$

$4\frac{1}{2}$

$\frac{5}{8}$

Tenon, $\frac{3}{4}$ in. by $\frac{3}{8}$ in.

$2\frac{7}{8}$

Double-wedged through tenon, 1 in. by $2\frac{1}{4}$ in.

Inside end view

Partial front view

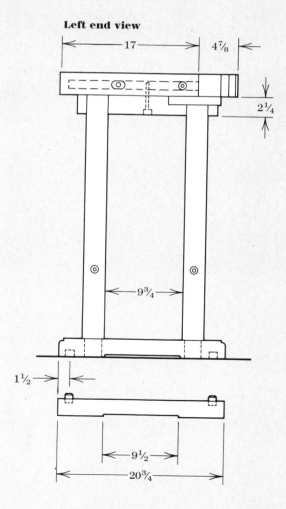

Left end view

17 $4\frac{7}{8}$

$2\frac{1}{4}$

$9\frac{3}{4}$

$1\frac{1}{2}$

$9\frac{1}{2}$

$20\frac{3}{4}$

Top view

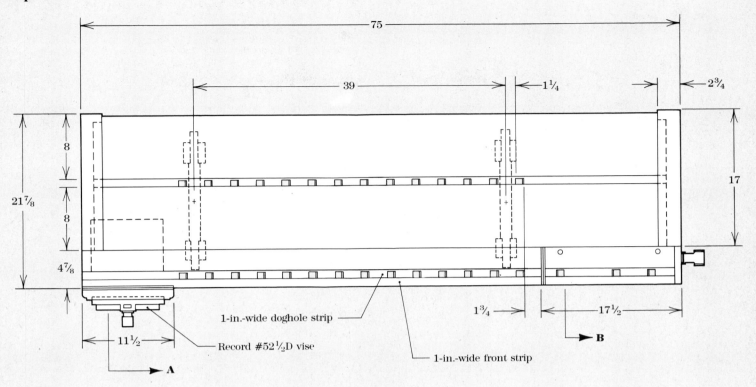

75

39     $1\frac{1}{4}$ $2\frac{3}{4}$

8

17

$21\frac{7}{8}$

8

$4\frac{7}{8}$

1-in.-wide doghole strip

$1\frac{3}{4}$ $17\frac{1}{2}$

B

$11\frac{1}{2}$

Record #$52\frac{1}{2}$D vise

1-in.-wide front strip

A

Front view

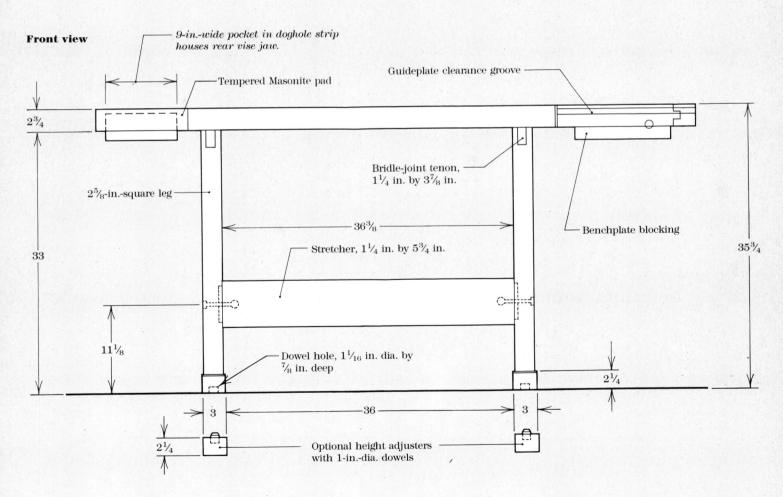

9-in.-wide pocket in doghole strip
houses rear vise jaw.

Tempered Masonite pad

Guideplate clearance groove

$2\frac{3}{4}$

Bridle-joint tenon,
$1\frac{1}{4}$ in. by $3\frac{7}{8}$ in.

$2\frac{5}{8}$-in.-square leg

$36\frac{3}{8}$

Benchplate blocking

33

Stretcher, $1\frac{1}{4}$ in. by $5\frac{3}{4}$ in.

$35\frac{3}{4}$

$11\frac{1}{8}$

Dowel hole, $1\frac{1}{16}$ in. dia. by
$\frac{7}{8}$ in. deep

$2\frac{1}{4}$

3   36   3

$2\frac{1}{4}$

Optional height adjusters
with 1-in.-dia. dowels

Section at A

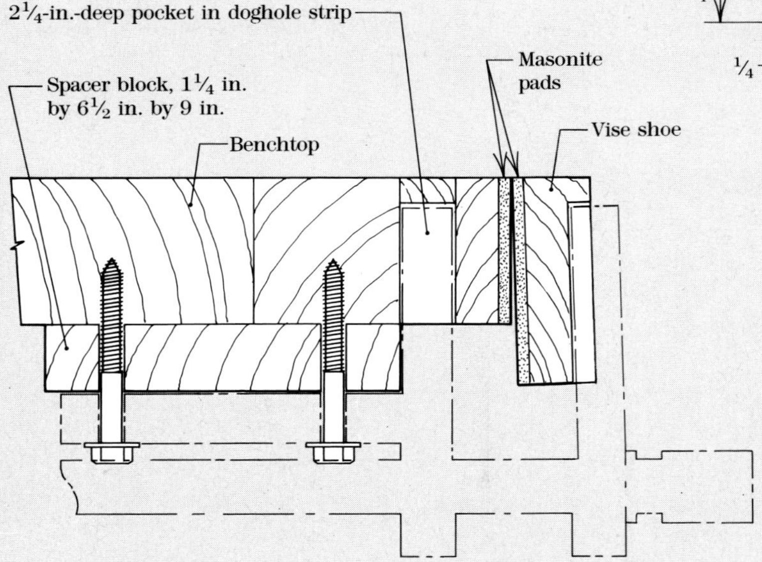
2¼-in.-deep pocket in doghole strip
Spacer block, 1¼ in. by 6½ in. by 9 in.
Benchtop
Masonite pads
Vise shoe

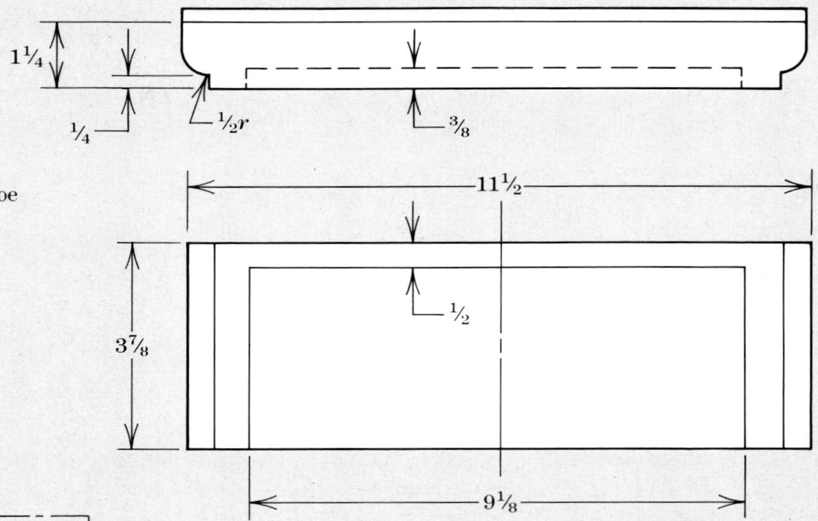
1¼
¼
½r
⅜
11½
3⅞
½
9⅛

Vise shoe

Tail vise

Section at B

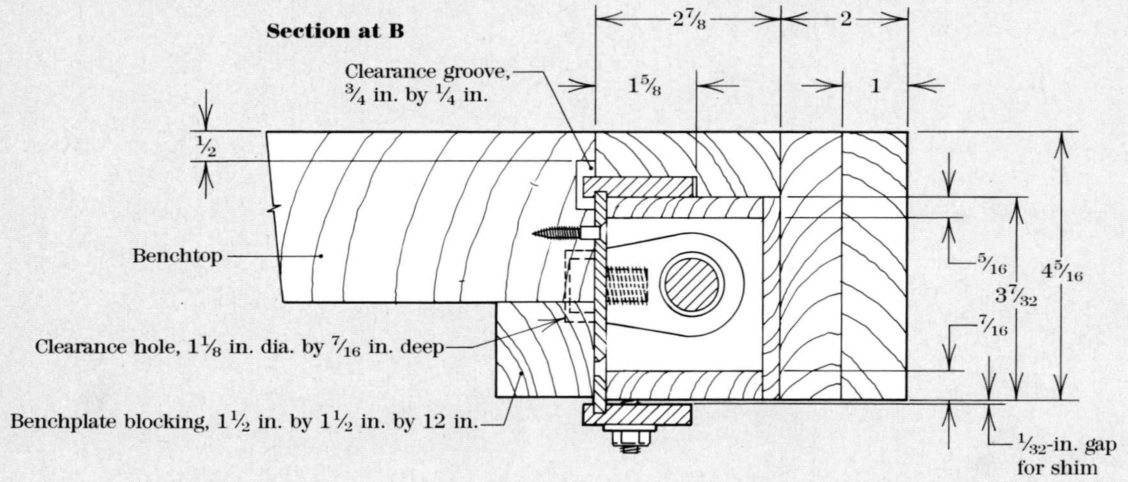
2⅞
2
1⅝
1
Clearance groove, ¾ in. by ¼ in.
½
Benchtop
5/16
3 7/32
4 5/16
7/16
Clearance hole, 1⅛ in. dia. by 7/16 in. deep
Benchplate blocking, 1½ in. by 1½ in. by 12 in.
1/32-in. gap for shim

Back view (benchscrew and benchplate removed)

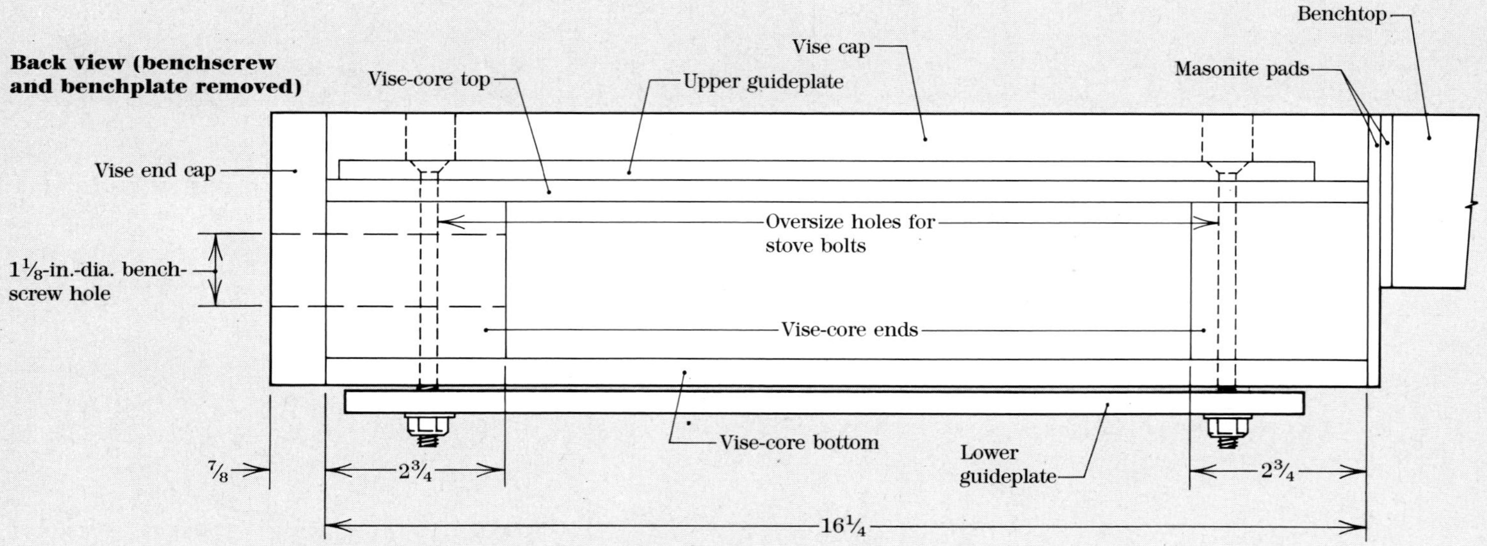
Benchtop
Vise cap
Vise-core top
Upper guideplate
Masonite pads
Vise end cap
Oversize holes for stove bolts
1⅛-in.-dia. benchscrew hole
Vise-core ends
Vise-core bottom
Lower guideplate
⅞
2¾
2¾
16¼

End-cap detail

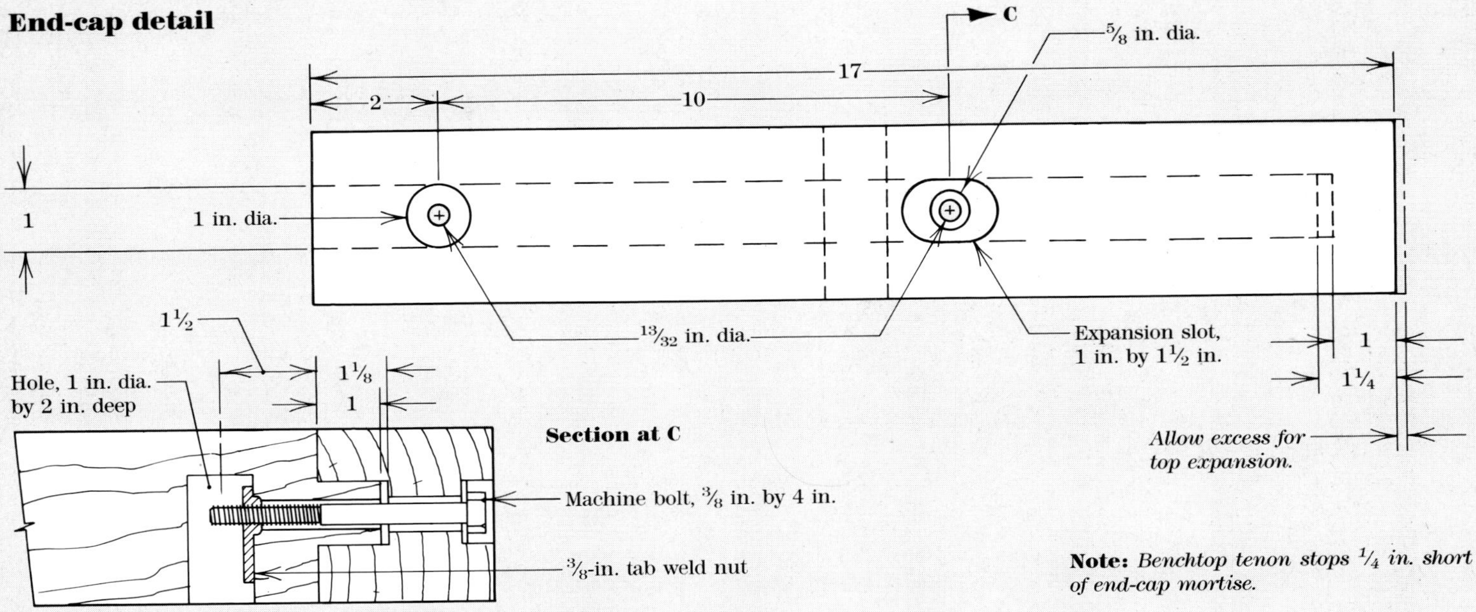
C
5/8 in. dia.
17
2
10
1
1 in. dia.
13/32 in. dia.
Expansion slot, 1 in. by 1 1/2 in.
1
1 1/4
Allow excess for top expansion.
1 1/2
1 1/8
1
Hole, 1 in. dia. by 2 in. deep
Section at C
Machine bolt, 3/8 in. by 4 in.
3/8-in. tab weld nut
Note: Benchtop tenon stops 1/4 in. short of end-cap mortise.

Top view

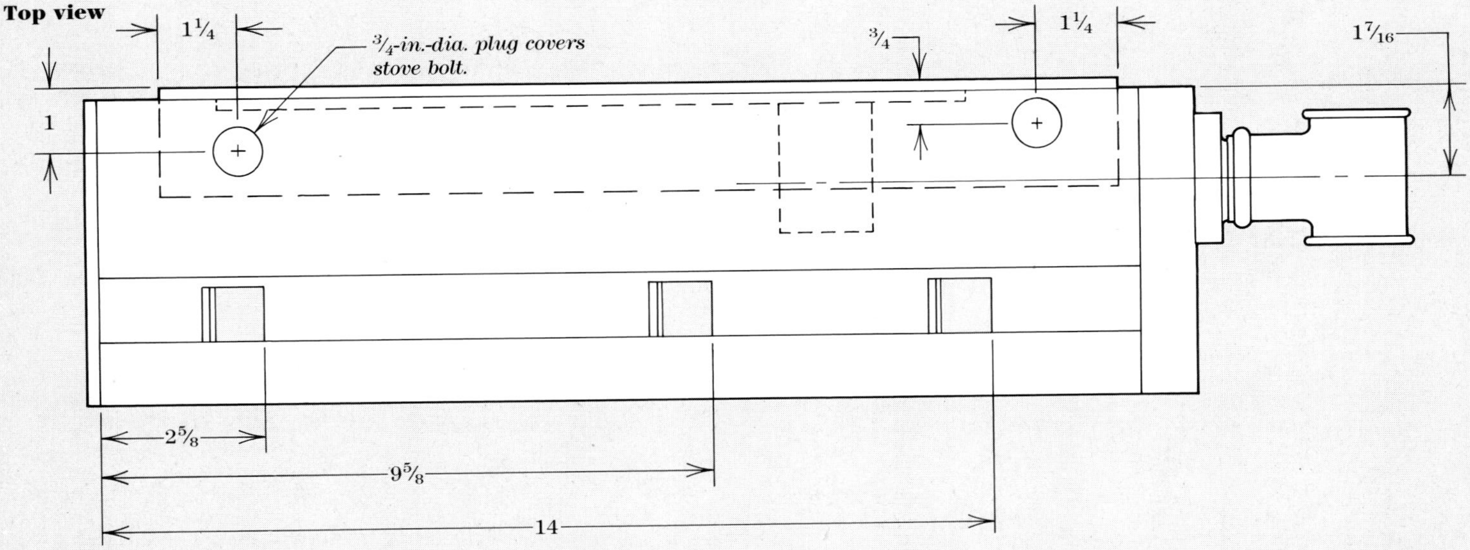
1 1/4
3/4-in.-dia. plug covers stove bolt.
3/4
1 1/4
1 7/16
1
2 5/8
9 5/8
14

Front view (vise core and front strip removed)

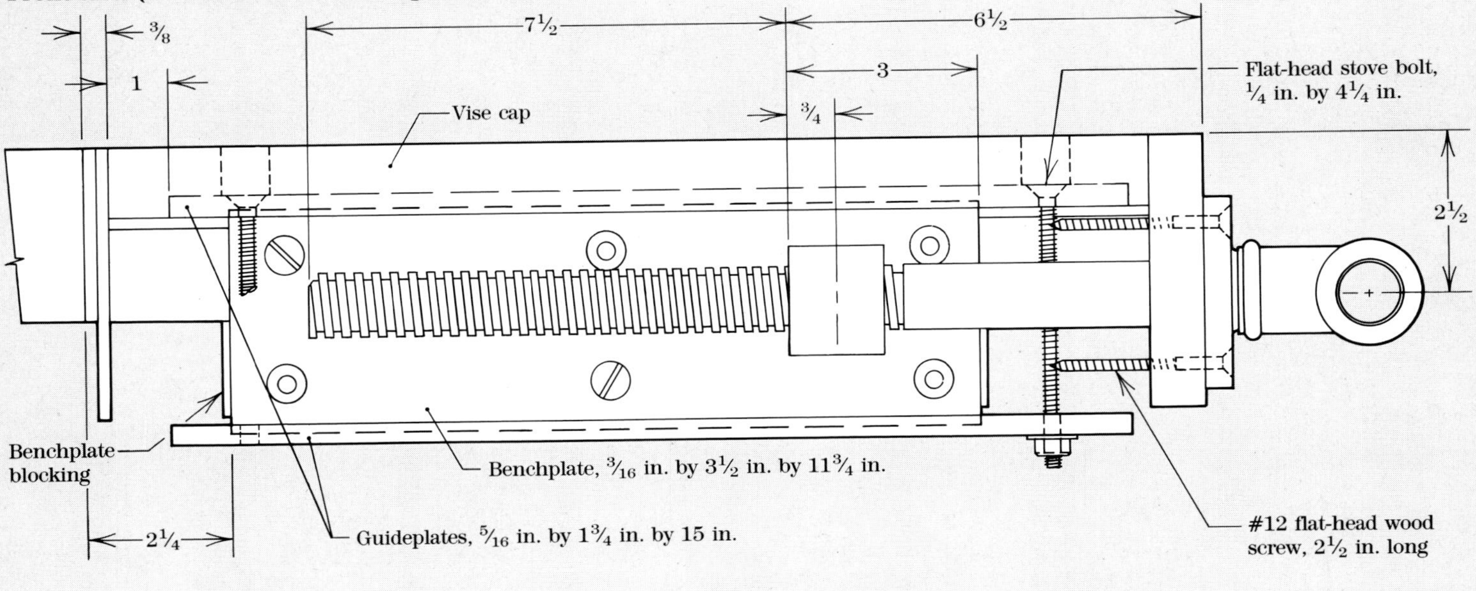
3/8
1
7 1/2
6 1/2
3
3/4
Vise cap
Flat-head stove bolt, 1/4 in. by 4 1/4 in.
2 1/2
Benchplate blocking
Benchplate, 3/16 in. by 3 1/2 in. by 11 3/4 in.
2 1/4
Guideplates, 5/16 in. by 1 3/4 in. by 15 in.
#12 flat-head wood screw, 2 1/2 in. long

Kirby bench

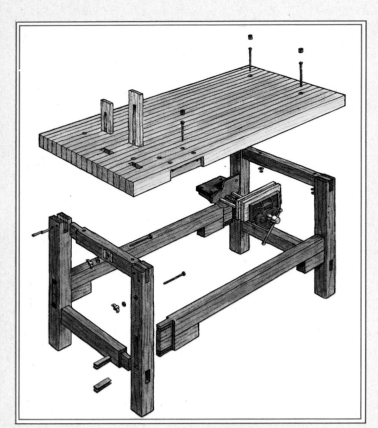

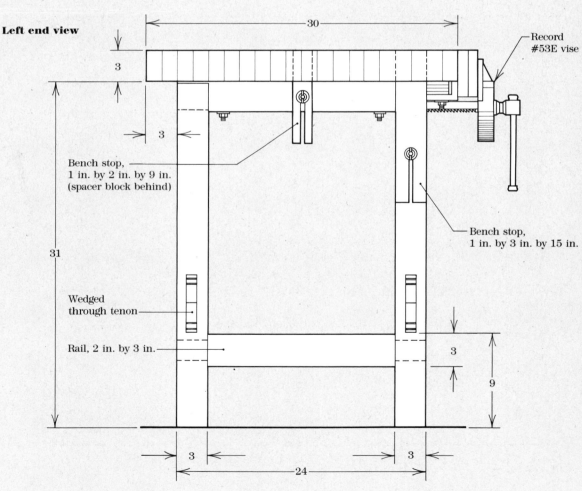

Left end view

30

Record #53E vise

3

3

Bench stop,
1 in. by 2 in. by 9 in.
(spacer block behind)

Bench stop,
1 in. by 3 in. by 15 in.

31

Wedged
through tenon

Rail, 2 in. by 3 in.

3

9

3

3

24

Top view

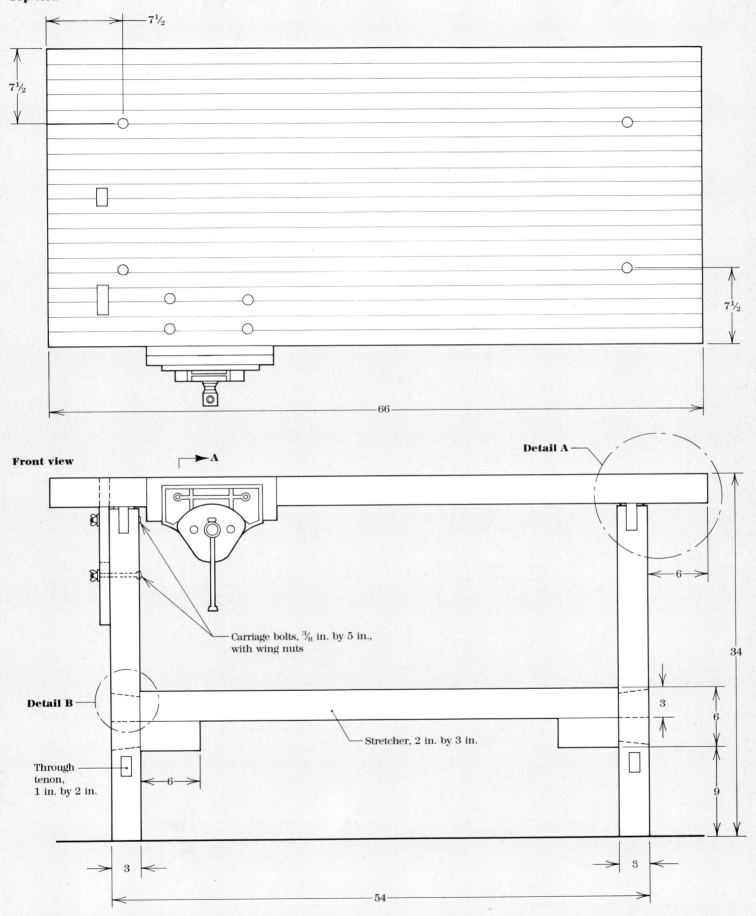

$7\frac{1}{2}$

$7\frac{1}{2}$

$7\frac{1}{2}$

66

Detail A

Front view

►A

Carriage bolts, $\frac{3}{8}$ in. by 5 in., with wing nuts

Detail B

Stretcher, 2 in. by 3 in.

Through tenon, 1 in. by 2 in.

6

34

3

6

9

3

3

54

6

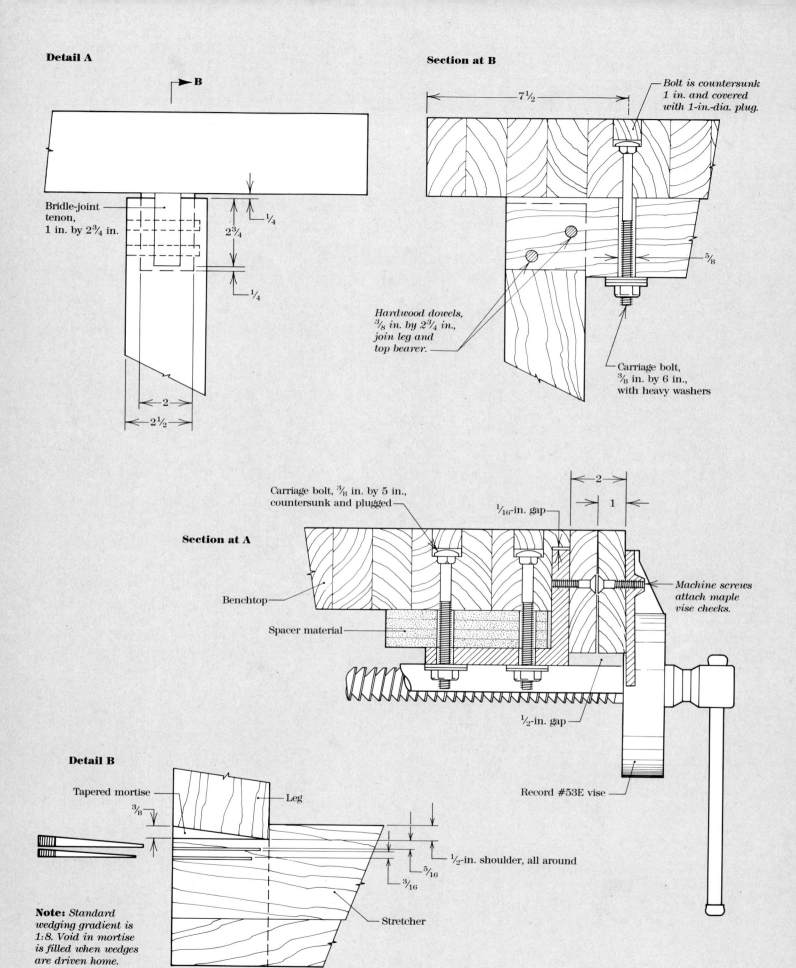

Detail A

→ B

Bridle-joint tenon, 1 in. by 2¾ in.

¼

2¾

¼

2

2½

Section at B

7½

Bolt is countersunk 1 in. and covered with 1-in.-dia. plug.

5/8

Hardwood dowels, ⅜ in. by 2¾ in., join leg and top bearer.

Carriage bolt, ⅜ in. by 6 in., with heavy washers

Section at A

Carriage bolt, ⅜ in. by 5 in., countersunk and plugged

1/16-in. gap

2

1

Benchtop

Spacer material

Machine screws attach maple vise cheeks.

½-in. gap

Record #53E vise

Detail B

Tapered mortise

Leg

⅜

½-in. shoulder, all around

5/16

3/16

Stretcher

Note: *Standard wedging gradient is 1:8. Void in mortise is filled when wedges are driven home.*

Bibliography

Books:

Alexander, John D., Jr. *Make a Chair from a Tree: An Introduction to Working Green Wood.* Newtown, Conn.: The Taunton Press, 1978.

Andrews, Edward Deming. *The People Called Shakers.* New York: Oxford University Press, 1953. Reprint. New York: Dover Publications, 1963.

Andrews, Edward Deming and Faith Andrews. *Shaker Furniture, The Craftsmanship of an American Communal Sect.* New Haven: Yale University Press, 1937. Reprint. New York: Dover Publications, 1964.

Barlow, Ronald S. *The Antique Tool Collector's Guide to Value.* El Cajon, Calif.: Windmill Publishing, 1985.

Blandford, Percy W. *Country Craft Tools.* Detroit, Mich.: Gale Research, 1974.

Drachmann, A.G. *The Mechanical Technology of Greek and Roman Antiquity.* Copenhagen: Munksgaard International, 1963.

Eaton, Allen H. *Handicrafts of the Southern Highlands.* New York: Russell Sage Foundation, 1937. Reprint. New York: Dover Publications, 1973.

Edlin, H.L. *Woodland Crafts in Britain.* London: B.T. Batsford, 1949.

Félibien, André. *Des principes de l'architecture....* Vols. 1 and 3. Paris: Chez Jean-Baptiste Coignard, 1676.

Frid, Tage. *Tage Frid Teaches Woodworking—Book 3: Furniture-making.* Newtown, Conn.: The Taunton Press, 1985.

Goodman, W.L. *The History of Woodworking.* London: G. Bell and Sons, 1964.

Greber, Josef M. *History of the Woodworking Plane.* Trans. by Seth W. Burchard for The Early American Industries Association. Forthcoming. Originally published as *Die geschichte des hobels* [Zurich: VSSM, 1956].

Greenhill, Basil. *Archaeology of the Boat.* Middletown, Conn.: Wesleyan University Press, 1976.

Hindle, Brooke, ed. *America's Wooden Age: Aspects of its Early Technology.* Tarrytown, N.Y.: Sleepy Hollow Press, 1975.

Hommel, Rudolf P. *China at Work.* New York: John Day, 1937. Reprint. Cambridge: MIT Press, 1969.

Hulot, M. *L'art du tourneur mécanicien.* Vol. 1. Paris: M. Roubo, 1775.

Hummel, Charles F. *With Hammer in Hand; The Dominy Craftsmen of East Hampton, New York.* Charlottesville: University Press of Virginia, 1968.

Jenkins, J. Geraint. *Traditional Country Craftsmen.* London: Routledge & Kegan Paul, 1965.

Jones, Bernard E., ed. *The Complete Woodworker.* Berkeley, Calif.: Ten Speed Press, 1980.

Kebabian, Paul B., and Dudley Witney. *American Woodworking Tools.* Boston: Little, Brown, 1978.

Kebabian, Paul B., and William C. Lipke, eds. *Tools and Technologies: America's Wooden Age.* Burlington: University of Vermont, 1979.

Kilby, Kenneth. *The Cooper and His Trade.* London: John Baker, 1971.

Krenov, James. *The Fine Art of Cabinetmaking.* New York: Van Nostrand Reinhold, 1977.

Lambert, F. *Tools and Devices for Coppice Crafts.* Young Farmer's Booklet, no. 31. London: Evans Brothers, 1957.

Langsner, Drew. *Country Woodcraft.* Emmaus, Penn.: Rodale Press, 1978.

———. *Green Woodworking.* Emmaus, Penn.: Rodale Press, 1987.

Mayes, L.J. *The History of Chairmaking in High Wycombe.* London: Routledge & Kegan Paul, 1960.

Meader, Robert F.W. *Illustrated Guide to Shaker Furniture.* New York: Dover Publications, 1972.

Mercer, Henry C. *Ancient Carpenters' Tools.* 5th ed. New York: Horizon Press for the Bucks County Historical Society, 1975. Originally published 1929.

Moxon, Joseph. *Mechanick Exercises. Or the Doctrine of Handy-Works.* 3rd ed. London: Printed for Dan Midwinter and Tho. Leigh, 1703.

Muramatsu, Teijiro. *A Hundred Pictures of Daiku at Work.* Tokyo: Shinkenchiku-sha, 1974.

Nicholson, Peter. *Mechanical Exercises.* London: J. Taylor, 1812.

Odate, Toshio. *Japanese Woodworking Tools: Their Tradition, Spirit and Use.* Newtown, Conn.: The Taunton Press, 1984.

Proudfoot, Christopher, and Philip Walker. *Woodworking Tools.* Rutland, Vt.: Charles E. Tuttle, 1984.

Pye, David. *The Nature and Art of Workmanship.* Cambridge: Cambridge University Press, 1968.

Roubo, Jacques-André. *L'art du menuisier.* 4 vols. Paris: L.F. Delatour, 1769-1775. Reprint. 4 vols. in 3. Paris: Léonce Laget, 1977.

Shea, John G. *The American Shakers and Their Furniture.* New York: Van Nostrand Reinhold, 1971.

Seymour, John. *The Forgotten Crafts.* New York: Alfred A. Knopf, 1984.

Sprigg, June. *By Shaker Hands.* New York: Alfred A. Knopf, 1983.

Starr, Richard. *Woodworking with Kids.* Newtown, Conn.: The Taunton Press, 1982.

Strandh, Sigvard. *A History of the Machine.* Gothenburg, Sweden: AB Nordbok, 1979. Reprint. New York: A&W Publishers, 1979.

Sturt, George. *The Wheelwright's Shop.* Cambridge: Cambridge University Press, 1923.

Tempte, Thomas. *Arbetets Ära.* Stockholm: Arbetslivscentrum, 1982.

Underhill, Roy. *Woodwright's Workbook.* Chapel Hill: University of North Carolina Press, 1986.

Viires, A. *Woodworking in Estonia.* Trans. by J. Levitan. Jerusalem: Israel Program for Scientific Translations, 1969. Reprint. Springfield, Va.: U.S. Department of Commerce.

Windsor, H.H. *The Boy Mechanic, Book 2.* Chicago: Popular Mechanics, 1915.

Yanagi, Soetsu. *The Unknown Craftsman.* Tokyo: Kodansha International, 1972.

Manuscripts:
Grant, Jerry V. and Douglas R. Allen. "Portraits of Shaker Furniture Makers, Their Life and Work." Pittsfield, Mass.: Hancock Shaker Village, 1985.

Tarule, Rob. "The Landscapes of England and New England in the Early Seventeenth Century." Unpublished manuscripts on British hurdle makers.

Western Reserve Historical Collection. "Journal of Passing and Important Events at Union Village Ohio," by Daniel Miller. [V:B-237]. Cleveland, Ohio.

———. Journal of Freegift Wells. [V:B-296]. Cleveland, Ohio.

———. Letter from Thomas Damon to George Wilcox, December 23, 1846. [IV:A-19]. Cleveland, Ohio.

Articles:
Blackaby, James R. "How the Workbench Changed the Nature of Work." *American Heritage of Invention & Technology* (Fall, 1986): 26-30.

Hoadley, R. Bruce. "Small Workbench." *Fine Woodworking* (No. 10, 1978): 86.

Kirby, Ian. "Laying Veneer." *Fine Woodworking* (No. 47, 1984): 41.

Klausz, Frank. "A Classic Bench." *Fine Woodworking* (No. 53, 1985): 62-67.

Langsner, Drew. "Making Wooden Buckets." *Fine Woodworking* (No. 40, 1983): 73-80.

———. "Body Mechanics and the Trestle Workbench." *Fine Woodworking* (No. 54, 1985): 82-85.

Lyman, E.D. "A carving/shaving bench." *Fine Woodworking* (No. 56, 1986): 58.

McKinley, Donald Lloyd. "Workbench: Ingenious ways to hold the work." *Fine Woodworking* (No. 16, 1979): 72-75.

Stanley, Mark. "Workench Design Ideas." *American Lutherie.* No. 1, 1985): 36-37.

"Traditional Workbench." *Woodsmith* (No. 50, 1987): 4-24.

Waterman, Duane. "Go-Bar Deck, Part 2." *Guild of American Luthiers Data Sheet #299.*

———. "Guitar-Body Vise." *Guild of American Luthiers Data Sheet #272.*

Credits

Sources of Supply

Key:

Benches **(B)**	Tops **(T)**
Vises **(V)**	Plans **(P)**
Hardware **(H)**	Kits **(K)**
Accessories **(A)**	Classes **(C)**

Acme Electric
Box 1716
Grand Forks, ND 58201
800-358-3096
(B)(V)(A)

Addkison Hardware Co., Inc.
PO Box 102, 126 East Amite
Jackson, MS 39205
601-354-3756
(V)(A)(K)

Alaska Wood Industry
RR #4
Ashton, Ontario,
Canada KOA 1BO
613-257-3597
(B)

Atlas Machinery Supply Ltd.
233 Queen St. W.
Toronto, Ontario,
Canada M5V 1Z4
416-598-3553
(V)(H)(A)

Busy Bee Machine Tools Ltd.
18 Basaltic Rd.
Concord, Ontario,
Canada L4K 1G6
416-738-1292
(B)

C.B. Tool & Supply
2502 Channing Ave.
San Jose, CA 95131
408-435-8810
(B)(V)(A)

Cambridge Tool Company Ltd.
131 Sheldon Dr., Unit 12
Cambridge, Ontario,
Canada N1R 6S2
519-623-4709
(B)(V)(H)

M. Chandler & Co.
6005 Milwee #709
Houston, TX 77092
800-247-9213
(B—custom)(V)(H)(A)(P)(C)

Clark National Products, Inc.
984 Amelia Ave.
San Dimas, CA 91773
714-592-2016
(V—Zyliss)

Craft Patterns, Inc.
PO Box 502
St. Charles, IL 60174
312-584-3334
(P)

The Cutting Edge
3871 Grandview Blvd.
Los Angeles, CA 90066
213-390-9723
(B)(V)(H)(A)(P)

Dallas Wood & Tool Store
1936 Record Crossing
Dallas, TX 75235
214-631-5478
(B)(V)(H)(P)(C)

The Emmert Vise Co.
PO Box 3553
Hagerstown, MD 21742
301-733-0730
(V—Emmert)

Fine Tool & Wood Store
336 NE 122 St.
Oklahoma City, OK 73114
800-255-9800
(V)(T)

The Fine Tool Shops, Inc.
PO Box 7093, 170 West Rd.
Portsmouth, NH 03801
800-533-5305
(B)(V)(A)

Frog Tool Co. Ltd.
700 W. Jackson Blvd.
Chicago, IL 60606
312-648-1270
(B)(V)(H)(A)(P)

Garrett Wade Company
161 Avenue of the Americas
New York, NY 10013
212-695-3358
(B)(V)(H)(A)(P)

Gaydash Industries, Inc.
1347 Middlebury Rd.
Kent, OH 44240
216-673-7054
(V—Versa Vise)

Highland Hardware
1045 N. Highland Ave. NE
Atlanta, GA 30306
404-872-4466
(B)(V)(H)(A)(P)

J. Philip Humfrey Ltd.
3241 Kennedy Rd., Unit 7
Scarborough, Ontario,
Canada M1V 2J9
416-293-8624
(V)(A)

Izhak's Woodworking Inc.
2324 Washington Blvd.
Venice, CA 90291
213-823-6110
(B—custom)

W.S. Jenks & Son
1933 Montana Ave. NE
Washington, DC 20002
202-529-6020
(B)(V)(H)

Kindt-Collins
12651 Elmwood Ave.
Cleveland, OH 44111
216-252-4122
(V—patternmaker's)

Kirby Studios Tools, Ltd.
710 Silver Spur Rd. #279
Rolling Hills Estates, CA 90274
213-544-7030
(B)

Laguna Tools
2081 Laguna Canyon Rd.
Laguna Beach, CA 92651
714-494-7006
(B)

Lee Valley Tools Ltd.
2680 Queensview Dr.
Ottawa, Ontario,
Canada K2B 8H6
613-596-0350
(B)(V)(H)(A)(P)

Leichtung, Inc.
4944 Commerce Parkway
Cleveland, OH 44128
800-321-6840
(B)(V)(A)

Peter Shapiro
2143 Canyon Dr.
Los Angeles, CA 90068
213-822-2299
(B/K—custom)

The Source
7305 Boudinot Dr.
Springfield, VA 22150
800-452-9999
(V)(H)(A)(P)

Stewart-MacDonald Mfg. Co.
21 N. Shafer St.
Box 900
Athens, OH 45701
800-848-2273
(B—for luthiers)

Techni-Tool
5 Apollo Rd., Box 368
Plymouth Meeting, PA 19462
215-825-4990
(B)(V)(T)

Trend-lines
PO Box 6447
Chelsea, MA 02150
800-343-3248
(B)(V)(A)

Waverly Woodworks
Rt. 5, Box 1204
Forest, VA 24551
804-525-7247
800-237-9328
(V)(H)(A)

Wood Carvers Supply, Inc.
PO Box 8928
Norfolk, VA 23503
804-583-8928
(V)(A)(P)

Woodcraft Supply Corp.
41 Atlantic Ave.
PO Box 4000
Woburn, MA 01888
617-935-5860
(B)(V)(H)(A)(P)

The Wooden Boat Shop
1007 NE Boat St.
Seattle, WA 98105
206-634-3600
(B)(V)(H)(A)

Woodline/Japan Woodworker
1731 Clement Ave.
Alameda, CA 94501
415-521-1810
(V)(H)

The Woodsmith Store
1063 So. Brentwood Blvd.
St. Louis, MO 63117
314-727-3077
(B)(V)(H)(A)(P)

Woodworker's Supply
of New Mexico
5604 Alameda Pl. NE
Albuquerque, NM 87113
505-821-0500
(V)(H)

The Wood Chop
PO Box 649
Ashland, OH 44805
419-289-2162
(B/P—for children)

Workbench Tool Co.
2833 Perry St.
Madison, WI 53713
608-273-0148
(B)(V)(H)(A)(P)

Notes:
There are many manufacturers of laminated benchtops and industrial suppliers of metal frame bases throughout the country.
The changing value of the U.S. dollar will affect the status of the imported benches.

Index

Publisher, Books: Leslie Carola
Managing Editor: Mark Feirer
Associate Editor: Roger Holmes
Design Director: Roger Barnes
Associate Art Director: Heather Brine Lambert
Production Editor: Nancy Stabile
Copy Editor: Victoria Monks
Illustrations: Mark Kara, Heather Brine Lambert
Pre-Press Manager: Austin E. Starbird
Coordinator of Production Services: Dave DeFeo
System Operators: Dinah George, Nancy-Lou Knapp
Scanner Systems Operator: Swapan Nandy
Production Assistants: Lisa Carlson, Mark Coleman, Deborah Cooper, Barbara Snyder
Pasteup: Marty Higham, Cynthia Nyitray
Indexer: Harriet Hodges

Typeface: Modern Medium, 9 point
Printer and Binder: New Interlitho, Milan, Italy